AF541579

Tea
Technological Initiatives

Tea
Technological Initiatives

Edited by

Dr. Niladri Bag
Associate Professor
Department of Horticulture
Sikkim University, 6th Mile, Tadong
Gangtok-737 102

Dr. Arundhati Bag
Assistant Professor
Department of Medical Biotechnology
Sikkim Manipal University
Tadong, Gangtok-737102

Dr. L.M.S. Palni
Professor & Dean
Department of Biotechnology
Graphic Era University
566/6 Bell Road, Clement Town
Dehradun-248 002, Uttarakdand

NEW INDIA PUBLISHING AGENCY
New Delhi – 110 034

NEW INDIA PUBLISHING AGENCY
101, Vikas Surya Plaza, CU Block, LSC Market
Pitam Pura, New Delhi 110 034, India
Phone: + 91 (11)27 34 17 17 Fax: + 91(11) 27 34 16 16
Email: info@nipabooks.com
Web: www.nipabooks.com

Feedback at feedbacks@nipabooks.com

ISBN: 978-93-85516-33-7

Cover Page Designed by: Dr. Niladri Bag

Composed, Designed and Printed in India

Preface

Tea is an ecologically driven agrofarming having huge potentials with high input. The high productivity is related to efficient plucking by pluckers, existence of good tea stands free from diseases, pests along with favourable climate, etc. Indian Tea has a 170 years old history and since its inception, it has occupied an important place in our national economy. After Independence, Indian Tea industry has achieved about 250 fold growths and still it is world leader in terms of quality and consumption. But, for the last few decades Indian Tea industry is passing through a bad phase and major problem is low yield.

There are crucial questions regarding the future of the Indian Tea Industry and people living on it. Therefore, it is an urgent need to reorient developmental planning for Indian Tea Industry in a manner that helps to revive the industry while meeting the aspirations of the people for a standard living.

Tea is intricately associated with Indian Society and today it is a common men's drink next to water. More than 70% of the country's population comes in contact with tea in one way or another including tea drinking, thus tea plays an important role in Indian society both directly and indirectly.

In this backdrop this is an attempt to bring together some young researchers work in the current volume entitled. The unique feature of the book is that it incorporates available knowledge on more operative aspects of tea cultivation, ranging from the history of this 5000 years old crop, nursery, planting materials, latest aspects of history of Darjeeling tea, role of small tea growers in Indian tea, insect pest management and pesticide residue, health benefit of tea, biotechnology of tea etc. All the chapters have been well compiled with great dedication and sincerity and authors own the authenticity of informations available in the book. We hope the book will be useful to a wide range of users starting from students, teachers, researchers, academicians, bureaucrates and policy makers etc. Authors have expressed their experiences in chapters and have taken the liberty to access to the research findings and other required informations from national and international literatures, which have been duly referred.

We acknowledge the help given by my research scholars, students, scientists in the laboratory, and field staffs. We are thankful to well wishers, colleagues and innumerable students who have made worth while contributions directly or indirectly by giving suggestions.

Finally, we wish to express our immense gratitude to the staff of 'New India Publishing Agency (NIPA)', for their quick response and interest regarding this important work, composer and copy-editor, for their job on manuscript, and production team for their valuable help to bring out the book in the present form.

Editors

Contents

Abbreviations

ACE : Angiotension converting enzymes
BF : Bio-fertilizer
BMPs : Better Management Practices
Chl : Chlorophyll
CP : Collar prune
Cu : Copper
DHT : Dihydrotestosterone
DS : Deep skiff
EC : Epicatechin
ECG : Epicatechin gallate
EGC : Epigallocatechins
EGCG : Epigallocatechin gallate
E : Transpiration
EW : Epicuticular wax
FAO : Food and Agriculture Organisation
FAOSTAT : Food and agriculture organisation Statistics
FFP : Frame forming prune
GAP : Good agricultural practices
GI : Geographical Indications
GIA : Global Industry Analysts
Green tea : GT
gs : Stomata conductance
GTP : Green tea polyphenols
LOS : Level-off-skiff
LP : Light pruned
LS : Light skiff
MP : Medium prune
MS : Medium skiff
NIOSH : National Institute for Occupational Safety and Health
OA : Osteoarthritis
Pn : Photosynthetic rate
PPFD : Photosynthetic Photon Flux Density
PSB : Phosphate Solubilising Bacteria
RA : Rheumatoid Arthritis
RP : Rejuvenation prune
SBGH : Sex hormone-binding protein globulin
Ta : Ambient temperature
Tb : Base temperature
UNEP : United Nations Environment Programme
UP : Un-pruned

VAM	:	Vesicular Arbuscular Mycorrhiza
VPD	:	Vapour Pressure Deficit
WHO	:	World Health Organisation
WUE	:	Water Use Efficiency
Ψ_L	:	Leaf water potential

List of Contributors

Dr. Ananda Das Gupta
Indian Institute of Plantation Management
Jnana Bharathi Campus
P.O. Malathalli
Bangalore – 560 056
email: anandadg06@gmail.com

Dr. Ananda Mukhopadhyay
Entomology Research Unit
Department of Zoology
University of North Bengal
Dist. Darjeeling , West Bengal–734 013
email:entoananda@gmail.com

Dr. Anil Kumar Singh
Darjeeling Tea Research and Development Centre
Tea Board A.B. Path
Kuresong – 734 203, Darjeeling, West Bengal
email: ssteaboard@gmail.com

Dr. Biswajit Bera
Tea Board of India
Brabourn Road
Kolkata – 700 001

Mr. J.S. Bisen
Darjeeling Tea Research and Development Centre
Tea Board A.B. Path
Kuresong-734203, Darjeeling
West Bengal, India

Miss. Jigisha Anand
Department of Biotechnology
566/6, Bell Road, Clement Town
Graphic Era University
Dehradun – 248 002

Mr. Kumar Basnet
Entomology Research Unit
Department of Zoology
University of North Bengal
West Bengal – 734 013

Mr. Mainaak Mukhopadhyay
Department of Botany
University of Kalyani, Nadia
West Bengal, India

Mr. Mrityunjay Choubey
Darjeeling Tea Research and Development Centre
Tea Board A.B. Path
Kuresong-734203
Darjeeling, West Bengal

Mr. N. Kumar
Darjeeling Tea Research and Development Centre
Tea Board A.B. Path
Kuresong – 734203, Darjeeling
West Bengal, India

Dr. Narayanan Nair Muraleedharan
Department of Entomology
Tocklai Tea Research Institute
Jorhat –785008
Assam

Dr. Nishant Rai
Department of Biotechnology
566/6, Bell Road, Clement Town
Graphic Era University
Dehradun – 248 002, Uttarakhand
email: nishantrai1@gmail.com

Miss. Sangmu Thendup
Department of History
Sikkim University
6th Mile, Tadong, Gangtok – 737 102
Sikkim, India

Miss. Sushma Rai
Department of History
Sikkim University, 6th Mile
Tadong, Gangtok – 737102
Sikkim

Mr. Roshan P. Rai
DLR Prerna, Darjeeling
email: rairoshan@gmail.com

Mr. R. K. Chauhan
Darjeeling Tea Research and Development Centre
Tea Board A.B. Path, Kuresong-734203
Darjeeling, West Bengal

Mr. R. Kumar
Darjeeling Tea Research and Development Centre
Tea Board A.B. Path
Kuresong-734203, Darjeeling
West Bengal, India

Mr. Prosenjit Biswas
Darjeeling Tea Research & Development Center
A.B. Path, Kurseong-734203
prjt.bsws@gmail.com

Dr. S. Baisya
Department of Agronomy
Tocklai Tea Research Institute
Jorhat - 785008, Assam
email: sbaisya@gmail.com

Miss. Soma Das
Entomology Research Unit
Department of Zoology
University of North Bengal
Dist. Darjeeling
West Bengal–734 013

Dr. Tapan Kumar Mondal
Division of Genomic Resource
National Bureau of Plant Genetic Resources
IARI Campus, New Delhi-110012
mondaltk@yahoo.com

Dr. Somnath Roy
Department of Entomology
Tocklai Tea Research Institute
Jorhat –785008
Assam, India
somnathento@gmail.com

Dr. Gautam Handique
Department of Entomology
Tocklai Tea Research Institute
Jorhat –785008
Assam

Tea: Technological Initiatives, pp. 1-18
New India Publishing Agency, New Delhi, India
Edited by Niladri Bag, Arundhati Bag and L.M.S. Palni

1

Story of Darjeeling Tea A Historical Background

Sangmu Thendup and Sushma Rai

"If you are cold, tea will warm you; if you are too heated, it will cool you; If you are depressed, it will cheer you; if you are excited, it will calm you."

-William Edward Gladstone

Abstract

The history of the development of the tea plantation industry in the Darjeeling district dates back to the early fifties of the nineteenth century when the English entrepreneurs took lease of extensive land on the mountain slopes of the Darjeeling Himalaya and started tea plantations. During the formative years of the introduction of the plantation industry the region was sparsely populated so, laborers from various parts of India and her neighboring countries were encouraged to settle in the fringe areas of the Tea gardens. Over the last 160 years many changes have taken place in their lifestyle. The main purpose of this paper is to reconstruct the history of Darjeeling tea and to explore the life of people in the Darjeeling region and also to learn how the local populace adjusted themselves with the given environment with the changing scenario of the area.

The rapid growth of Darjeeling can be traced back to the introduction of Tea Plantation Industry in the hills which was followed by the introduction of Railways (1881) and other Missionary activities like establishment of various educational institutions and other construction works. Therefore, it would not be wrong to say that the history of Darjeeling is incomplete without the study of history of the plantation Industry in the region.

Key words: *Darjeeling Tea, Plantation, History, Railways, development*

Introduction

According to the English dictionary 'Tea' is define as the dried leaves of a tropical evergreen shrub *Camellia sinensis* and a hot drink prepared by infusing these leaves in boiling water is called 'Tea'. The Latin name for tea is *Camellia sinensis*, literally meaning the Chinese Camellia. Besides being an agricultural crop, it also has an industrial base. The cultivation, maintenance, harvesting and processing of tea is labour-intensive and provides regular employment to millions. Tea industry is one of the chief foreign-exchange earners in most of the developing countries.

Tea is known to be as the 'Green Gold' of India. It is not only a much liked beverage of the people in the country but also fetches a good amount of foreign exchange. Being one of the world's largest producer, consumer and exporter of tea, India plays a significant role in the world tea trade.

Historical Background

Although tea has been grown as a plantation crop for nearly two centuries now, its origin is still a big mystery. There is still considerable speculation about the place of its origin. It is generally believed that it originated somewhere in South-east Asia. However current distribution patterns of tea types or varieties suggest that tea possibly originated in the vicinity of the Irrawaddy basin from where it dispersed to South-east China, Indonesia, and Assam. Tea varieties specific to these three principal regions have characteristic biological features, but it is rarely that only one particular type is to be found in a particular locality (Banerjee, 1993).

Another widely accepted theory regarding the origin of tea is that it was from China, that tea came and its origin there is lost in the mist of legends. Its legendary origin can be traced back to around 2737 B.C. when tea was ascribed to have medicinal properties. It was called a "divine healer" by the then Chinese emperor Shen Nung. It was a chance discovery when Shen Nung while camping in the countryside was resting against the trunk of a tea-tree and watched the cauldron boiling. Some leaves, disturbed by the breeze fell into the boiling-water. Intrigued by the aroma, he tested the concoction and discovered the most popular beverage called Tea (Subramaniam, 1995). That is how tea was discovered, according to the legends. It was believed that tea was certainly drunk in China well before the beginning of the Christian era. At least one container marked with the pictogram ch'a or tea, has been discovered in a tomb of the Han dynasty (206 B.C. – 220 A.D; Wright, 2011).

Tea as a wild plant might have grown from time immemorial but its original home is still a big mystery to the researchers. It is believed by some scholars that tea travelled to China from India, where wild tea bushes used to grow in Himalayan valleys since time unknown. The plants travelling to China were attributed to the Indian Buddhist Missionaries. Although credit goes to China for starting cultivation of tea for the first time, some authorities also believe that tea was known to some tribes in Shan state in Burma and Thailand as early as it was made use of in China (Adikesavan, 1975). In Columbian Encyclopedia, tea plant or *Thea Sinensis* is clearly described as 'Indigenous to Assam'. Chinese credit lies only in taking up the cultivation first and popularizing it. The commercial cultivation of tea started in China in about 8th century A.D. It is said that for centuries, Chinese people purposely kept the processes of tea manufacturing a secret to the world. Some of the Chinese Emperors even issued strict edicts for not revealing the secrets of tea production to any European. But some of the corrupt Chinese were bribed by the Britishers, to spell out the secrets of tea plantations (Lama, 2008).

So it is likely that in both China and India tea bushes grew from time immemorial but gradually other countries like Japan, Sweden, America, Brazil, Java, Sri Lanka, Africa, France, Mauritius, Jamica, Sumatra, Mexico, Kenya, Peru, Nepal, Bangladesh etc. started cultivation and production of tea.

Darjeeling Tea

In India the oldest record of tea is found from the writings of Mandebelo in the year 1640 (Chakraborty, 1997). Between 1818 and 1834, several private individuals and Government Officials took an initiative for the cultivation of tea in North East India. Discovery of wild tea plants in Assam around 1823 is a major landmark in the history of the introduction of tea industry in India. In 1821 Major Robert Bruce and in 1824 Mr. Scott, discovered the tea plants growing wild in Assam, but much expense and considerable delays were consequently incurred in bringing plant and seeds from China, and Chinese men to teach the people of India the art of growing and manufacturing tea. Because of the profits to be earned in future, the British Government itself undertook the formation of experimental plantations in Upper Assam and the districts of Kumaon and Garhwal; in 1839 private speculation took the field, and the Assam Tea Company was formed. But the Tea plantation in the real sense was first started in India about 1834 (Adikesavan, 1975).

Tea plantation is a highly specialized commercial enterprise employing wage labour. It has many characters of modern factory except the fact that it is land based. In India, tea plantation as an organized industry came up around the middle of the nineteenth century. Tea is cultivated extensively in Assam and the two districts of Jalpaiguri and Darjeeling of West Bengal. It is also grown in some parts of south India.

From the eighteenth century onwards, the British were drinking more tea than ever. By the year 1720, the British had acquired such an appetite for tea, that the East India Company imported one million pounds of tea in that year from China. Gradually the demand for the beverage grew to such phenomenal heights that five years later five million tons of tea was imported (Sharma and Das, 2008). In 1773, British Parliament passed the Regulating Act which gave the East India Company full monopoly of Tea trade with China. Again in 1813 the Charter Act was passed which although deprived the East India Company of its monopoly of trade in India but it was still enjoying its monopoly of trade with China and trade in tea. This monopoly of the company finally came to an end by the Charter Act of 1833, which totally abolished the company's monopoly of trade in tea and trade with China.

Finally in 1834, considering its demand; Lord William Bentick, the then Governor-General of India appointed a committee to enquire into the prospects of tea cultivation in India. And in 1835, the British Government showed serious interest to start the tea cultivation in India. Historically, the district of Darjeeling was acquired by the Britishers from neighboring states of Sikkim, Nepal and Bhutan. By 1835, Darjeeling was ceded from the Sikkim Raja and Mr. Campbell became the Superintendent of Darjeeling. The credit of establishment of the tea industry in Darjeeling goes to the enterprise of Dr. Campbell, a civil surgeon of the Indian Medical Service in 1841 (O'Malley, 1907). Along with him Dr. Withecombe, another Civil Surgeon and Major Crommelin also started experimental tea plantation in Lower Valley called Lebong (Lepcha word meaning, the tongue – like spur, a corruption of ali, a tongue and abong, a mouth). The results of these experiments were very satisfactory. But according to Dr. Hooker and few others, too much moisture and too little sun of Darjeeling at a height of 7,000 feet was not in favour of the large scale cultivation of tea in the area (Dash, 1947). This was not the case of lower sites of Pankhabari and Kurseong (Kurseong – it is a Lepcha word. It has been suggested that this name is a corruption of Kurseon-rip, the small white orchid, which used to grow plentifully round Kurseong and that it means the place of white orchids. Another suggestion is that it refers to a cane which used to grow there in rich profusion and which the Lepchas in their "Rong-Ring", as they term their own language, call 'Kur' and that 'Seong'. There are still a few of these canes to be found in the forest behind Eagle's Cras.), where plantation of both tea and coffee was established by Mr. Martin. The British Government also established tea nurseries during the period, but commercial exploitation began during the 1850s. Before the transfer of Dr. Campbell to Darjeeling in 1839, the authorities had already given some consideration to the possibility of developing the cultivation and manufacture of tea in the region under the East India Company.

In 1840, Dr. Campbell was transferred from Kathmandu to Darjeeling, and soon he started the experimental growth of tea in Aloobari (Taken from a Nepali word meaning potato garden) area at an altitude of 7000 feet. Within a period of twenty years, several tea gardens appeared in Darjeeling. It is a matter of some wonder as to how huge pieces of machinery for tea gardens were manhandled up the hill when a simple journey was so difficult at that time.

The history of Darjeeling presents a late chapter in the extension of British rule. The Britishers selected hill areas of Darjeeling for both bodily comfort and control over hill resources. The rapid growth of Darjeeling can be traced back to the introduction of Tea Plantation Industry in the hills which was followed by the introduction of Railways (1881), roads, ropeways and other Missionary activities like establishment of various educational institutions and other construction works.

There is a debate regarding the reason why the East India Company selected North Bengal (Terai and Darjeeling District) for the plantation industry. The conventional theory is that the ecology and climate of this region were favorable for the plantation industry. But the post colonial researchers have raised certain question in regard to the selection of this region. According to them the ecology and climate of other hilly areas of India were also favorable for the plantation industry. They believe that the abundance of land was one of the major reasons for the selection of this region for plantation industry. Besides, these regions were sparsely populated; naturally the planters did not have to face any resistance from the local inhabitants. However the issue is not free from controversy.

Importance of Darjeeling Tea

Unlike most Indian tea, Darjeeling tea is normally made from the small-leaved Chinese variety of *Camellia sinensis var. sinensis*, rather than the large - leaved Assam plant (*Camellia sinensis var. assamica*). Among the different kinds of teas cultivated in India, the most celebrated one comes from the Darjeeling Himalayas. The best of India's prized Darjeeling Tea is considered the world's finest tea. The region has been cultivating, growing and producing tea for more than the last 160 years. The Darjeeling tea industry at present employs over 52 thousand people on a permanent basis while additional 15,000 persons are engaged during the plucking season which lasts from March to November (Bomjon, 2008).

Darjeeling tea not only occupies a place of pride for Darjeeling but for the whole of India. The aroma and taste of Darjeeling orthodox tea is unparallel in the world. There are at present a total of 87 tea estates in hills which have been accorded the status for its produce, as 'Darjeeling Tea' by the Tea Board of India.

Introduction and consolidation of the tea plantation industry in Darjeeling during the later half of the nineteenth century was geared to a demand abroad for an exotic drink. The tea gardens of Darjeeling Hills are one of the only industries which have survived for more than one and a half century. Although most of the tea bushes are more than a hundred years old they are still producing tea with '*Muscatal Flavor*' and '*Exquisite Bouquet*'. Tea from the Darjeeling region has traditionally been prized above all other 'Black Teas', especially in the United Kingdom and the countries comprising the former British Empire.

The credit for the introduction of tea plantation in Darjeeling entirely goes to Dr. Arthur Campbell, who was appointed as the first Superintendent of Darjeeling in 1839. For about a year he spent his time in Kurseong next to 'Constantian' and experimented with the planting of tea saplings obtained from the Calcutta Botanical Garden (Lama, 2008). But in Darjeeling the first trial of tea plant was made only in 1841. By this time Dr. Campbell was also shifted to Darjeeling permanently and experimented with tea cultivation in his Beechwood Estate just below the Municipality Building of Darjeeling Town and proved that the area was ideal for the cultivation of tea on a large scale (Lama, 2008).

Around the same time, a British army officer Captain Samler who had betrayed the crown was hiding in Kurseong along with his men occupied the present Makaibari (Nepali word meaning Maize farm) Tea Estate area and planted the saplings which they stole from Campbell. Samler was finally granted amnesty after he helped the British Crown during the Indian Rebellion of 1857 (Sepoy Munity). Later he became an agent of 'Darjeeling Tea Company' and also the legal owner of Makaibari Tea Estate. It is believed that Samler pioneered the cultivation of tea in Makaibari area which probably showed the real potential of tea in Darjeeling. But there is no recognition of official records to establish the fact. In 1859 before he died, Samler sold the garden to G.C. Banerjee, his assistant. The Makaibari Tea Estate continued to be run by the Banerjee family, and was the only Garden in Darjeeling district which had a resident landlord until 2014, when it was sold off by the family.

Meanwhile Dr. Campbell after shifting out from Kurseong to Darjeeling kept his experimentation on. It was soon found that the plant thrived at the altitude where Darjeeling was situated (about 7,000ft). Having heard of this, many other tea planters from England started flocking into Darjeeling to make a fortune out of it.

The British government soon started distribution of seeds to those who desired to cultivate the plant. Writing in 1852, Mr. Jackson says in his Report on Darjeeling that "I have seen several plantations in various stages of advancement, both of Assam and China plant, and I have found the plants healthy and vigorous, showing that the soil is well adapted for the cultivation. In the garden of the Superintendent, Dr. Campbell, in Darjeeling, in more extensive plantations of Withecombe, the civil Surgeon, and major Crommelian, of the Engineers, in a Lower Valley called Lebong, the same satisfactory result has been obtained: the leaves, the blossom and the seeds are full and healthy; the reddish clay of the sides of the hill at Lebong seems to suit the plant better than the black loam of Darjeeling. This has been the result at and about Darjeeling itself, at a height of 7,000 feet ; but the opinion of Dr. Hooker and of others competent to judge seems to be that there is too much moisture and too little sun at Darjeeling to admit of the cultivation on a large scale becoming remunerative : this objection, however, does not apply to the lower sites of Pankhabari and Kurseong, where plantation of both tea and coffee has been established by Mr. Martin and the plants are now in a highly – thriving condition. In this tract of country, between the Morung and Darjeeling, every variety of elevation and aspect is to be found, and there seems to be little or no doubt that tea cultivation in that tract would answer" (O'Malley, 1907).

These plantations appear to have been merely experimental plots, but by the year 1856 the industry began to be developed on an extensive scale, especially on the lower slopes, as it was believed that the elevation of Darjeeling was too high for the plant to be very productive. By January 1857 tea has been raised from seed at Takvar (A corruption of Lepcha word tak, a hook-thread, and vor, a fish-hook, a name suggested by the curve of the land) by Captain Mason, at Kurseong by Mr. Smith, at Hope Town by a company, on the Kurseong flats by Mr. Martin and between Kurseong and Pankhabari by Captain Samler, agent of the Darjeeling Tea Concern. At this stage development was preceded at a rapid rate. In 1856 the Alubari tea garden was opened by the Kurseong and Darjeeling Tea Company and another garden by the Darjeeling Land Mortgage Bank on the Lebong spur. In 1859 the Dhutaria garden was started by Dr. Brougham and between 1860 and 1864 gardens at Ging (Taken from a Tibetan word meaning the stretched-out slope), Ambutia (Taken from the Nepali word meaning the place of mango trees), Takdah and Phubsering (The word is taken from the Tibetan word. The name is said to be that of a Bhotia Sardar, who first opened out the tea-garden now known by this designation. Properly it is Phurpusring i.e., Sring who was born on Thursday (Phurpu), It being a common practice to name Tibetan children after the day on which they are born) were established by the Darjeeling Tea Company and at Takvar and Badamtam (The word is taken from a Lepcha word. The padam bamboo, the giant bamboo which furnishes the Lepchas with their milk jugs, water-vessels, etc. There was formerly a forest of these bamboos at the place known by this name) by the Lebong Tea Company. Tea gardens like Makaibari, Pandam and Steinthal were also opened up during this period. In the Terai also experimental plantation was started and in 1862 the first garden in the Terai was opened out at Champta near Khaprail by Mr. James White who had previously laid out one of the largest gardens of the District at Singell near Kurseong. Other gardens had been opened in the Terai by 1866 (Dash, 1947).

The earlier planters, owing to want of experience, made many mistakes, and their ventures did not meet with success. But these mistakes were remedied later and in the next ten years steadily increasing prosperity was noticed.

A meticulous study of the plantation industry of India and its spectacular growth shows that in the first phase the tea industry was pre-occupied by the Europeans Entrepreneurs. Amongst the European entrepreneurs it was the Scottish and British planters who have shown their extraordinary courage in founding the plantation industry in the region.

It is an obvious thing to question as to why would Britishers come all the way from England to a place like Darjeeling where no infrastructure was developed as such at that time. But one must notice a fact that in the early days, only those Englishmen who failed to make it as soldiers, sailors, clerks and by default, with nothing else to loose and nowhere to go, took up life as a "tea planter" (Lama, 2008). But at the same time there were many British planters who did take a deep interest and pain in growing the gardens and genuine concern for the welfare of the labourers actually became the real pioneers of the Darjeeling Tea Industry.

Around 1841, a small band of Moravian (German) missionaries had come to Darjeeling to spread Christianity. Later they settled around the Tukvar area with the intention of funding their mission with the 'Sweat of their brow' but they just could not compete with local labour. After the death of their British sponsor Rev. William Stuart, a Baptist, who brought them to Darjeeling, these missionaries, in order to sustain their lives, began planting tea and unwittingly became pioneers in the Darjeeling tea industry. Their gardens are still known today as the "*Padre Kamans*" (Garden of Priests).

It is believed that the Wernicke (Moravian missionaries) family was the pioneers to introduce tea in the Darjeeling district on a commercial basis. One of the original ancestors of this family had personally visited China to obtain the seeds and plants (Banerjee and Banerjee, 2007).

In 1919, after retirement from the Army Lt. Col. Hannangan began his career as a tea planter's assistant to the late Mr. E.A Wernickle, managing Proprietor of the then Bannockburn Tea Estate. Slowly other missionary people also entered Darjeeling, Rev. W. Start was one of them, who actually made the first attempt to reach the hill people through English education. In 1840 he started his mission on the site where now lies by the manager's bungalow of the Takvar Tea. Co. Ltd. Later he also opened a school for Lepchas in Darjeeling (Banerjee and Banerjee, 2007). Soon other missionary groups like Niebels, the Stolkes and others also joined the Wernickes in the noble cause. Thus we can attribute the introduction of educational institutes in Darjeeling to the incoming of the plantation industry in the area.

Among the Indian planters Bipra Das was one of the pioneers named in the plantation of tea. Tea gardens like Gayabari, Tindharia and Mohurgang were entirely his own achievements. Bhagat Bir Rai also planted tea in 1845 in Samripani, which is a division of Dhootriah Tea Estate at present. In 1950 he secured the proper ties of Saurani and Phuguri and planted tea. But his descendants sold all the gardens to European Planters before 1910. Kamal Krishna Haldar of Barrackpur who was the first "Tahasildar" manager of Maharjadhiraja Bahadur Mahatab Chand of Burdwan in Darjeeling also planted Kamalpur Tea Estate near Bagdogra in Terai in late 1950s.

The role played by the pioneers of tea prior to independence of India in 1947, is a saga of courage, entrepreneurship and determination. Sir Percival Griffiths, in his 'History of the Indian Tea Industry' (London, 1967) - describes the first planters as having had 'to hew their way through track less Jungles to cope with disease and the ravages of wild beasts, to recruit and maintain the morale of the workers from distant provinces, and last, but not the least, to learn the technique of tea cultivation and manufacture'.

The following table shows the number of gardens, extent of land cultivated with tea, together with the out-turn, etc, for each of the five years from 1866 to 1870 inclusive, for 1872 to 1874 and for 1885, 1895, 1905 and 2013.

Table 1: Comparative Table of Tea Operations in Darjeeling District, For Years 1866 – 1870, 1872 – 1874 and in 1885, 1895 and 1905 and 2013

Year	Number of Gardens	Extent of Land Under Cultivation in Acres and Hectors	Out – Turn of Tea in Lbs. and kg	Number of Labourers Employed
1866	39	10,392	4,33,715	Not Known
1867	40	9,214	5,82,640	Not Known
1868	44	10,067	8,51,549	6,859
1869	55	10,769	12,78,869	7,445
1870	56	11,046	16,89,186	8,347
1872	72	14,503	29,38,626	12,361
1873	87	15,695	29,56,710	14,019
1874	113	18,888	39,27,911	19,424
1885	175	38,499	90,90,298	Not Known
1895	186	48,692	1,17,14,551	Not Known
1905	148	50,618	1,24,47,471	Not Known
2013	87	17,818 (in hectares)	8.56 (million kg)	Not Known

Sources: W.W. Hunter: A Statistical Account of Bengal (for the year 1866 - 1874) and O' Malley (1907): Bengal District Gazetteers Darjeeling (for the year 1885, 1895 and 1905 figure), and I.T.A. Report on Darjeeling Tea (Indian Tea Association Darjeeling Branch: for the year 2013)

It is evident from the above table that, from 39 tea gardens in 1866, the numbers kept on increasing and went up to 186 gardens in 1895. But by 1905, the number of tea gardens started decreasing however, total production was increasing in the region. This was mostly because of the fact that a numbers of gardens were merged together while the production was not affected adversely. By this time, tea trade became the staple means of livelihood of the people (Sharma and Das, 2008). It is also noticed from the above table that after 1885 the workers were not taken into account on the ground that by then the industry did not require more labourers as they started to get sufficient workers from the local areas also. Secondly, the planters started introducing the system of *Hattta Bahira* (Taken from a Nepali word meanint oust the workers after their superannuation). Since 1940 production increased considerably in spite of difficulties with management, transportation and costs. In 1942 the output was 26,478,500 lbs of black tea and 1,24,200 lbs of green tea. In 1943, the amount of black tea produced was 25,593,000 lbs and 2,572,500 lbs of green tea (Dash, 1947).

Table 2: Distribution of Tea Gardens in Darjeeling 1940

Sl. No.	Thana	No of Tea Estates
1.	Darjeeling	19
2.	Jorebungalow	16
3.	Sukhiapokri	9
4.	Pulbazar	2
5.	Rangli Rangliot	9
6.	Kurseong	25
7.	Mirik	5
8.	Siliguri	27
9.	Kharibari	11
10.	Phansidewa	13
11.	Kalimpong	0
12.	Gorubathan	6
	Total	**142**

Source: A.J. Dash: Bengal District Gazetteer, 1947.

Only in the Kalimpong Subdivision (which was taken from Bhutan in 1866) land was withheld from development under tea, as Government's policy was to reserve that area for forest and ordinary or other cultivation.

Table 3: Valley Wise List of Gardens Producing Darjeeling Tea (2013)

Darjeeling East	Darjeeling West	Kurseong North	Kurseong South	Mirik	Rongbong	Teesta Valley
Singtom	Badamtam	Rungmook	Longview	Gayabaree	Gopaldhara	Tukdah
Orange Valley	Ging	Oaks	Makaibari	Singbulli	Chamong	Rangli Rangliot
Reeshihal	Bannockburn	Ringtong	Goomtee	Phuguri	Selimbong	Namring
Chongtong	Phoobsering	Margaret's Hope	Jungpana	Soureni	Sungma	Geille
Lingia	Pandam	Singcll	Mahaldiram	Thurbo	Nagri	Teesta Valley
Tumsong	Tukvar	Monleviot	Mohan Majua	Okayti	Nagrilarm	Samabeong
Mim	North Tukvar	Amboolia	Malootar	Seeyok	Avengroove	Ambiok
Pussimbing	Vah Tukvar	Springside	Sivitar		Dhajea	Mission Hill
Dootcriah	Soom	Castleton	Nurbong		Turzum	Kumai
Kalej Valley	Happy Valley	Dilaram	Nabarda Majua			Upper Fagu
Poobong	Rungneet	Edenvale	Rohini			Glenburn
Moondakotee	Rangaroon	Balasun	Giddapahar			Peshok
Stienlhal	Aloobari		Tindharia			Lopchu
Liza Hill	Barncsbeg		Sepoydhura			
Arya			Selim Hill			
Mary Bong						

Source: Status Report on Darjeeling Tea by J.P. Gurung, conducted by Gorkha Territorial Administration (G.T.A).

At present there are 87 tea gardens spread across roughly 19,000 hectares (46930 acres) of land area, employ about 52,000 permanent workers, and 15,000 contract employees who are mostly of Gorkha origin. The gardens collectively produce about 10 to 11 million kilograms *i.e* 22 million Pounds of tea every year.

The largest tea concern in Darjeeling District was that of the Darjeeling Tea Company Limited which used to own about four gardens, established between 1860 and 1864. These gardens were Ambutia, Ging, Takda and Phubserang. The headquarters of the company was in London, its local management was vested in the hands of superintendent, with five European assistants. The total area held by the Company in 1872 was 8547 acres, of which 1300 acres were under plantation. The number of labourers employed on the Company's garden was on an average, one to every acre of cultivated ground. This was the average for the year, a larger number of hands were employed during the manufacturing season, from March to November, and a smaller number during the months when no tea was made. The labourers were paid at the rate of about Rs.3 per month for children, upto Rs.5 or Rs.5.8 for able-bodied men. As a rule the labourers were readily procurable; the majority was Nepali immigrants, the remainder being made up of Lepchas, Bhutias, and tribals/ adivasis from the plains. These people were encouraged to settle down permanently on Company's gardens, by assigning to them small plots of land unsuited for tea, for the cultivation of cereal crops, such as maize, millet etc. (Hunter, 1874).

The following table indicates the prices of Darjeeling tea from 1910 to 1940. But at present the prices of Darjeeling tea varies from one tea estate to another, prices ranging from Rs.600 to Rs. 8,000 per kg approximately, based on I.T.A (Indian Tea Association) Report, 2013. It was reflected from reports that prices of Darjeeling tea from 1910 to 1940 were fluctuating (Table 4).

Table 4: Prices of Darjeeling tea from 1910 to 1940

Sl. No.	Year	Prices at Calcutta Auction (Rs./ Pound)
1.	1910	8.9 (6.5 to 10.3)
2.	1915	10.9 (7.11 to 11.10)
3.	1920	7.5
4.	1925	16.0
5.	1930	14.9
6.	1935	12.2
7.	1940	16.0

Source: A.J. Dash, Bengal District Gazetteers Darjeeling, 1947

Most of the tea plantation areas in Darjeeling have the China variety, which was for many years considered the only kind suited for the production of fine tea. Of late, the variety known as the "Assam indigenous" has been much in favour, and is certainly capable of producing the very finest tea; but is very delicate, and with anything like rough treatment soon becomes so weak as to be not remunerative. A good hybrid from these two varieties later proved most suitable all round. Some fields were even planted with the "Manipur indigenous" which was the most hardy of all the varieties, and gave a good yield, but the tea produced was almost invariably coarse and rank in flavor (O'Malley, 1907).

Recruitment of Workers

One of the biggest problems faced by the planters at the time of the introduction of the tea industry was the recruitment of large numbers of labourers, as plantation industry is a labour intensive industry. During that time, Darjeeling region was sparsely populated. So the Britishers had to recruit labourers from various parts of India and the neighboring countries. Another reason for this was that the Sikkim Rajah had forbidden his subjects to work for the British and Lepchas were far too independent to bother about jobs as their needs were few their wants even fewer and were happy with their simple way of living. So, under compulsion Britishers forcibly rounded up natives from the Chotta Nagpur hill forest areas of the Deccan Plateau and brought them to Darjeeling as bonded labour to work on the tea plantations. As they came from the same altitude; the Britishers presumed that they could adapt to the weather, but they could not cope with the cold and damp of the Darjeeling hills so finally left from the hill areas and ultimately settle down in Terai areas and many of them went as far as Dooars region. Again the Britishers had to face labour problem. This time they turned to Nepal for Nepali workers who were known to be famous for their cheerful and hardworking nature.

But again because of poor sanitation, improper water supply and inadequate medical treatment, plantation workers used to run away from the gardens and managers went around the villages with money in bags to allure the workers and discourage them from running away from the gardens. This was the annual occurrence during the winter months as then having no bridges in the rivers, workers used to run away in winter times from the gardens (Sharma and Das, 2008).

Recruitment system in the tea plantation of North Bengal region was different from that of Assam. The Labourers were never placed under any contract and in that sense they were free. Recruitment in this area was done mainly through garden *Sardars*. The *Arkattis* who were the intermediaries (recruiters) between the planters and the laborers played an important role in the recruitment of laborers. The Sardars on the other hand were tea garden labourers and not local recruiters like Arkattis. They were sent to the recruiting grounds in the recruiting season which generally began after the rains in October or November and ended in February.

Each garden had a team of *Sardars* and *Gallawalas* (labour recruiters) who went to Nepal and Sikkim to recruit laborers during the winter season. The Darjeeling Planters Association even had to bring out a standing order to stop enticement of workers of one garden for another garden. The recruiters used to get Rs.10 each respectively as recruiting bonus. After being brought to the gardens, the workers were tied down in such a way that they became almost like bonded labor. In order to achieve this end a para-military force known as North Bengal Mounted Rifles was kept at the plantations (Sharma and Das, 2008).

Around 1907, the average rate of wages for men was Rs.6, for women it was Rs.4 to 8 and for children Rs.3 to Rs.2 per month; but in addition to wages, they used to get living quarters in the garden, often with water laid on and free medical attendance and medicine (O'Malley, 1907). But the present scenario is very different with workers being denied of even the basic facilities like plastic shoes (Gumboot), umbrellas made out of bamboo (*Ghum*), canvas to wrap their clothes during heavy rain (*Barsati*), basket to carry tea leaves (*Doko*) etc.

Today more than 30% of Darjeeling's land is under tea plantation. The industry provides employment directly and indirectly to about 50% of the population in the district. And 50% of the directly employed workers are women (Rai and Chakroborty 2012). Next to Assam, West Bengal is the second largest tea growing state accounting for 22% of the total area under tea cultivation and 22% of total production of the country (Mitra, 2010).

Conclusion

Tea was one of the most important items of the British consumption. And it is for this reason that the British government took extra effort in the development and promotion of it. But over the years especially after 1990s, Darjeeling tea is not only losing its position in the global market but also its colour and flavour. Various reasons have been attributed to it by Miss Smritima Diksha Lama (Lama 2012) in her article 'Darjeeling Tea Industry: Implications of Globalisation Triggered Fair –Trading'. They are as follow:

- Emergence of new growers like Vietnam, Indonesia and Kenya resulting in an over supply of tea in the international market.

- Indian tea is loosing its position in the export market as a result of high production cost and poor quality.
- Higher production cost as a result of expenditure on fixed expenses like fuel, power and labour.
- Tea is mostly sold through auction in which price realization is doubtful as the brokers are said to be in cooperation with the big buyers to keep prices low.
- Existence of higher percentage of ageing bushes leading to decrease productivity and degradation in quality.
- There also appears to be a lack of sufficient and up-to date statistics regarding the tea sector without which proper planning and fund utilization is not possible.

Keeping this critical crisis situation of the Indian tea industry in perspective, when we look at the Darjeeling tea industry, we find that the Darjeeling tea industry has been facing much varied and localized crises that need very specific intervention and specialized redressal (Lama, 2012).

The other problems faced by the tea plantation in the Darjeeling hills pertains to the tea garden workers who have been marginalized in many aspect of their social and economic life. During the British period, the tea plantation workers of Darjeeling hills were treated as *'Kamanee* or *Coolies'* (Kamanee or Coolies: Nepali term used to refer the Tea Garden Labourers) living under the complete sovereignty of 'Gora Sahibs' (Gora Sahib: Nepali term used to refer to the European White Tea managers) assisted by ignorant 'Sardars' who use to lure their respective caste and tribal workers from the agrarian belts of Darjeeling, Dooars, Terai, other places of Eastern Himalayas and even from Nepal. The Gorkha Sardars felt fortunate to get the post of Sardars at the plantations as they got '*Bakshis'* (Bakshis: Nepali term used to denote reward in terms of money or kind) for recruiting workers/labourers for the management. Even today the management indoctrinates the workers with the idea that the tea plantations are their colonies and estates where workers are taught to be the slaves of the managers (Gurung, 2014). In addition to this, the population in the plantation industry started growing and in order to meet the situation the planters started implementing 'Contract Clause' of Bengal Act III of 1915 which provided not only for the recruitment of labourers but also incorporated provisions of labour retrenchment. Thus from 1930 onwards, many workers were retrenched from tea plantations in the name of '*Hatta Bahira'* system (Sharma, 2003). The system of *Hatta Bahira* might have thrown the labourers into more precarious condition. Although Darjeeling Plantations were not characterized with indentured labour like Assam but there existed to some extent, coerced labour (Rasaily, 2003).

Theoretically Darjeeling tea industry is the mainstay of the economy in the hills and therefore promises to provide a rewarding way of life which provides to its workers a steady livelihood, housing, statutory benefits, allowance, incentives, crèches for infants of working mothers, children's education, integrated residential medical facilities for employees and their families and many more. But in reality this is not the case. The labourers, who are the heart and the soul of the plantation industry still do not have the basic facilities like housing, education, allowance, incentives, and medical facilities, which they deserve. Although these people have been living in the areas for more than hundred years they still do not have any documents of their land holdings. So, if any worker goes against the management, they have the right to throw them from their land holdings. In 1951 government of India had tried to take an initiative by passing the Plantation Labour Act in order to address the sufferings of the laborers, but it hardly had any effects on them. It is high time now that both central and state governments should work on the various issues of the plantation laborers and try to solve them.

Introduction of the Tea plantation industry in the later half of the nineteenth century was one of the major events in the history of the Darjeeling region. It was soon followed by the other major activities like introduction of Railways, construction of roads, ropeways, bridges etc and finally the establishment of missionary schools and colleges. The construction of the railways and roads in Darjeeling, as in other parts of the country was introduced by the British in order to enhance their economic benefits by being able to transport raw materials, cash crops and plantation crops from the remotest villages to the presidencies of British India. Thus we can say that the development of Darjeeling in the nineteenth century was largely due to the introduction of the plantation industry in the region by the Britishers, though they were only interested in the profit rather than the growth and development of the area. However, the present scenario reflects a sad picture, tea plantations which was the very reason to have initiated the development and modernization process to Darjeeling is now lagging behind as underdeveloped and marginalized in the Darjeeling region. The wages of the tea garden labourers remains a meagre Rs 122.50 per day even after many efforts by the workers' union and political parties to increase the wages of the workers. It is high time that the state and the central government arise from their slumber and give Darjeeling Tea, the area and the Tea Garden workers the due that they deserve.

References

Adikesavan A.G. (1975). Economic Products Gallery, The Director of Meseum, Government meseum- EGMOPE, Madras.

Annual Report of I.T.A. (Indian Tea Association, Darjeeling Branch) for the year 2013.

Banerjee B. (1993). Tea Production and Processing, Oxford and IBH Publishing Co. Pvt.Ltd, New Delhi.

Banerjee G.D. and Banerjee S. (2007). Darjeeling Tea – The Golden Brew, International Book Distributing Co, Lucknow.

Bomjon O.S. (2008). Darjeeling – Dooars People And Place Under Bengal's Neo – Colonial Rule, Bikash Jana Sahitya Kendra, Darjeeling, 2nd ed (Revised Updated and Enlarged Edition).

Chakravorty R.N. (1997). Socio – Economic Development of Plantation Workers in North East India, N.L. Publishers, Dibrugarh, p28.

Dash A.J. (1947). Bengal District Gazeteer, Alipur, Darjeeling, Bengal Government Press.

Griffith P. (1967). The History of Indian Tea Industry, Routledge, London.

Gurung J.P. (2014). Status Report on Tea, G.T.A. (Gorkha Territorial Administration), Darjeeling (unpublished)

Gurung P. (2014). 'Marginalization of Tea Workers in Darjeeling', In: Ray D.C. and Chhetri B. (eds.) Discourses on Darjeling Hills, Gama Publication, Darjeeling, p157.

Hunter W.W. (1874). Statistical Account of Bengal, Government Printing Press, Calcutta.

Khawas V. (2006). "Socio-Economic Condition of Tea Garden Labourers in Darjeeling Hills", Council for Social Development, Sangha Rachana.

Lama B.B. (2008). Through The Mists of Time: The Story of Darjeeling, The Land of Indian Gorkha, Kurseong.

Lama S.D. (2012). Darjeeling tea industry: implications of globalisation triggered fair trading. In: Lama M.P. (Eds.), Globalisation and cultural practices in mountain areas: dynamics, dimensions and implications, Indus Publishing Company, New Delhi.

Mitra D. (2010). Globalization and Industrial Relations In Tea Plantations – A Study On Dooars Region of West Bengal, Abhijeet Publications, Delhi.

O'Malley, L.S.S. (1907, Reprint 1999). Bengal District Gazetteers Darjeeling, Logos Press, New Delhi.

Rai R. and Chakroborty S.R. (2009). 'Two Leaves and a Bud – Tea and Social Justice in Darjeeling' In: Bose P.K. and Das S.K. (eds.) State of Justice In India- Issues of social Justice, Volume 1, SAGE Publication, India Pvt. Ltd. New Delhi.

Rasaily R. (2003). Labour and Health in Tea Platations: A Case Study of Phuguri Tea Estate, Darjeeling, Unpublished PhD. Thesis submited to the Jawaharlal Nehru University, Centre for Social Medicine and Community Health, School of Social Sciences, New Delhi.

Sharma K. and Das T.C. (2008). Agony of Plantation Workers in North-East India, Kalpaz Publications, Delhi.

Sharma K. (2003). Tea Plantation Workers: In A Himalayan Region, Mittal Publications, New Delhi.

Wrigh G. (2011). The Darjeeling Tea Book, Penguin Enterprise, New Delhi.

Tea: Technological Initiatives, pp. 19-38
New India Publishing Agency, New Delhi, India
Edited by Niladri Bag, Arundhati Bag and L.M.S. Palni

2

Small Farmers: Locating them in "Darjeeling Tea"

Roshan P. Rai

Abstract

"Darjeeling Tea" as a global brand brings attention to the Darjeeling District and creates a global imaginary which obliterates the dark colonial history, existing exploitation of the workers and the natural resources of the region. The global brand name is protected under the certification trademarks which has a geographical indication. Thus only tea from the 87 tea plantations can be called Darjeeling Tea. There is an increasing trend towards Darjeeling Tea going organic and fairtrade. This trend is critiqued with questions of how green and fair can a monoculture plantation get. Within this imaginary, small farmers' tea offers an alternative paradigm based on equity and diversity. This paper explores the alternative by highlighting the (United Development Organization) journey of a small farmers collective Mineral Spring SanjuktaVikashSanstha.

Key Words: *Darjeeling Tea, Tea Plantations, Small Farmers' Tea, Organic and Fair trade.*

Introducing "Darjeeling Tea" a Conscious Euphemistic Approach

"Darjeeling Tea" brings global attention to Darjeeling, a tiny district in the Himalaya, tucked in the north-east of India. This construct around Darjeeling Tea overrides socio-political complexities of the region and commodifies the lives of the people and Darjeeling in this global imaginary. Locating small farmers in this global imaginary is a daunting task as it is not inclusive of small farmers'.

Darjeeling is the only hill district in the state of West Bengal. It was inhabited by the Lepchas and Nepali-speaking residents who also lay claims to the landscape (Chettri, 2013; Golay, 2006).But the ones who have left an indelible mark on the region are the British who came into the region after the East India

Company obtained the lease of "a small strip of country in the south of the Sikkim Himalaya for the purpose of a sanatorium and an outpost of strategical importance on the northern frontier of India" (Darjeeling Gazetteer, 1907). The British influence in the region's socio-economic profile still continues today.

Darjeeling is socio-ecologically different from the rest of Bengal and coupled with regional marginalisation, has led to demand for separation from Bengal which begins before Indian Independence with the demand first placed in 1911. The demand continues today and with the recent agitation, has resulted in the creation of the Gorkhaland Territorial Administration, 2011, an autonomous governance institution with West Bengal. It differs in physical boundaries of the Darjeeling District and comprises mostly of the hill sub-Divisions of the district.

Dr. Arthur Campbell and Lieutenant Robert Napier "set to work to fell the forest and lay the foundations of the hill station of Darjeeling". Dr. Campbell, who later became the Superintendent of Darjeeling, and Major Cromellin was particularly influential in establishing the experimental tea stations with seeds from China.(Darjeeling Gazetteer, 1907) By 1866, Darjeeling had 39 tea plantations producing a total crop of 21,000 kg of tea. In 1870, the number of plantations increased to 56 producing 71,000 kg. By 1874 there were 113 plantations in approximately 6000 hectares. (Khawas, 2011). The Darjeeling Gazetteer, 1907, notes that "natives of the surrounding country quick to avail themselves of the blessing of life" under the new administration of the East India Company provided the major work force in the tea plantations. According to the Tea Board of India, at present Darjeeling produces 10 million kilograms tea year, spread over 17,500 hectares of land in 87 registered tea plantations.

The Tea Board of India, the sole official authority for tea in India, uses a description of Darjeeling Tea which constructs the imaginary to be "as exotic and mysterious as the hills themselves. First planted in the early 1800s, the incomparable quality of Darjeeling Teas is the result of its locational climate, soil conditions, altitude and meticulous processing. The tea has its own special aroma, that rare fragrance that fills the senses. Tea from Darjeeling has been savoured by connoisseurs all over the world." This narrative has been evolving over the years and is an attempt to gloss up, hide and recreate a unique story of Darjeeling Tea resulting from a complex relationship of the tangible and the non-tangible forces. "A product of the environment which goes beyond human control and management yet at the same time crafted by the women pluckers whose lives are dedicated only to make the tea." In this description there is also a connotation of linking it to a religious belief but I will not go to that length and quote it. The final part of the construct is the way only a special few "connoisseurs" enjoy it which relegates others to the ordinary, uninitiated or not good enough to partake. The official description is part of the narrative which creates the myth and mystique of Darjeeling Tea.

The author intends to deconstruct this "Darjeeling Tea from the 87 tea plantations" and pose the question of the relevance of small farmers' tea within it. This question has a relevance with background of "Tea Board allocates Rs. 300 crores for small growers in 12th Plan (Sundar, 2012). Small tea growers break new ground with major allocation in 12th plan (Dutta, 2014)." These news items show that the think tanks of the country not just feel that small farmers' tea needs to be promoted but has allocated financial investment for the promotion. At a personal level, the question of the relevance of small farmers Darjeeling Tea is based on the author's relationship with small farmers' tea collectives in Darjeeling since the late 1990s through his work with DLR Prerna (an NGO, Darjeeling, www.darjeelingprerna.org) and one of these relationships with Mineral Spring is used as a case to argue for small farmers' tea. Some of the analysis on Darjeeling Tea is built on the authors' dissertation on Plantation Labour Act Implementation (1995) and a co-authored paper on Tea and Social Justice in Darjeeling (2009). The paper proposes that small farmers' tea presents a socio-ecological paradigm shift and offers more equitable and sustainable regional development pathways than the existing Darjeeling Tea paradigm.

Romance Hides Reality – Deconstructing "Darjeeling Tea"

Deconstructing the official language of "Darjeeling Tea" entails removing the romance and conducting a reality check. "A tradition steeped in history" actually means that the history of Darjeeling Tea is steeped in colonialism, a product of the exploitative and extractive relationship that the colonial masters exerted on the native slaves. The introduction and expansion of Darjeeling Tea in the mid-1800s (The Darjeeling Gazetteer, 1907) questions the indigeneity of the tea to Darjeeling. Darjeeling Tea has a history of about 100 years only but the portrayal projects indigeneity. The romance hides the fact that large tracts of indigenous forests were cleared for plantation tea, taxonomically described as *Camelia sinensis* syn. *Thea chinensis,* which was brought from China.

More than half a century after independence the quality of life of the over 52,000 permanent workers (Workers means a person employed in a plantation for hire or reward, whether directly or through any agency, to do any work, skilled, unskilled, manual or clerical, having wages less than Rs. 750/- p.m., but does not include – medical officer, managerial staff and temporary workers.(Hind Mazdoor Sabha-2002) and 15,000 engaged temporarily in plucking (Tea Board of India) has not improved. 60% of the workers are women who form the bulk of the lowest rung in the worker hierarchy. The workers have remained in pretty much the same unjust situation since the beginning of the tea plantations over six generations ago. The estimated USD 7.5 million (Tea Board of India) annual turnover of Darjeeling Tea does not necessarily imply that the wages and lives of the workers are good or fair. In the recent years an increasing amount of literature documents the unjust and exploitative conditions of plantation tea in general and includes Darjeeling Tea (Besky, 2008; Columbia Law School Human Rights Institute, 2014).

The tea workers in Darjeeling get paid a wage even lower than the Rs. 120 daily wage under Mahatma Gandhi National Rural Employment Guarantee Act. Thus we have a glossy annual turnovers and "booked orders at a record price of $1850 per kg" (Tea Board Announcement, 2014) juxtaposed with below minimum daily wage of tea workers. The portrayal of "an idyllic existence close to nature's heartbeat" hides the culture of violence and exploitation.

The fact that large amounts of tea not grown and produced in Darjeeling are being sold as Darjeeling tea has led to its protection.The "Darjeeling Logo" and "Darjeeling" have been registered under the Certification Trade Marks with Geographical Indication as being tea from the 87 tea gardens in Darjeeling above 2000 metres above mean sea level. This protectionism is non inclusive of small farmers, even though they might share boundaries and have the same tea stock as the plantation. So one can forget about the larger Darjeeling hills who have the similar geo-physical and ecological features to be included in the Darjeeling Tea. I use the term 'similar' and not the same as constantly communicated that every plantation and every block of the plantation is unique.

Along with the trade protection policies, there is a romantic representation of Darjeeling tea as a traditional craft, "The crafting of Darjeeling Tea begins in the field. Where women workers begin plucking early in the morning, when the leaves are still covered with dew. The spirals of walking women gradually twist, then unfold to form a line. The tea is picked fresh every day, as fresh as the crisp green leaves can make them. The tea bushes are mystic messages on the Earth's canvas. A tale of excellence, brewed cup by cup, produced with the loving care lavished by the workers. Caressed to state-of-the art perfection by unchanging tradition." This representation presents a picture of hand crafting and enhances the role of the women, when most of the processing that gives the tea its characteristics is done in a large factory. It also creates an illusion of the ownership of the process by these women who have owned the tea and the process since time immemorial when the fact is that tea expanded rapidly in the mid-1800 after it was introduced by the colonial masters. To add insult to injury the romance continues in this description of, "The women pluckers smile and, with the radiance of their joy, the sun rises over the plantations. Every morning, as the mist rises from the mountains, women tea pluckers make their way up the steep mountain paths towards the 87 fabled plantations that have been producing the highly prized black teas of Darjeeling." Plucking is an extremely back breaking job especially in steep mountain paths and this is where most of the women are employed. The industry projects itself as employing more than 60% of the workforce to be women but conveniently fails to mention their job description and difficulties. The bulk of the picking is in the monsoon when the sun does not shine on most days and it rains days on end. Even if the sun does shine, picking tea is a physical hardship and mentally repetitive and unchallenging and as stated above financially without remunerative justice. The smile that a women worker has is for the camera or as the famous Nepali song goes, "*mu to mathi dhungarakhi hasnu parya cha*" (I smile with a rock on my heart).

Fig. 1: A typical tea plantation in Darjeeling
Photo: Author

Increasingly these plantations are being referred to as gardens. "Located on grand estates veiled in the clouds, the gardens are in fact plantations that, at times, stretch over hundreds of acres. But, they are still 'gardens', because all tea grown here carries the name of the estate, or garden, in which it is grown". Plantations and their dark history of colonial exploitation is being removed with usage of "gardens". It is recreating a history minus its dark side. Reference to "gardens" also creates a picturesque imaginary of colour which is snug and personal where as a plantation is an industrial monotone.

"A melody of greenness surrounded by blue skies and the sparkle of the mountain dews. And tied to the circle of life, the tea bushes sustain themselves day in day out, season after season, through the years. Life on a plantation is a completely natural, refreshing state of being." This hides the environmental history of the clearing of natural forest of Darjeeling to plant tea from China. It also attempts to greenwash large stretches of industrial mono culture plantations as being natural. It does not talk about the extensive toxic synthetic agro-chemical use in the tea industry as has been highlighted in the "Trouble Brewing, Pesticide residues in tea samples in India, Greenpeace, 2014[1]. Monoculture plantation tea is the least sustainable of natural systems with low productivity per area at an ecosystem level and high potential risks whether it be pests or climate variability.

Mineral Spring *Sanjukta Vikash Sanstha* – a Journey of a Small Farmers' Tea Collective

Darjeeling Tea, the single largest employer in the district and grown in 17,500 hectares of landis controlled by a handful of people. Access to the 87 plantations club is extremely limited even if one falls within Darjeeling.Thus even if one grows tea but one is not part of the clique of "Darjeeling Tea" and the club of 87 plantations, it is extremely difficult to be a part of the institution[2]. Hence, in spite of the increasing popularity of small farmers nationally and internationally, Darjeeling Tea merely pays lip service as there are very few examples of small farmers' Darjeeling Tea.

These few small farmers occupy the shadowy spaces of "Darjeeling Tea" and the grey areas of the so-called illegal hand-made tea that is sold locally. Some who have become a bit successful took the initiative to organize themselves

[1]An investigation carried out by Greenpeace India has found residues of hazardous chemical pesticides in a majority of samples of the mainbrands of packaged tea produced and consumed in India. Over half of the samples contained pesticides that are 'unapproved' for use in tea cultivation or which were present in excess of recommended limits. A total of 34 pesticides were found, with 46 samples of branded tea – or 94% - containing residues of at least onepesticide. 59% (29 of the samples) contained 'cocktails' of more than 10 different pesticides, including one sample which contained residues of 20 different pesticides." This report has been rubbished by the Tea Industry. Tea Board has said all teas from India are within the stringent laws of the country and has come out with a Plant Protection Code, 2014 which aims to achieve sustainability through Good Agricultural Practices (GAP) including integrated pest management, promotion of alternative control strategies (Biological control etc.) to gradually reduce the dependence on chemicals. PPC shall focus on responsible chemical management that includes proper selection, judicious usage, safe storage and proper disposal, occupational health and safety and green chemistry. PPC is committed to minimising the possible negative impact of pesticides.

[2]Similar examples of regional protection is quoted by Darjeeling tea and the examples that are used are Champagne and Scotch. In both the instances the geographical indication is not limited to large corporate houses and the trademark is based on quality. Thus one has family owned, co-operative and small owners who produce Champagne and Scotch as they are able to maintain the quality required. Within the Darjeeling context, only the 87 tea plantations have access to the trademark so even if a small farmer grows the same Chinese stock of tea right next to a tea plantation, the small farmer is unable to call it Darjeeling irrespective of quality.

into a collective which gave them a stronger voice and more bargaining power.

A large number of tea plantations closed in Darjeeling post-independence due to various reasons. This included the war effort and independence movement which saw the departure of British companies. The new Indian companies were speculative without long-term developmental goals and operated with extremely centralised management systems. The changing political, financial and legislative environment and worker awareness and demands for rights also played a crucial role in the closure of tea plantations. The closed plantations were converted to agriculture but retained some of the tea. Besides the tea from these closed plantations, there has been some interest towards growing tea amongst the small farmers in Darjeeling for its steady cash flow, lower maintenance efforts as well as a strategy to manage increasing human wildlife conflict[3].

Fig. 2: Bird's eye view of Dabaipani and Yangkhoo photo: DLR Prerna

The evolution of Mineral Spring *Sanjukta Vikash Sanstha* or Mineral Spring United Development Association (MSSVS) is one example of a small farmers collective that evolved after the closure of Harrison's Tea Company, and last registered as Lebong and Mineral Spring Tea Company.

MSSSVS is made up of Harsing, Dabaipani and Yangkhoo *Busties*(villages) which lie on the Lebong Spur of the Darjeeling-Jalapahar Range, one of the great hill ranges radiating northwards towards Darjeeling town from the central point, a saddle at Ghoom. Harsing *Busty* is located 10 km away from

[3]Tea in the Darjeeling Hills come in 4 spurts of growth in a year called 'flushes' and this is spread around nine months thus giving a steady year-long cash income. Once a tea bush settles into a soil, it has shown to last over a 100 years in the Darjeeling Hills and especially in a small farmers context, nutrient flow from annual crops provide the tea bush with its requirements and most maintenance work of pruning is done in the few winter months only thus the work load in comparison to annuals is high reduced. The hills of Darjeeling have been experiencing increased influx of wild animals into agricultural land and destroying crops. Many are resorting to planting tea as none of the wildlife that are engaged in the conflict feed on tea thus planting tea ensures crop survival and protection from this conflict.

Darjeeling town. Yangkhoo and Dabaipani are further away across the valley. The extremity of Dabaipani would be 15 km from Darjeeling town. These *Busties* form the major portion of what constituted the Harrison's Tea Estate owned by the Harrison's Tea Company established by Mr. Harrison. The tea plantation was large by Darjeeling standards with 1200 workers. The plantation in its last registration a gross acreage of 575, as Lebong and Mineral Spring Co. Ltd, – Registration Number 1973 had acres under tea. Documents show that the plantation had income from cardamom and forest resources too. Thus the tea plantation was well endowed with both natural resources and a fair income.

The name Lebong and Mineral Spring has a special significance. The Mineral Spring face of the plantation is known in Nepali as *Dabaipani* or 'Medicine Water'. Legend has it that an Englishman with festering sores washed himself in one of the natural springs and was cured. Therefore Mineral Spring or *Dabaipani.*

Between 1952 and 1960 the Lebong and Mineral Spring tea plantation had three closures, 1952-53; 1955-56; 1957-60. An interesting point is that all the closures were after the plucking season. The closures can be attributed to mismanagement of the plantation on listening to the stories by the people in Mineral Spring who were workersor young adults in the tea plantation at the time of the closures.

The workers of the tea plantation survived by selling tea leaves to the neighbouring plantations during the plucking season as well as working in there. At other times, they felled trees in the tea plantation reserve forest and sold firewood and charcoal. The people lived in extremely difficult circumstances and hoped that the plantation would reopen. But, by 1962, the reserve forest had exhausted and the hope of the tea plantation re-opening had receded. The people began to partially uproot the tea bushes and started to cultivate other crops.

Post 1965, people started "Land Grabbing[4]" in spite of their fear of appraisals from the management. The people distributed the land among themselves which gave birth to new settlements. The distribution was done arbitrarily, brain and brawn being the only criteria for the size of the land grabbed. The elders remember only the verbal disputes at the time of distribution. This could be because there was land enough for all.

The people of these *busties* were descendants of daily wage labourers in the

[4]The term "Land grabbing' is being used here as is officially described, as tea plantations are registered for a lease from the government for 33 to 99 years and the workers have no land registered under them including their living quarters.

tea plantation. They were engaged throughout the year in the plantation at various stages of tea production. They depended solely on the tea plantation for their livelihood and were not engaged in any other productive economic activity. Thus, agriculture was an alien lifestyle for them in which they failed miserably. They began to grow maize and millet. The production was very low. The lack of knowledge of cultivation and the infertility of the soil from long tea monoculture were the causes of such low production.

Cultivation of monoculture tea over a long period renders the soil unsuitable for agriculture. This is because tea an exotic[5] plant, was primarily grown as a monoculture plantations by the colonists. Even after uprooting the tea bushes it takes years before the fertility of the soil is regained as monocultures plantations tend to be extractive on a system. Tea is cultivated in slopes whereas for settled tillage agriculture to be viable in the hills, one has to practise terrace cultivation. It takes expertise to cut terraces in the hills otherwise one has to pay heavy losses due to soil erosion. Traditional knowledge on terracing had been lost during the plantation labourer days.

The people supplemented their subsistence income by selling milk. Most of them had bought cows with the loans obtained from middlemen at very high interest. The interest rates ranged from 72 percent to 120 percent per annum. The milk was bought by the same middlemen. The price paid for it was a mere 44 paisa per litre. The annual income per family of Harsing *Busty* only was approximately Rs. 600/- (National Social Service, 1971).

No government help was forthcoming because of the tea plantation status. The Plantation Labour Act, 1951 a Central Government Enactment governs plantation life with West Bengal Rules 1956 for implementation.The preamble to this Act aims at providing for the welfare of labour and to regulate the conditions of work in the plantations. But the Act does not have provisions for action after closure or lock-out of the plantation. Every aspect of tea plantation life falls under the purview of this Act where responsibility of social benefits fall on the owner of the tea plantation, so, the government development and social welfare schemes under the Panchayati Raj Institutions were not undertaken in the plantations till the year 2000 in the Darjeeling Hills[6]. Mineral Spring, in spite of the closures of the tea estate in 1952-53; 1955-56; 1957-60, legally designated as tea plantation, was deprived of government development and welfare

[5] Exotic plant is a plant introduced to an area from outside its native range, intentionally or accidently. The tea in the Darjeeling Hills were introduced by the British from China.

[6] There is limited Panchayati Raj Institution schemes in tea plantations as most of them are land based and land is not in the hands of the workers in a tea plantation.

schemes. The people, therefore, had been rendered virtual destitute in their own land, bonded to the middlemen and resigned to their fate.

Through the Kissan Sabha in 1977, official measurement and distribution of land among the people were initiated. On 31May, 1983 the first Panchayat election took place.

The transformation into agricultural communities in the unorganised sector with individual initiative and operation has been a difficult proposition. From a mindset of a cog in the wheel of the lowest order in a plantation, to owners of land and agricultural entrepreneurs, the journey has required behavioural changes and skills that were not easily acquired. From a lowly worker to the owner of land assets, deciding upon it and drawing sustenance has been a major challenge. To decide and plan for oneself and the community from being a group of workers taking orders brought about responsibilities that were not there previously.

The transition period was a tumultuous one for the people of Mineral Spring and this has never been fully documented. One can gauge the difficulty of circumstances when one hears of the SOS call that led to the first civil society intervention by the National Service Scheme students and teachers of St. Joseph's College and Hayden Hall Institute Darjeeling. The initial intervention were relief camps that had to be organised to address issues of malnutrition and ill health. Teachers recall the pitiable conditions of ill health, poor economic condition and exploitation. A stark reminder of the dehumanising process is the inability of the people to communicate with outsiders in the days when the relief programme was being undertaken. While civil society intervention ameliorated the conditions of the community, but one also notes the inability of the government machinery to address the situation.

In 1973 with the intervention of Hayden Hall Institute, Darjeeling, a dairy union was established and supported with medical and adult education programmes. In 1996 with the intervention of DLR Prerna, a Darjeeling-based NGO, the people of Mineral Spring initiated the proposed[7] *Sanjukta Vikas* Co-operative with milk as its first product. DLR Prerna has been actively partnering with the collective since 1996 and at present engaged primarily in knowledge management.

Mineral Spring *Sanjukta Vikas Sanstha* is a registered organisation today and has a membership of 456 families. The governance structure is based upon

[7] The term "proposed" is used as the officials in the co-operatives department were reluctant to register a non-government milk co-operative as they already had HIMUL a state sponsored milk co-operative functioning in the district.

member families with a vote each resulting in hamlet committees being formed at the hamlet level. Executives from the hamlet committees form the apex Board that manages and develops policies of the entire collective. The board also has functional working committees and employees from within the members to implement its activities. An advisory body of members along with a DLR Prerna representative forms an additional monitoring body. The collective is engaged in savings and credit union, drinking water collection and distribution; milk, vegetables, horticulture and spices and tea marketing; knowledge management; communication and newsletter and welfare activities.

MSSSVS with its plantation past had retained some of the tea bushes, so today, continues its relationship with tea through a memorandum of understanding with Tea Promoters India, Private Limited(TPI). TPI processes and sells the tea exclusively as Mineral Spring Small Farmers tea. Systems have been developed where strict measures of quantity produced, processed and sold are maintained[8]. Mineral Spring small farmers' tea is thus sold internationally as a partnership between the collective and TPI. Mineral Spring has been making 10, 000 to 11, 500 kg of made tea annually in the last five years.

Green leaf tea[9] at Mineral Spring in 2014 was sold to TPI at Rs. 52 per kg[10]. The breakdown is as follows: the primary producer- Rs. 38 per kg; the collector per hamlet - Rs. 2 per kg; *headload*, the person carrying the tea to the central point - Rs. 2.50 to Rs 3 per kg depending on the distance; Internal Control Fund – Rs. 2 per kg; MSSVS administration Rs. 6.25 - Rs. 8.25 per kg. The variation on the administration sum is because hamlets further away from the central collecting point pays a higher wage to the *headload*. Only from the central point does a jeep/s transport the tea, the rest are carried on foot in a bamboo

[8] Tea that is being sold is primarily of the stock from the old tea plantation days so is of the original Chinese stock. New saplings have been sourced through TPI from the Darjeeling stock. Tea management expertise is sourced from TPI and sustainable agriculture practices from DLR Prerna. At every stage from the hamlet level to the central level the tea plucked is checked for quality and lower quality tea plucked is sent back. On the way to Selimbong the tea is further aired and cooled to stop fermentation.

[9] The price is for freshly picked tea leaves and not processed tea. MSSVS picks it tea based on the timings decided by the Tea Committee which decides days of the week for picking. On designated picking dates, each hamlet has a time deadline by which the individual members need to take their tea to their collector who weighs and inspects it for quality. This is taken on the backs of 'headloads' in bamboo woven baskets or tea net bags to a central point where it is re-weighed and inspected. From this central point jeeps are loaded and sent off to Selimbong, TPI, 2 to 3 hours depending on traffic where it is processed. Members are paid on a weekly basis.

[10] The price for green tea quoted here is maintained throughout the year. In other circumstances the price fluctuates during the year based on the flushes and the quality of tea the different flushes produce.

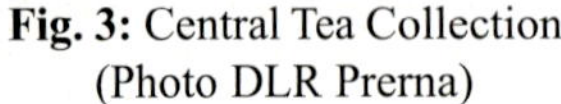

Fig. 3: Central Tea Collection (Photo DLR Prerna)

Fig. 4: 'Headloads' photo: DLRPrerna

woven basket or a tea net bag. All the people who financially benefit from the sale of tea belong to the MSSVS including the office secretary. The price that MSSVS gets for their green leaf tea is one of the best in the Darjeeling Hills but the price that the tea is ultimately sold internationally is not available to the author.

MSSVS is certified organic producer under national and international standards since 2001. Internal Control and Internal Regulation Systems of organic farming with internal organic inspectors has enabled the organic certification to be renewed annually till date (2014), which shows the commitment of the members towards an organic way of life. Every member signs a memorandum with the MSSVS stating their commitment to go organic and are then provided with terms of reference for going organic. Based on this, agricultural practises are defined and regulated, which includes farm management, seeds sourcing and saving, integrated pest management and documentation at a farmer and collective level. These terms of references are strictly adhered to with repercussions at individual or hamlet level for non-adherence. Every farmer documents their daily intervention in their farm diary and afarm map which delineates the farm outline and major land use pattern. At a collective level, every farmer is coded and their data centralised. Cases of land fragmentation and new codification is done once a year at the collective level. Based on the systems, Internal Inspection Officers conduct a 100 percent inspection of all the farmers of MSSVS. The internal inspection officers' skills are constantly upgraded. The External Inspector conducts at least 10 percent random field inspection and full documentation inspection annual before the collective is certified organic on an annual basis.

Since 2003 the collective is Fairtrade Labelled as a producer organisation of small farmers' tea by the Fairtrade International. The labelling guarantees defined international recognised standards and conventions through inspection and certification. This ensures that the product fetches a Fairtrade minimum price.

The appropriate minimum price being, "covers cost of production and also leaves enough to plan for the future of the business. The pricing of the product can be over the minimum price depending on the bargaining power". Over and above the minimum price, a premium price of USD one for higher grades and USD fifty cents for lower grades is quoted which is channelled directly to improve the lives of the communities and families of the primary producers. This premium price has been utilised by MSSVS to expand their physical infrastructure for tea like tea saplings, weighing sheds, and office; social infrastructures like small paths, bridges and drinking water; welfare activities like landslide relief, medical assistance and knowledge enhancement programmes.

What Sets Mineral Spring Apart?

MSSVS stands apart from the "Darjeeling Tea" in that, land is owned by its members and the land holding pattern is between an acre to 12 acres per family. The agro-ecology of the collective is based on diversity. A large number of food and non-food crops are grown in the land of each member with some tea in a section, at the edges of the terrace or between other crops within the terrace. Thus, depending on the altitude, one sees tea bushes among Darjeeling mandarin, ginger, potatoes, greens, lentils, turmeric, millets, carrots, radish, squash and maize. Agro-forestry is also part of the agro-ecology in Mineral Spring. Even though the market sees only tea that is sold as an organic product, the entire agro-ecology of MSSVS does not use synthetic agro-chemicals and its diversity is one of the key elements which enables this tea to stand apart from the plantation monoculture organic Darjeeling Tea.

Fig. 5: Diverse agro-ecology of Mineral Spring with tea as one of its components (Photo: Author)

Fig. 6: Rolling tea (Photo: Author)

The Mineral Spring experience challenges Darjeeling Tea and offers an alternative paradigm.

The people own their primary production asset, land, which is never a possibility in tea plantations. This ownership of land enables the decision making of its use by the people. Thus, one sees that the use of land assets in Mineral Spring is highly evolved and diverse when compared to tea plantation workers. Only 85% of the members actually grow tea, the rest engage in other forms of agriculture. The ones who do have tea do not cover their entire land with tea, but have grown it along with other food crops. Importantly each member decides for themselves when they would start selling their tea to TPI. Thus some of the best tea is kept for hand-made tea[11] for domestic and local consumption.

The decision making and the structure of MSSVS, is democratic with elected member representatives taking on various roles of governance. The participation spaces for the members including regular election and meetings of the board and hamlets furthers equity and collective decision making. The Board has also taken special initiatives to include and nominate women within it. The ownership of primary assets by the members, the democratic institution and processes for further inclusion promotes human dignity and rights of the members whereas a plantation dehumanises workers.

MSSVS tea stands apart from the monoculture Darjeeling Tea of the 87 plantations. Tea is one of the many crops in the complex agro-ecology of Mineral Spring and this complexity and diversity makes the system more sustainable, resilient and provides greater eco-systems services than a plantation. This diversity allows natural nutrient flows within the system to maintain and improve fertility and allows natural relationships to deal with pest and disease[12]. The diversity enhances food and nutrition security of members of the collective. This equation is pretty simple in that one cannot survive by drinking tea nor does it provide all the nutrition needs one needs however much it is touted to be "Better to be deprived of food for three days, than tea for one – Ancient Proverb"(Tea Board of India)

In a world with an increasing number of tea plantations gaining organic certification, Mineral Spring is the only organic tea which comes from a multi-cropping system. While acres and acres of monoculture tea in a plantation can

[11]*HateyChia* or handmade tea occupies rural homes and the shadow markets in the Darjeeling Hills. It is hand-rolled and sun dried or over fire which might give it a smoky aroma.

[12] Mineral Spring is not pest and disease free, taking the case of tea, a number of diseases and pests are prevalent but the diversity contains and limits it reducing the need for pest repellents. Eg. In the lower reaches of Mineral Spring one finds the presence of *heliopeltis*or the commonly tea mosquito but the diversity has kept it in check with no intervention needed till date since its appearance 10 years ago.

be called organic, this is organic farming in its most limited interpretation. It simply means chemicals are below a "maximum residue level". Certified organic tea is a step above tea contaminated with chemicals, especially in the light of the Greenpeace 2014 report, but it still has a long way to go before it can claim to address issues of bio-diversity conservation and equity. A more meaningful interpretation of organic and sustainable agriculture requires more than a reductionist approach to synthetic agro-chemicals and needs to incorporate an ecosystems approach based on diversity. To paraphrase plantation literature, organic is a way of life of live and let live, but organic tea plantation only allows a single crop to grow and essentially does not allow the vast workers to live a life dignity and equity.

MMSVS brings about an experience in the certification process, where a collective of small farmers is certified organic in national and international standards. Certification is a cumbersome and expensive process, but with proper systems and support, MSSVS shows that small farmers can access organic certification and the market. But the key to this discussion is the ecosystems approach to organic farming of MSSVS which is a philosophical and a pragmatic approach and not just a market led intervention. A proof of this is that no member of MSSVS till date has converted their farm into a mini-monoculture tea plantation even though there is a guaranteed market. This decision to have a limits to the amount of tea planted against all market logic is an understanding of the true path of an organic way of life based on agro-ecology of the region and working within it. It is an ecosystems approach that uses invisible and visible values to define the development needs of the collective.

With the international trade of tea adding social concerns to the environmental concerns, Fairtrade tea from Darjeeling is rapidly increasing. One sees a range of Fairtrade tea from the Fairtrade International, Rainforest Alliance to Ethical Tea Partnership all of which have a mix of social and environmental standards. But, at the end of the day, all are Darjeeling Tea from plantations governed by the same Plantation Labour Act paying pretty much the same daily wage and within same the plantation paradigm. This is contradictory in nature as Fairtrade has an ideology, philosophy and standards of participatory planning, accountability and transparency which are conspicuous in their absence in the plantation system. This raises the question of "Can a Plantation be Fair?" (Besky, 2008). Fairtrade Darjeeling Tea is contradictory."Fair Trade is a trading partnership, based on dialogue, transparency and respect, that seeks greater equity in international trade. It contributes to sustainable development by offering better trading conditions to, and securing the rights of, marginalized producers and workers – especially in the South. Fair Trade Organizations, backed by consumers, are engaged actively in supporting producers, awareness raising and in campaigning for changes in the rules and practice of conventional international trade." (A Charter of Faitrade Principle 2009). Tea plantations are based on exploitation that has continued from its colonial roots and cannot be called fairtrade. In the context of Darjeeling Tea, by labelling plantation tea to be faitrade, Fairtrade International is doing exactly the opposite to its belief of "securing rights of workers" and perpetuating the existing exploitative system of plantation tea. Fairtrade has diluted its ideology and created standards of hired labour which enables trading and labelling plantation Darjeeling tea. It is also a market driven initiative as there are no alternatives to existing Darjeeling Tea as yet. Mineral Spring in spite of its inability to process and trade directly, is still the best example of faitrade in Darjeeling Tea.

Mineral Spring, thus, proves that tea can be grown in models more equitable and sustainable outside of the existing paradigm of Darjeeling Tea, which is synonymous to the monoculture plantations. Importantly, it is out of the box of a singular imaginary, "Monocultures of the Mind" (Shiva 1993), which decides that the existing plantation tea is the only way to cultivate tea. The imaginary construct greenwashes Darjeeling Tea and certification processes like organics and fairtrade gives it a market legitimacy. While this legitimacy requires immense time and effort to uncover, understand and act upon, a conscious consumer, the driving force of the organic and fairtrade market, usually just has enough time to read the broad labelling of the product in the socially and environmentally conscious section of a store. The critique of organics and fairtrade Darjeeling Tea is so difficult to come by, that a consumer will not be privy to it even in a niche store. Thus continues the trading of a green washed and so-called socially responsible tea with all its contradictions.

Mineral Spring Tea is not without its contradictions. It completely depends on TPI for the processing and marketing of its tea. The registration process[13] of its members in the Tea Board of India still continues so that till date Mineral Spring is justified based on the old records of Lebong and Mineral Spring Tea Estate rather than as a small farmers' collective. Even when all the registrations are done, there is no clear provision within the Darjeeling Tea certification trade mark and geographical indicator for a small farmer collective outside of the 87 tea plantations. The legacy of the trade of Darjeeling Tea is monopsonic, which starts with a small group producing and selling tea and an apex small set of buyers in a buyers' market that decides and protects the industry and business.

[13] Registration in Tea Board of India requires each farmer to apply individually with land papers which are verified by an accredited land surveyor. With the history of Mineral Spring, the land documents are not exactly what the Tea Board prescribes and the accredited land surveyor is far from Darjeeling and expensive. It also calls for soil testing and soil testing facilities are not available in Darjeeling.

This system does not easily allow a new player to enter. A small farmers' collective today has all the right reasons to succeed in the existing market trend towards ethical and environment friendly tea. But it has miniscule capacity and access to this group, so the collective faces an extremely uphill task to enter and succeed in it in the present state of affairs.

Importance and Implications of Mineral Spring

The market for Darjeeling Tea globally is guaranteed, "The Darjeeling district has 87 certified tea gardens, as they are locally known, producing about 20 million pounds of tea every year. That is why local tea growers grew annoyed that as much as 88 million pounds of tea were being sold as Darjeeling on the global market each year." (Yardley, 2012). This means that the international market has the capacity to absorb four times the production capacity of the tea plantations of Darjeeling. To change this market practice, the certification trade marks with geographical indication was brought in to check the selling of tea from other regions as Darjeeling. "European Union to drink 100% pure Darjeeling Tea; gets PGI status" states that "tea produced only in Darjeeling can be sold as Darjeeling tea in the European Union countries. Along with this, a section of blenders selling the brew with a certain percentage of the commodity as Darjeeling tea was given a five-year time to shift to the new business. Some blenders in the EU countries generally mix 49% of any tea with 51% of Darjeeling tea and still sell it as Darjeeling tea. The blenders were handed out a caveat that only those people whose products were in the market five years before October 14, 2009, could continue selling their blended product as Darjeeling tea for the next five years. At present, EU imports nearly 3-4 million kg of Darjeeling tea, which accounts for nearly 60% of Darjeeling's tea export." (Ghosal 2014).

This essentially means that the global markets requirements for Darjeeling Tea from a supply perspective is reducing dramatically. Logically this should bring about looking at ways of increasing production. Since the 87 plantations have tea in all possible land and produce to their fullest capacity, the only way to do it is to promote small farmers who are from the Darjeeling hill sub-divisions. In all the three subdivisions of Darjeeling, there are tea plantations registered under the geographical indication which means all of the sub-divisions can potentially grow Darjeeling Tea. This can happen only if the small farmers' are promoted to grow tea based on the experience of Mineral Spring. Yet, the existing strategy is to restrict it to just the 87 tea plantations rather than use a criteria of stock and quality and expand to small farmers nearby. Mineral Spring has time and again shown that even with the difficulties of transport[14] small farmers can

[14] Increase in transfer time from plucking to processing reduces quality and MSSVS tea travels 2 to 3 hours to reach Selimbong by jeep, plus the time for it to be carried from the hamlets to the central collection point.

produce quality Darjeeling Tea which is socio-ecological safe and equitable.

Discussing global organic markets, "the global trade is expected to cross US $104 billion in 2015 at an estimated annual rate of 12.8 per cent" (Jishnu and Sood, 2012) shows the upward trend in green business and this goes hand in hand with increasing fairtrade market models. This is primarily fuelled by increasing conscious consumers who want to lead healthier lives and contribute towards improving lives of producers so purchase ethical products. This has meant that there is dramatic market opportunities for small farmers' organic Darjeeling Tea like Mineral Spring. Yet, an exclusionary rather than a proactive approach has been adopted when it comes to small farmers' Darjeeling Tea.

Within the context of Darjeeling Tea and small farmers there is a need for the Tea Board of India to take a visionary stand and really promote small farmers' Darjeeling Tea. This can be done only if the certification trademarks and its geographical indication categorically states that small farmers within the parameters of stock and quality can be called Darjeeling. Likewise there is a need for registration processes to be made appropriate and accessible to small farmers in terms of process and expense. For the 12th five year plan vision of promoting small farmers' tea and Rs. 300 crores investment to make a difference in Darjeeling, a strong political will is required which as of now is lacking. Existing debate continues within the framework of plantation monoculture tea and has yet to imbibe the rich and diverse agro-ecological experience of Mineral Spring. Technical discourses stay within the boundaries of the plantation so fail to capture the potential and opportunities of a diverse agro-ecology. Political debates on equity still remain within the limitations of the Plantation Labour Act and fail to ask deeper questions of ownership, equity and decision making thus remain caught up in wrangles of wage rates and bonuses.

Civil society intervention has been an important aspect of the Mineral Spring experience. This partnership of sharing life's experience between the community and civil society has been instrumental in the developmental journey of Mineral Spring. Key to this experience is the role of knowledge management and a sustained partnership as dramatic increments are needed in knowledge and skills to go beyond being a primary producer. Thus, for small farmers to succeed there is a need for them to organise under a collective umbrella and gain knowledge and skills beyond planting, nurturing, picking and supplying. External agencies', government, non-government or corporate, long-term partnership enables the essential access to the knowledge and skill that is needed.

In the world tea trade based on monopsony, the role of corporate bodies is of critical importance. There is a need to relook at tea plantations and their labour practices as a continuation of the colonial exploitative past and bring about a much needed change based on present day values of rights and dignity. But this can happen only if the monocultures of the mind changes in the corporate sector of tea beyond exploitation, protectionism and greed to inclusion, diversity and shared living.

Ultimately, promoting small farmers' Darjeeling Tea means regional development, one of the factors that is lacking which sees Darjeeling time and again agitating for more autonomy and regional development.

References

Besky S. (2008). Can a Plantation be Fair? Paradoxes and Possibilities in Fair Trade Darjeeling Tea Certification. *Anthropology of Work Review* 29.1: 1-9.

Chettri M. (2013). Choosing the Gorkha: At the Crossroads of Class and Ethnicity in the Darjeeling Hills. *Asian Ethnicity* 14.3: 293-308.

Columbia Law School, Human Rights Institute (2014) The More Things Change…The World Bank, Tata and Enduring Abuses on India's Tea Plantations.

Darjeeling Planters Association Records (2000).

Dekens J. (2005). Livelihood Change and Resilience Building: A village study in the Darjeeling Hills, Eastern Himalayas, India. Diss. University of Manitoba.

Department of Labour, Government of West Bengal (1994) Special Issue, Plantation Workers in West Bengal, Labour Gazette, Government of West Bengal.

DLR Prerna (1996 – 2014). Annual Reports and Papers.

Dutta I. (2014). Small tea growers break new ground with major allocation in 12th plan. The Hindu, retrieved from www.thehindubusinessline.com

Fairtrade International (2011a). Fairtrade Standards for Small Farmers' Organisation. Retrieved from www.faitrade.net

Fairtrade International (2011b). Fairtrade Standards for Hired Labour. Retrieved from www.faitrade.net

Fairtrade International (2011c). Fairtrade Standards for Tea for Hired Labour. Retrieved from www.faitrade.net

Ghosal S. (2014). European Union to drink 100% pure Darjeeling Tea; gets PGI status. The Economic Times, retrieved from www.articles.economictimes.indiatimes.com

Golay B. (2009). Rethinking Gorkha Identity: Outside the Imperium of Discourse, Hegemony, and History. *Indian Nepalis: issues and perspectives* 73-94.

Jishnu L. and Sood J. (2012). Organic Universe. Down to Earth. www.downtoearth.org.in

Khawas V. (2011) Status of Tea Garden Labourers in the Eastern Himalaya: A case of Darjeeling Tea Industry, In: *Mamta Desai and SaptarshiMitra (eds), Cloud, Stone and the Mind: The People and Environment of Darjeeling Hill Area.*

National Service Scheme, St. Joseph's College (1971). Darjeeling Survey Report.

O'Malley L.S.S. (1907, reprint 1999). *Bengal District Gazetteers Darjeeling,* Logos Press, New Delhi.

Rai C. B. and Sarkar R. L. (1986). Indian Institute of Hill Economy, Darjeeling – 1986, Development of Human Resources in Darjeeling Hills – a Case Study of Hayden Hall Programmes on the Weaker Section Population An Evaluation Report sponsored by CARITAS India.

Rai R. and Patrick (1995). A study of the adequacy of the Plantation Labour Act, 1951 in protecting the interests and promoting the welfare of the tea plantation workers in West Bengal, with special reference to North Tukvar Tea Garden, Darjeeling. Dissertation: Center for Research in New International Economic Order, Chennai.

Rai R. and Chakraborty S.R. (2009). Two Leaves and a Bud: Tea and Social Justice in Darjeeling, *State of Justice in India: Issues of Social Justice, VOl Ied. Ranabir Sammadar.*

Shiva V. (1993). Monocultures of the Mind: Perspectives on Biodiversity and Biotechnology. Palgrave Macmillan.

Sukhtankar A. and Rosenblum P. (2014). Can there be a 'socially responsible' tea? www.kafila.org

Sundar P. S. (2012). Tea Board allocates Rs. 300 crores for small growers in 12thPlan. The Hindu, retrieved from www.thehindubusinessline.com

Tea Board of India (2014). a. Brand India Tea, About Darjeeling Tea, b. Darjeeling Tea, the best. A quest for intellectual property rights, c. Plant Protection code (November 2014, Ver. 3.0), Policy on usage of Plant Protection Formulations in Tea Plantations of India. www.teaboard.gov.in

Yardley J. (2102). Good Name Is Restored in Terrain Known for Tea. Retrieved from The New York Times, www.nytimes.com

Tea: Technological Initiatives, pp. 39-61
New India Publishing Agency, New Delhi, India
Edited by Niladri Bag, Arundhati Bag and L.M.S. Palni

3

Human Resource Strategies for Sustainable Development of Tea Plantation Workers in India

Ananda Das Gupta

Abstract

Ensuring the provision of safe and healthy working conditions to all workers and promoting the availability to them of certain minimum welfare amenities are the constitutional obligations of all employers as well as the Indian state. Enlightened employers tend to provide for these facilities on their own, as part of their performance and retention strategies. Others, however, largely consider these obligations as burden. Some recent researches reveal that the pursuit of numerical flexibility policies is becoming rampant amongst employers in the advanced world; they are doing so to cut costs and become more competitive. This means more use of labour-cost-saving devices and neglect of some basic rights of workers. Of course, intense competition amongst companies puts pressure on them to attempt cost-cutting and ignore these legal provisions. It should be noted however, that the law relating to working conditions is perhaps the most basic of labour laws in any country; therefore, its sanctity has to be maintained with an added responsibility by all concerned and as a societal value.

Introduction

The discovery of the tea bush in Assam by Robert Bruce inspired the colonial capitalist to make large-scale investments in it. The availability of suitable land and thin population were favourable conditions for growing tea in Assam, so was the climate of Assam. Harler (1964: 33) pointed out that the Brahmaputra Valley is perhaps the best tea growing area of the world with favourable soil, climate ad topography. Once the problem of land was over the planter had to manage necessary capital. To attract the investors the colonialist enacted many law in their favour. Within two decades many more companies with British capital made their entrance in different parts of Assam (Nag 1990: 51-52). Between 1859 and 1866, the British Authority cleared the hills of Assam for new tea gardens and tried to attract huge investments for the industry. Within a few decades, tea manufacturers in Assam had covered 54% of the market in the United Kingdom and had outstripped China (Fernandes, Barbora and Bharali 2003: 2).

Faced with labour shortage the planters had to get workers from other sources. That is when they began to recruit workers from other parts of India mostly present day Jharkhand, Bihar, Uttar Pradesh etc. as indentured labourers in slave like conditions. This class of people was uprooted from their land and livelihood by the *Permanent Settlement* 1793 meant to ensure regular tax collection for the colonial government. Impoverishment was the consequence. So they had no choice but to find other sources of livelihood. In the absence of other alternatives, they were forced to follow the labour contractor and become indentured labourers on the land that the Assam indigenous communities had lost under the same colonial processes to the tea plantations. Their initial recruitment was done through professional contractors who were notorious for abuses and exploitation. The tea garden community folk songs have passed details of such exploitation down from one generation to another (Gupta 1990: 51-53). This labour force has been popularly called as the 'tea tribes' and 'ex-tea tribes' and there is debate about this nomenclature but we shall not enter into this debate. In 1997, a total of 5.9 lakh labourers along with their dependants were working in 1012 registered gardens spread in an area of 2.32 lakh hectares, which is 2.9% of Assam landmass (Sagar 2002: 1).

The role the trade union is important. The emergence of trade unions is recent in the tea industry of Assam. As late as 1946 the planters recognised them on a differential basis. They identified the ones with whom they could negotiate. But the planters were well organised from the beginning of 1879. Because of their poor organisation the workers are dissatisfied with them. It also results in unrest from time to time, for example the killing of managers that one has witnessed in recent years. One sees two main tendencies among the workers. One is their sudden outbursts and the second, an organised and protracted struggle. One saw the first type in the case of the Victoria Jute Mill workers in West Bengal in 1993 and the second by the Kanoria Jute Mill (Debnath 2003: 34).

Social Security services generally mean the basic facilities that are necessary for the mental, physical and intellectual development of a person. They should include food, shelter and health care. Carl Wellman (1996: 268) defined social benefits as some form of assistance provided to an individual in need. Thus "welfare" or "social security" is the collective name for all social benefits, especially for groups that need protection to grow into better citizens. A welfare state has a moral obligation to ensure the good of all its citizens, particularly the weaker sections. If it cannot provide all the facilities, it can take the help of other agencies. (Madan and Madan 1983: 163). A human being can lay an ethical claim from his/her society on the minimum livelihood in the event that he or she lacks the means of sustaining life because of circumstances beyond his or her control (Wellman 1996: 268). Though the other welfare measures in their narrow sense do not include education particularly of the working class, PLA makes an exception to it and includes it among the amenities to be provided to the workers. We can, therefore, justifiably include it among their social security provisions.

The tea industry in India employs more than 1.5 million workers. Every seventh worker in the organised sector (out of the total estates in India, about 65 percent estates are owned by the Corporate giants and around 35 percent of the estates belong to the Proprietors) industries is a tea worker. The tea plantations of Assam and West Bengal together account for the employment of more than a million workers as well as about 80 percent of the total production in India. The tea plantation workers, mostly tribals and lower castes, are the backbone of the tea industry in eastern India. They are fourth generation descendants of indentured immigrants brought by the colonial planters 150 years back from the tribal tracts of Bengal, Bihar, Orissa and Madhya Pradesh. Literatures abound depicting the inhuman living and working conditions of the tea workers during the colonial period. Living and working in isolated plantation enclaves they were as good as bonded labourers. Laws existed to penalize the workers not to protect them. The planters were the omnipotent authority. It was their raj.

With increasing job opportunities for good material, that too in Urban areas, the problem that tea gardens are facing now and which may accentuate in future lies in management of tea garden, in that no talented person shall like to make a career in secluded life of tea garden, with problem of good education to children, job-opportunities for spouses, limited chances of promotion, lack of cultural/social activities and cut-off life from civilised world. Furthermore, there is hardly any career planning or motivation. Obviously, tea garden may have to reconcile with second class talent to run the garden even at attractive pay scales and perks. Most of the gardens are also facing shortage of Work-Force to harvest green leaf in peak seasons. There is an attitudinal change in the workers and less number of workers is willing to do manual jobs. The welfare measures are a major factor as far as productivity is concerned. The labourers need to be housed well and provided with good medical facilities and amenities. The plantation labour act stipulates that a plantation with more than 1000 workers should have a full fledged hospital and that the medical facilities should be provided free of cost. This is vital because the plantation setup is geographically treacherous areas. The workers have to work in high altitudes and during heavy rains. Besides this they have to walk long distances from home to the work place. Transportation is difficult as vehicles cannot ply to all parts of the tea gardens. The plantation is an out door oriented work in the forests and is highly physically demanding for the worker. Moreover the nutrition factors are important. The survey revealed that 100% of the workers in the south took normal rice and dhal based food. While non vegetarian food (red meat) is consumed at a minimum of once a week and not more than twice a week. Fresh vegetables are not available in village shops; in addition the workers also complain that they have no time to go to the nearest town to make any real purchases. An interconnected aspect to this is the rest time for the workers. While they have to leave home early so as to reach the work site in time, breakfast is not taken or the left over food is taken in the morning. Rest time during the working hours came up as an important factor. Absenteeism due to laziness have be reported by all the field officers and the managers and supervisors alike. The same has been presented in a different angle by the male workers. They complain of lack of rest during working hours and they say that the real time work is more than eight hours a day. Mental disorientation was also noticed amongst all the male and female workers. The whole lot reported that they are working just for a living and they do not find any happiness or pleasure or satisfaction in working in the plantation under the present wages, housing and other welfare measures provided, and also add that it is out of financial compulsion.

The Scenario

In spite of the sporadic efforts being made to improve labour conditions during the pre-independence days, certain appreciable changes in the condition of plantation workers took place only after Independence. The first and foremost step in this direction was the enactment of the Plantation Labour Act of 1951. This Act is a very comprehensive document and its provisions extend to the health, housing, education, social welfare, leisure, recreation and working conditions etc. of plantation workers. The Act also intended to bring about uniformity in all matters concerning plantation labour and, improve their quality of life in all spheres. It provides safeguards for them against exploitation by regulating their hours of work, rest intervals, minimum age of employment of children and adolescents and provides for annual leave with wages etc.

Wages

The most striking feature of the Northern and Southern gardens is the difference in wages. While average monthly wage of the workers in Assam and West

Bengal figured at around Rs. 48 per day plucking, in Tamil Nadu they were around Rs. 81 per day plucking. The only difference between Tamil Nadu and the two northern states (Assam and West Bengal) is that workers in the latter are given a part of their wages in kind. They are given rations of 2.25 Kg of rice and wheat per week at a subsidised rate. Workers in Tamil Nadu are given rations at rates, which are slightly below market rates, but the quantity is more. The subsidy worked out to be around Rs. 5 per day.

Housing

While comparing the housing conditions in Assam, West Bengal and Tamil Nadu, we find that there are similarities between the two former states and the differences with the latter state. It gives a picture that had the employers been adhered to the Plantation Labour Act 1951, all houses would have been made permanent in Assam and West Bengal by 1969. In Tamil Nadu Plantations, however, the houses meet the specifications of the Act.Again in terms of sanitation and water, Tamil Nadu plantations have adequate sanitary facilities where each house has a toilet or two houses are provided with a common toilet, the labour lines in Assam and West Bengal do not have these facilities. Barring a few gardens maintained and owned by the corporate groups, access to health services in the gardens in Assam, West Bengal and Tamil Nadu is not adequate, because, by and large, the requirements of the Plantation Labour Act have not been met. The ill-maintained creches, barring Tamil Nadu, where a better record is available in maintaining creches, are also a factor to be contributing to the welfare of the plantation labourers.

Women Workers

Women workers, in general, are agitated over their harsh working conditions. During the plucking seasons, they are forced to work for more than 8 hours - either late into the night or start very early in the morning. Lactating mothers face a lot of inconvenience. Even pregnant women are forced into deep hoeing. Women, who join after maternity leave are not given light work and the planters do not care a fig about their health conditions. Subharani, a permanent worker in Baishahabi tea estate, developed complications immediately on joining work. She could not work for the next few months. Finally her job was terminated.

Lack of women supervisors is also a source of trouble for women. Women workers get apprehensive and work amidst constant fear of sexual harassment and assault, especially when working in secluded areas or in bungalows of the management staff. The workers of Baishahabi estate alleged that two girls, Durga and Mani were raped by, the manager, in 1996. Durga committed suicide and Mani is still missing. The two families have been denied any justice thereafter.

Casualisation of the work force is on the rise in the tea industry. Through the decades of 1950s and 1960s, the number of permanent workers in the Assam tea plantations was reduced by a staggering 25.61 per cent. Within a short span of seven years, between 1984 and 1991, the temporary labour force increased from 170,495 to 268,450 and has almost touched the three-lakh figure. Majority of them are women. Women workers have been gradually shifted to the temporary labour category. According to a tea worker herself and also the member of State Women's Commission, the planters have resorted to this ploy 'primarily to deprive the women workers of their basic rights as enshrined in the Plantation Labour Act (PLA), such as maternity benefit and medical benefit.' "Many of us are thereby denied housing, subsidised food grains, provident fund and bonus," she says.

Medical Facilities

Health and medical care of tea garden workers has been a very controversial topic mainly because o(lack of uniformity. This has invited sharp criticism in the case of some gardens and appraisal for others. It was only after the passing of the Plantation Labour Act of 1951 that a uniform Medicare format for tea plantation labour was worked out. According to the Plantation Labour Act, 1951, every tea plantation, employing not less than 1000 workers is required to maintain a hospital or have a lien on a certain number of beds available at the hospital on the nearest neighbouring plantation. Basic first aid kit and medicines should always be there even if it cannot provide a hospital or dispensary. Tea estates must employ necessary qualified staffs who work under the supervision of a medical officer in the hospital of the nearest tea garden. For every thousand workers the estate hospital must possess at least 15 beds. Such estates are required to have a full time medical officer, a nurse, a midwife, a health assistant, a *compounde*r (paramedical staff) and a dresser. These hospitals have to be equipped with an operation theatre; a delivery room, an outdoor patient consultation room, a kitchen, and other necessary equipment such as oxygen cylinders must be there. There should be a separate ward for female patients and one isolation ward has to be provided as well. The hospital should be equipped with basic and necessary medicines.

Some small tea estates especially in Darjeeling Hills do not have resident Medical Officers but they usually have ad hoc arrangements with outside doctors who visit on fixed weekly intervals. But this is not a very suitable solution because sometimes for serious and emergency cases it becomes difficult to shift patients to hospitals and lack of timely attention can be quite dangerous. Often female patients suffering from complications during delivery have died because no doctor could be contacted in time to handle such cases. Hence the government and the industry should find some better alternatives. The District Medical Centres and government dispensaries should be located near such tea estates because there is often a high concentration of resident population on plantations, many of whom are not workers or direct dependants of workers. Such people often have no access to any medical facilities, hence both workers and non workers of such far flung estates could benefit if such facilities were provided. Each estate is also required to have at least one ambulance for sending patients to district hospitals. Now almost all tea gardens have acquired ambulances.

As in the case of other amenities, the quality of medical facilities and hospitals in tea gardens differ from estate to estate. Some well run tea gardens are maintaining very decent, neat and clean and well equipped hospitals with proper buildings and compound, whereas there are some where medical facilities are nothing but a sham. This was the case in the early days as well. Some hospitals were just temporary sheds without even a qualified doctor. In those days it was quite common for pioneer planters and managers to treat tea garden patients themselves. However with time, there was gradual improvement and very soon the native 'doctor babus' appeared on the scene.

Case Study

Condition of the Workers-Assam

Across the many plantations in Assam, most of which are situated in the upper parts of the state, the condition of the tea garden workers is nothing short of abysmal. Adivasis brought in as indentured slave-labour from Central India by the British form the vast majority of the workers, with the rest consisting of other local tribal communities, as well as Nepalis, Bengalis, Oriyas and so on. During the initial decades from the 1850s till the 1920s under the British, the working conditions were akin to harsh slavery, with flogging, rape, torture and even the throwing of dead workers in rivers. While certainly not comparable to earlier times, the working conditions today are still far from being the well-regulated environment that functions according to the Plantation Labour Act brought out in 1951 to protect the interests of workers in plantations, who form the single largest organised sector workforce in Assam and the entire Northeast region numbering anywhere between 8 to 10 lakhs depending on the season.

The North Eastern Social Research Centre based in Guwahati conducted a comprehensive study across 172 tea gardens in Assam along with numerous interviews and group discussions with workers and families. The study brought to light numerous violations of the Act, including inadequate or completely non-existent provisions for drinking water, creches, schools, proper health facilities, sanitation for women workers and shelter. Even a cursory observation of the plantations today confirms these findings. Upon further investigation and discussions with workers, one learns that wages paid are much lower than prescribed minimum wage rates, no over-time payment is made, and occasional physical abuse occurs.

Babloo, Signus and Ranjit (last names withheld upon request), all workers in Mornia Tea Estate in Lower Assam, complained that they had to drink bitter-tasting, hard water from pre-existing wells, when in fact they're supposed to receive drinking water either through taps or tankers from a public water source. Late wage payments were another huge problem, with some workers receiving their wages as late as 3 to 4 months after the due date. Garden workers received around Rs. 1400 per month on paper, but portions were cut from that for shelter repair (which hadn't been conducted in over 10 years), canteen facilities (non-existent), and educational facilities (again non-existent). This translated to a real wage of about Rs. 45 per day, far lesser than the prescribed daily minimum wage of around Rs. 54. They further said that Provident Fund had been cut on a monthly basis from their salaries, yet since 2000 no retired worker had received gratuity from PF. When asked about this, management simply shifted the blame to their predecessors. The school was in a decrepit condition and the only education the children received, when they weren't working, was from the local church.

Further up north in Nagaon, this author was privileged to attend a few wide-ranging discussions with workers in various tea gardens in the area (whose identities have been protected due to their worker-mobilising activities) as well as with Arup Mahanto, a rural workers movement leader. All the workers said that the little benefits they did receive in earlier times were rapidly getting eroded over the years. This included rations, free medicines at the hospital in Kandoli Tea Estate (which has now been downgraded to a dispensary), money for firewood at Sagubhai Gardens and many others, all of which have disappeared with further and further deregulation measures in favour of capital in the post-liberalisation era. Mahanto further pointed to the nexus between management, police and corrupt union leaders as one of the crucial reasons for the deteriorating situation. Indeed, 4 of the 5 workers interviewed had been suspended and dismissed due to their attempts at mobilising workers, and all 4 now try to eke out livelihoods by working in the even more exploitative stone-quarry industry or selling firewood, while trying to fight a legal battle to get reinstated. Women, who are the backbone of the tea industry and the large majority of the workforce, face even harsher working conditions. In all the tea estates visited, one couldn't spot a single creche for infants and toddlers. Sanitation facilities were either inadequate or completely non-existent. And while nothing explicitly was mentioned, there have been many instances of verbal, physical and even sexual abuse. Women are in fact preferred as labour because most managers feel that they are particularly suited for garden work and easier to exploit. Thus, while women labourers for the most part get the same as their male counterparts, not a single woman can be spotted in the plantation factories where the wages for workers are marginally higher than their garden counterparts.

It must also be added that tea garden workers are caught between the proverbial rock and hard place, and forced to accept increasing labour exploitation due to harsh material conditions and lack of choice. Some who have access to cultivable land tend to be better off and more self-sufficient, at times working in the gardens only for short durations of time out of temporary necessity. Those possessing no or uncultivable land, and who leave the gardens, often end up as informal labour in earby towns and cities. Education levels, health indicators and poverty levels for the workers are among the lowest in Assam. Many families find it difficult to get their children into educational institutions and later on in finding proper employment. Thus the oppressive environment of the tea garden is often the only recourse for many of these families.

The provisions of the Plantation labour Act of 1951 have several provisions for the uplift of plantations workers. These include provisions relating to housing, sanitation, drinking water, medical facilities, canteens, creches and primary schools for children of workers. These provisions, if implemented properly, can play a major role in improving the working conditions of labour.

Human Right is a multi-faceted concept. While production is the actual output, human resource development with an accent on human rights is only a means to achieving the output. It is concerned with an effective and efficient utilization of resource-capital, material, energy, information and human. Factors affecting productivity may be (i) Technology (ii) Capital (iii) Labour Quality (iv) Economies of scale and (v) Resource allocation.

Some definitive Action Plans should now be chalked out for workers to face the challenges of the millennium in the plantations:

1. Continuous and periodically-held counselling regarding absenteeism, alcoholism, productivity, performance etc and educating womenfolk about small family, child-care, good housekeeping and health care.
2. Encouraging Thrift by the way of deposits in Post Office, Co-operative and Nationalised Banks. At the same time Life Insurance Policies can also be introduced as a part of self-awareness programme.

3. Stopping Badli system by providing opportunities/financial support to the children of the workers and hiring better workers by spotting the right person.

4. Enhancing the role of the Government in setting up Training and Motivational Centres in plantations.

5. Ergonomic studies are to be taken up for preventing fatigue and enhancing productivity.

6. Making provisions for facilitating the Women workers (constituting the major portion in the work force) with equal rights on the economic front, stoppage of sexual harassment in the work place and providing a cleaner environment for the general well-being.

7. The "Mothers' Club" Approach: The original plan was to have one mothers' club member for every 20-25 households in the garden. The only qualifications thought for membership were that the women concerned should be literate, active in community affairs and enjoy the respect and confidence of the community to which they belong. Among the many responsibilities and duties of the mother club the important ones are:

    i. Each member is to be responsible for 30-35 households

    ii. Teaching mothers on weaning food

    iiii. Education and motivating workers on family planning programmes.

    iv. Motivating through line meeting and group meetings among women.

    v. Arranging immunization programmes for children

    vi. Looking after pregnant women and arranging antenatal check-ups in garden hospital. Chlorinating drinking water sources etc.

It is imperative that the Human Resource Development (HRD) perspectives along with Human Rights Initiatives (HRI) in the Plantation Sector are of utmost importance in the direction of workers' participation in Management, their promotional scopes, the welfare measures and the fatigue study to enhancing productivity. The role of government and the positive attitude of corporate sector are to be initiated in tandem for ushering their sector into a new dimension, which is but compulsory and nevertheless, challenging along the lines suggested below:

1. Management Sensitivity

While tea industry management readily perceives a direct linkage between the nutrition and productivity of tea bushes, they exhibit a relative lack of appreciation of the input-output relationship in respect of the industry's major asset — *labour*. Even from the standpoint of purely commercial considerations, therefore, there is a convergence between the health and welfare of workers and the interests of management. Better health leads to higher labour productivity, which in turn justifies viewing health and welfare outlays as investment rather than consumption expenditure. Enhancing the health and welfare of workers calls for a good deal of coordination between the operational and welfare departments, besides close interaction with the trade unions. It also follows that, in order to maintain the major plantation asset at an optimum productivity level, the factors influencing it, such as the workers' life at home, their life in the community and their relationships at the workplace must be kept in good order.

2. Worker Motivation

In the present situation throughout South Asia, the wage structure and incentive schemes may have become an inflexible element, particularly in view of their applicability across the board for all workers within a given region or locality. It is, therefore, possible that the workers - especially, the women pluckers - may have calculated that extra work at the incentive rate is not worth the effort. This is all the more likely if the valuation women workers place on domestic chores or on leisure time at the end of the day is greater than the income derived from additional plucking. In that regard, it may be that motivational, rather than monetary factors would help to augment productivity, particularly among the women pluckers.

3. Optimizing Welfare Investments

In most cases, plantation managements are already incurring welfare expenditure for labour, although in varying degrees. If, through operational research and other means, such amounts are better spent and made more cost-effective, there will be both a quantitative and qualitative improvement in the returns on these investments and hence an augmentation in worker productivity. For this to happen, it is necessary to integrate welfare into the normal functioning of the estate.

4. Sexual Equality

In the male-dominated society of the South Asian region, the tea industry presents a refreshing contrast, being almost wholly devoid of sexual discrimination. In fact, the greater part of production activity on tea plantations centres on women workers who not only share a separate labour identity but also form the key element in the family basis of estate employment. Macro-level data from Assam in North India, for instance, suggest that the division of labour and cooperation between men and women, as well as the care of children are based on reciprocal relationships rather than on domination and exploitation. This is markedly different from the situation that exists in most rural households where, despite all the hard work put in by the womenfolk, they do not get paid for their labour. The independent economic status accorded to women tea workers and the nearly equal terms under which they operate should form the platform from which management can embark on a forward-looking personnel policy that will bring out the best in the women workers. In fact, the industry's major asset is not labour as such, but female labour. It is therefore important for management to take into account the socio-economic needs of women tea workers.

5. Controlling Common Illnesses

It has been observed that the two categories of illnesses - respiratory and water borne - account for 60-70 per cent of the diseases prevalent among tea workers in the sub-continent. These diseases are also the major contributors to absenteeism, sickness benefit costs and expenditure on drugs on estates. By and large, these illnesses are controllable through ensuring a protected water supply, proper disposal of human and animal waste, better personal hygiene and improved living conditions. By placing the emphasis on a preventive rather than a curative approach, not only will there be a reduction in costs to management but, more importantly, there will be a lower incidence of illness, with its attendant positive impact on worker productivity.

With increasing job opportunities for good material, that too in Urban areas, the problem that tea gardens are facing now and which may accentuate in future lies in management of tea garden, in that no talented person shall like to make a career in secluded life of tea garden, with problem of good education to children, job-opportunities for spouses, limited chances of promotion, lack of cultural/social activities and cut-off life from civilised world. Furthermore, there is hardly any career planning or motivation.

Obviously, tea garden may have to reconcile with second class talent to run the garden even at attractive pay scales and perks. Most gardens are also facing shortage of Work-Force to harvest green leaf in peak seasons. There is an attitudinal change in the workers and less number of workers are willing to do manual jobs. The welfare measures are a major factor as far as productivity is concerned. The laborers need to be housed well and provided with good medical facilities and amenities. The plantation labor act stipulates that a plantation with more than 1000 workers should have a full fledged hospital and that the medical facilities should be provided free of cost. This is vital because the plantation setup is geographically treacherous areas. The workers have to work in high altitudes and during heavy rains. Besides this they have to walk long distances from home to the work place. Transportation is difficult as vehicles cannot ply to all parts of the tea gardens.

The plantation is an out door oriented work in the forests and is highly physically demanding for the worker. Morcover the nutrition factors are important. The survey revealed that 100% of the workers in the south took normal rice and dhal based food. While non vegetarian food (red meat) is consumed at a minimum of once a week and not more than twice a week. Fresh vegetables are not available in village shops; in addition the workers also complain that they have no time to go to the nearest town to make any real purchases. An interconnected aspect to this is the rest time for the workers. While they have to leave home early so as to reach the work site in time, breakfast is not taken or the left over food is taken in the morning. Rest time during the working hours came up as an important factor. Absenteeism due to laziness have be reported by all the field officers and the managers and supervisors alike. The same has been presented in a different angle by the male workers. They complain of lack of rest during working hours and they say that the real time work is more than eight hours a day. Mental disorientation was also noticed amongst all the male and female workers. The whole lot reported that they are working just for a living and they do not find any happiness or pleasure or satisfaction in working in the plantation under the present wages, housing and other welfare measures provided, and also add that it is out of financial compulsion.

Development of educational facilities, right from the primary level onwards, is an important issue which both trade unions and employers must urgently address. Second, greater stress has to be laid on the development of the areas outside the plantations. This will create more avenues of employment near plantations. It will also reduce the isolation of the plantations. Here, too, trade unions and other development agencies, including the state governments, can play important role in pushing this idea through. Finally the cultural development of plantation workers is a long-neglected area. Their living conditions need to be improved.

Fresh capital inflow is needed right at this moment for the tea industry of India. Investment in new plantations and production machineries must come immediately to compete in the international market. Since tea industry has to compete globally, it is necessary that they should have access to global capital at competitive rate. This can bring life to the industry and those who live on it, especially workers.

Recognizing the fact that the tea industry's crisis in India has multiple causes, which require a variety of solutions–one of the most important steps from the government part shall be to introduce a stronger competition law to curb the misuse of corporate buying power and promote social objectives at the garden level. We believe that focusing on the role of the larger tea companies, which hold a great deal of power in Indian tea market can have a significant influence over conditions for workers on plantations and small growers.

Observations from the Fields

Observation 1: There is a huge difference in labour conditions in West Bengal tea industry when compared to south and Assam as well. This is true with respect to most of the factors deciding socio-economic conditions like wages paid, social amenities provided, living conditions, education rate etc.

Observation 2: The cause for this is multi-faceted. Number of issues can be identified to get a holistic look at the cause of the miserable condition of plantation workers. However major problem lies in politicised trade unions in tea gardens. Trade union movement in West Bengal has completely misled the labour. The Unions do not allow the promotion of a healthy work culture in the labour. The root of the problem can be identified with left front politics, which have utilised labours as their poll pawns. That is why there is no work culture. In South India, though wages paid is higher than that of West Bengal, they have a work culture and are not mislead by the trade union leaders.

Observation 3: Our survey conducted in tea growing areas of West Bengal revealed that, the labours are paid below statutory wages. This is true for some of the proprietary gardens. The Government is also equally responsible, which has completely neglected the plantation labours. It is not making use of statutory tools present to protect the workers. Apart from this, enough funds are also not being diverted to develop this sector. However, it is also the global economic slump and reduced tea consumption over years that are also responsible for the miseries of the plantation workers.

Case Study: Health Aspects-West Bengal

Kalchini's hospital was found to be the best equipped among all others. It is an old building in a reasonably good shape, with two 8-bed wards - separate ones for males and females. Besides that it has a maternity ward, TB ward and dysentery ward. There are two doctors residing near the hospital and a compounder. The compounder said that the most common diseases with which the workers suffer is malaria, dysentery and leprosy. The workers however rate TB as the most common disease. Says 19-year-old worker, "My father died of TB while my mother and elder brother suffer from this disease." However, TB cases are not treated in the company hospital, but the company provides transportation to take the patients to a nearby hospital.

Agrees a nurse in Rheabari Tea Estate, Hospital, "TB is increasing as the workers do not maintain hygiene and have close contact with the patient which only spreads this further. The children usually suffer from diarrhoea, mumps and glandular fever."

The government provides medicines for treating leprosy in the Kalchini hospital. Regular camps are held for, postnatal care and for cataract operations. There is a labour room and a well- maintained operation theatre equipped with freezers and cupboards to store the medicines and equipment for performing minor surgeries. However, the mat covering the operation table was splattered with dried bloodstains, proving that hygiene is still compromised with. The exterior of the hospital was very shabby, with naked bricks peeping from a thin coat of paint. The toilets were very dirty, with an overpowering stench of urine and moisture laden walls.Hygiene amongst the plantation workers is a cause for concern. The management insists that they are very poor in keeping themselves and their premises clean. Their way of living, quality of the cooking and clothing is kept dirty. A related problem was also highlighted by the managers and field officers. The workers have at least one set of relative living in their premises at all times. This is a constraint on the recourses made available at the labour quarters. The working area within the house is thus reduced per labour and there are constrains on the bedding and toilet facilities available. The chances of diseases spreading are very high in such situations. The relatives coming over to live in the estates is more of a social problem than a problem related directly to the estates. Sexually transmitted diseases have been reported from all the estates. Within the past two years cases of HIV have also been reported in all the estates. Most large estates in the south have the facility for screening people for HIV/AIDS. With regard to awareness of HIV all the people interviewed are aware of the disease, but the actual details of the diseases are not known to them. The reasons some cases of HIV and STD in estates is attributed to the low knowledge of the diseases as well as poor hygiene conditions. Besides this the social setup extra marital affairs, illegal relationships, multiple relationships and mental frustrations of the workers are the reasons.

Alcoholism

100% of the married female workers claimed that their husbands were regular drinkers. While 98% of the males claim that they were not. This is a contradicting situation. All the estate doctors visited that the males were alcoholic and that this would affect the productivity. They were physically weak, disoriented and frustrated which ended up in family feuds and further affected the mindset of the women workers.

Loans and Advances

During the study one more factor that came out was the aspect of loans and advances. Labourers take loans and advances from the company and from banks and LIC. They pay the premium for LIC for a period until which they can avail of a loan. This money is then used for the purchase of capital items such as land, household furniture, refrigerator, television. etc., Besides this the labourers also pawn their jewels and other belongings and borrow money from the moneylenders. Special mention has to be made to the Valpạrai area in South India. The minimum interest rates charged by the moneylenders there is 10% per month. This exorbitant rate of interest is collected from the labourers on wage day with the help of *goondas*. While this question was placed before the labourers they claimed that the wages have reduced in the recent past along with price rise. Hence they were unable to fund their children's education. There was truth in their statements as every family had their children studying in convents.

Debt-anxiety

If the density of population in a household (including dependent relatives and unemployed sons) exceeds 5, it causes stress and affects performance. Sleeplessness and intermittent claudication have been a common complaint for being absent.The recent findings on absenteeism conducted by the author on behalf of the Indian Institute of Plantation Management on the tea garden managers from Assam and West Bengal by way of administering structured questionnaire, summarize the following points:

Causes of Absenteeism

General Observations

a. Lured by higher wages in other sectors

b. Proximity to Townships brings in better job opportunities, particularly on contractual basis

c. Lack of motivation/laziness

d. "Self-satisfaction" [complacency]

e. Seasonal ailments and general ill-health during cultivation – period (Monsoon)

f. Maternity casual-leave

Personal Reasons

a. Sleeplessness/Claudication

b. Big family (Exceeding 5 members)

c. Debt-anxiety

d. Strained relationship at home

Special Observations

a. "A wet morning plays a role for absenteeism"

b. "A late night after a pay day attributes to absenteeism"

c. "Lack of good living-conditions, lack of good entertainment facilities an dill-health contribute to absenteeism".

The Indian Tea Association (ITA) and its Branches continue to undertake many projects for the people living in areas surrounding the tea estates. Such projects include financial assistance in building or extension of school and colleges, construction of Sports Stadiums and Cultural Complexes etc.

In Assam some of the corporates and the industry as a whole, under the umbrella of the Assam Branch ITA (ABITA) have undertaken a number of Agricultural Projects at different locations to provide help and support to local farmers to go in to multi cropping and yield improvement. The problem of low agricultural productivity in Assam needs to be urgently addressed. Though its soil is rich and water resources are abundant, productivity level of Assam's agriculture is handicapped by out-moded agri-practices and its farmers remain trapped in the vicious cycle of low yields-low supplies and low investments in the field. Many experts have pointed out that the North Eastern region as a whole and Assam, in particular, has tremendous potential for growth in agriculture and allied sectors.

What is needed is for the farmers to be aware of the benefit of modern methods of multi-cropping, use of high yielding varieties of seeds and along with it the use of pesticides, fertilizers etc. Some of the major companies on their own and Assam Branch of ITA , on behalf of all its member companies, stepped in to provide a support service through their agriculture projects at various locations in Assam - the services include supply of high yielding varieties of seedling at free or concessional prices along with fertilizer and pesticides, and technical and advisory through professionals.

Action By Objective (ABO) Model: Strategic Leveraging

This theory exemplifies a method of self-control and the way of fixing the target with reference to one's own strength and weakness to execute the projects. It comprises the sequential conduct of the following four functions:

- Reviewing and renovating strategies;
- Proving a job-improvement plan and congenial work environment;
- Using present and potential performance review in a systematic way; and
- Strengthening the ability and enriching the skills of the leaders through effective training and development.

In ABO approach, the objectives are handed down with no spirit of authoritarianism and as a tool for self-appraisal and self-development, it moves towards self-reliance, which is having four components: (i) Self confidence, (ii) better understanding, (iii) team spirits, and (iv) excellence in work.The whole process increases the credibility of the review results and performance levels in the participatory approach under the "bottom-up" theory; under which 'processing' in guided by ABO objectives.

The concept of ABO is basically based on five principles:

1. Caring for others
2. Helping them to solve the problems
3. This makes the team morally-driven
4. The team becomes more independent
5. This leads the team to self-realization

Thus, the ABO team members are those who:

1. Produce performers
2. Set personal examples; visible accessible on the move
3. Find ways to overcome obstacles and
4. Manage them.

Therefore with the technique of ABO, the work-culture an organization can be developed which is pointed towards Harmonization paradigm which embraces two aspects:

a. Immediate goal-achievement factors and;

b. Ultimate goal-achievement factors

The immediate goal achievement factors comprise:

i. Direction setting

ii. Resolution of conflicts and

iii. Team spirit.

These three factors culminate ultimately into ultimate goal-achievement i.e. the self-reliance -the centre of a halo from which the following four results –oriented factors are being diffused:

i. Knowledge–based performance

ii. Value-based management

iii. Development forward excellence and;

iv. Shared goals.

Finally, the Work-Culture dynamics of the HRD (Human Resource Development) depends upon two major factors, namely,

A. Selection of ABO criterion.

B. Value based mission –which propagates:

   i. Involvement of people for creating opportunities among the selves;
   ii. Innovation and growth for building-up the future
   iii. Developing inspired leadership as an integral part of strategic planning and
   iv. Forming an "enlightened" team which comprises

Operation team (for planning and execution)

Facilitator team (for day-to-day management like):

i. Review of working-style

ii. Review of problem-solving

iii. Formation of steps to be taken

iv. Addition of value to work and;

v. Job-re-arrangement for short-term and long-term strategic actions)

Vision team (for overseeing changes coming and formulating steps to be taken accordingly)

Some key issues and proposed action plans towards Workers' Welfare in Tea Industry:

Key Issues

1. Inadequate or poor facilities of housing
2. Poor facilities of drinking water
3. Lack of communication among the workers
4. Interference of belligerent unions for vested interests and not linking wage increase with productivity
5. Influence of outside agencies (like local political parties to woo votes)
6. Increase in incidence of theft (particularly green leaf) and general rowdiness due basically to unemployment
7. Insurgency problem
8. Malaria-prone areas
9. Proper implementation of PLA and Factories Act
10. The disproportionate increase in workers' family population.

Proposed Action Plans

1. Workers' education training and motivation programmes for productivity and work culture development
2. Periodically held health awareness programme
3. Implementation of PLA and Factories Act in full
4. To bring about an openness approach in the Management style and involve more people in the decision-making process
5. Population control measures and creativity awareness amongst the work-force
6. Adequate medical facilities for the work-force
7. Day-to-day functioning may be discussed in a group involving a "leading" group of workers
8. Encouraging thrift among the workers

9. Children/adolescents of the workers are to be motivated in a positive way.

Conclusion

There is need to develop suitable devices and attachments to convert their muscle power to useful mechanical power to useful mechanical power. Providing them with tools and implements capable of operating comfortable posture could enhance the productivity of workers. The tools may be push and pull type weeders, trolley sprayers, etc. these mechanical devices generate less fatigue during operation and thus result more work output. A man, on an average spends 65 to 75 percent of his working time with manually operated devices. In conventional tools time spent in use of them, is high in proportion to amount of work accomplished. During continuous work, a man can develop horsepower equivalent to about $1/10^{th}$ of his own weight (0.5 hp). There is need to improve existing tools and devices for efficient utilizations of energy. The desirable features of a good hand tool should be as under:

1. Ergonomic design to enable operator to work in least tiring posture,
2. Lightness of weight of tool for easy transportation,
3. Easy construction and comfortable operation with tool,
4. High quality of its components,
5. Low cost and facilities for local production of tools.

The working part of a tool should be designed for efficient performance of job. The handle is important as it determines position of worker and method of doing work. It has been experienced that drudgery and physical exertion are main features of Indian agriculture, which affect adversely man's intelligence and ability.

Success Case: Kanan Devan Hill Plantations (KDHP), Munnar, Kerala

The consequences of the new strategy were truly noteworthy. Within a year of the formation of new company not only were employees able to wipe off losses, KDHP Co. Ltd. earned an enormous profit. Productivity and quality too improved. New approaches have been tried out at the plantation like diversification, involvement of workers in company decisions, recruitment of business development managers etc. The birth and growth of the KDHP model has opened up a new corporate thinking in the trouble torn plantation sector in India. It is a success story of vision, courage and leadership, blended with innovation, teamwork and collective effort, and driven by hard work. The concept of employee ownership and participation in management practiced as in KDHP for the first time in the history of plantation sector. The Employee Buyout (EBO) is seen as a breakthrough in dealing with both the issue of employee entitlements and for corporate rescue. It can also address a key problem of globalization, by helping to anchor capital and jobs locally and preserve communities and skills.

The birth and growth of the KDHP model has opened up a new corporate thinking in the trouble torn plantation sector in India. It is a success story of vision, courage and leadership, blended with innovation, teamwork and collective effort, and driven by hard work. The concept of employee ownership and participation in management practiced as in KDHP for the first time in the history of plantation sector. The Employee Buyout (EBO) is seen as a breakthrough in dealing with both the issue of employee entitlements and for corporate rescue. It can also address a key problem of globalization, by helping to anchor capital and jobs locally and preserve communities and skills (Jensen, A., 2006). EBO in Tata Tea is a perfect example of the same. World Bank had recently sent a team to study the participative management in the 'KHDP model' and has plans to back similar initiatives in other major tea-growing areas of India. The Bank is not the only one to take notice of the 'quiet revolution' in the tea gardens of Munnar. Government of India isexamining the ''KDHP Model'' to revive nine other defunct tea gardens.

So successful has the 'Munnar story' of participative management been that the World Bank (WB) recently sent a team to study it, and has plans to back similar initiatives in other major tea-growing areas of India, especially the famed Darjeeling district of West Bengal and in Assam state. The Bank is not the only one to take notice of the 'quiet revolution' in the tea gardens of Munnar. Paton and Holmstron (as cited in Jensen, 2006) outline three key issues for the success of a worker buyout programme. The successful EBO needs a cohesive and homogenous group of skilled workers. This reduces the number of disagreements that arise and contains those that do. It also seems to encourage employees with the latent ability and confidence to take on managerial roles. Leadership is needed which is totally committed and able to balance the short and long term social and economic goals of the enterprise. An individual or a group as in the elected assembly board of experienced workers can provide leadership. Finally, the presence of effective and consistent advisory support mechanisms is needed. We can see that the employee buyout model in the company hastaken care of all the three aspects by Paton and Holmstron and hence remain successful.

The key design features suggested for an ideal EBO by Jensen are:

1. **Equity Participation** – collective and individual equity holdings for all employees ensures individual motivation and the long term stability of the organisation
2. **Corporate Governance** – a two tier board is recommended involving key stakeholders, enabling the legitimisation of management and the ability to deal with complexity more effectively than the Anglo Saxon unitary board.
3. **Employee Participation** – the emergence of worker self management and a team based model, restructuring the insolvent enterprise from command and control to maximise employee engagement and performance improvement
4. **Trade Union Involvement** – enabling collective bargaining over social and remuneration issues and acting as a catalyst for improved communication and training practices, resulting in high performance workplaces
5. **Networked Companies** – companies working in a cluster to leverage resources and tendering opportunities, also as a buffer against the market offering greater employment security and hence a willingness of employees to engage in workplace change and the improvement of corporate performance.

[This Paper is based on a Field Study done by the author on behalf of Indian Institute of Plantation Management Bangalore sometime back in Assam and West Bengal.]

References

Debnath K. (2003). 'West Bengal: The Neo-Liberal Offensive in Industry and the Workers' Resistance' *Revolutionary Democracy*, 9(1): 31-37.

Fernandes W., Barbara S. and Bharali G. (2003). *Children of The Plantation Labourers and Their Right To Education.* Guwahati: North Eastern Social Research Centre.

Gupta and Das R. (1990). 'From Peasants and Tribesmen to Plantation Workers', In: Karotemperal S. and Dutta Ray B. (eds). *Tea garden Labourers of North East India: A Multidimensional Study on the Adivais of the North East India*: Shillong: Vendrame Institute. pp. 1-34.

Jensen A. (2006). In solvency, Employee Right and Employee Buyouts -A Strategy for Restructuring - A Report. Ithaca Consultancy, London. KDHP News, Volume 1, No.3, January 2006

Nag A.K. (1990). 'The Condition of Tea Garden Labourers in North- East India and Its Background', In: Karotemperal S. and Dutta Ray B. (eds). Op. cit. pp. 51- 57.

Sagar R. (2002). *Plantation Labourers: Know Your Rights*. Guwahati: North Eastern Social Research Centre.

Tea: Technological Initiatives, pp. 63-89
New India Publishing Agency, New Delhi, India
Edited by Niladri Bag, Arundhati Bag and L.M.S. Palni

4

The Science and Practice of Tea Pruning and Tipping in North East India

S. Baisya

Abstract

Pruning is the most important cultural operations in tea cultivation. Time of pruning affects health of bush, incidence of pests and disease, yield and distribution of crop. Root starch reserve at the time of pruning affects the recovery of bush and hence considered as the most crucial factor for deciding time of pruning. Downward movement of photosynthates was found maximum in December and January, thereby maximizing the starch reserve in roots. Maximum recovery of tea bush and minimum loss in annual crop was found when pruned during December – January. Pruning in remaining months (February – November) may result in crop loss to the tune of 5 per cent – 37 per cent in plain areas. Clonal varieties differ in root starch accumulation during winter months hence pruning time of individual clone needs to be decided accordingly to enhance recovery of plants. Clones like TV -17, TV-18, TV- 19, TV- 20, TV- 22, TV - 23 should ideally be pruned in January while clones like TV-1, TV-7, TV-14, TV-25, TV-26, TV-30 can be pruned in December. Different types of pruning and skiffing that practiced in north east India have a direct bearing in yield and distribution of crop. The magnitude of crop loss is directly proportional with the severity of pruning and skiffing. The loss in crop in the year of prune may be as high as 60-70 per cent in rejuvenation prune compared to unprune. The quality of tea increased with the severity of pruning and skiffing. In general, about 25-33 per cent of tea bush are light pruned, 17 to 33 per cent of tea bush are skiffed and rest of the bush are kept as un prune / light skiffed in each year under commercial cultivation. Seasonal distribution of crop is important for uniform flow of green leaves in processing unit and also for proper utilisation of human resource. In addition to time of pruning, Agro - climatic condition influence the distribution of crop. Tipping affects the health and yield of tea bush. Careful selection of pruning cycle is important for economic viability and profitability of tea plantations in north east India.

Introduction

Tea as a perennial crop, if allowed to grow, would develop in to a tree producing flowers, seeds and less of green leaf to harvest as an economic crop. Pruning converts a naturally growing plant to a bush of manageable height, desired spread of productive frame and healthy branches to support a vigorous system of green leaf production.

As the age from pruning increases, the weight of pluckable shoots decreases, larger number of banjhi shoots appear on the plucking surface and more and more apical buds fail to develop in to pluckable shoots thereby making cultivation unproductive and uneconomic (Barua, 1956a).

In mature tea, periodical pruning is required to renew the branches with new set of productive foliage to keep the bush in a vigorous vegetative phase for production of succulent shoots having desired quality attributes. The objective of pruning are (Barua *et al*., 2008):

1. Renew fresh branching system on bush frame for more and more vegetative growth
2. Increase the flow of metabolites to growing shoots
3. Improve the quality of tea
4. Maintain the height of 'plucking' table for convenience of plucking
5. Reduce the incidence of pests and diseases
6. Mitigate the adverse impact of weather on the tea plant

Time of Pruning

Pruning in tea is a practice done to remove certain amount of top growth along with productive leaves. The amount of removal is proportional to the severity of pruning and bushes have to remain leafless for a certain period of time till the new layer of foliage is established. The time of pruning has a profound effect on the degree of recovery (Dutta, 1961), total crop in a year (Dutta,1956, Rahman & Mitra, 1973), seasonal distribution of crop (Dutta, 1969, Rahman and Mitra, 1973), incidence of pests & diseases (Sarma, 1956, Macalpine, 1956, Das,1957, Dutta, 1961), and abiotic stress on new growth. The time of pruning is decided by health of bush, weather parameters (Macalpine,1956, Dutta, 1957), spatial need of green leaf *etc*.

In north east India, it has been a long practice to prune teas during winter season when growth is slow and tea plants are in a state of dormancy.

Physiological Basis to Decide Time of Pruning

Following pruning, the dormant buds on pruned branches are stimulated to grow in to new primary branches, foliage is renewed and the system of plucking is re-established to support economic production, similar to a pre pruning stage. It was a matter of great curiocity that how a pruned bush without any foliage could survive and produce new growth while some bushes fail to recover. In early years of tea cultivation in north east India, the reason for mortality was thought to be due to disease infection. (Tunstall, 1938).

Starch Reserves

In tea bush there is a general tendency for die back of frames and branches after pruning and even mortality of pruned bush. Examination of uprooted dead plants indicated that in most of the cases no fungal association was observed but roots were devoid of starch reserves while the healthy plants contained enough starch (Tunstall,1938).

It has also been observed that bushes with low root starch reserves have poor and delayed recovery after pruning, manifested by thin and weak growth, dieback of branches and were subjected to severe infestation by secondary fungi such as *Colletotrichum camelliae,* Masse (Brown blight), *Pestalozzia theae,* Saw (Grey blight). In severe cases bushes die out completely (Tunstall, 1938). In addition to fungal infection on leaves, roots were severely infected with secondary fungi such as *Botryodiplodia theobromae*, Pat and *Macrophomina phaseoli* (Maubl) Ashby (Sarma, 1956).

Pruning causes a few layer of tissues on cut surface to dry up, exposing the cut surface to the infection by air borne spores of wound parasites like *Poria hypobrunnea*, *Aglaospora aculeata*, *Nectria* spp. etc. Under favourable condition, the parasites gradually extend downwards, killing individual branches in the beginning and after few years, the infection spread to main stem, finally killing the entire bush. It was also observed that, due to higher surafce area, larger cuts on thick stem found more prone to fungal infection (Sharma, 1956).

In pruned stick, callus formation take place just below the cut surface which heal up the wound prior to infection of disease pathogens into live tissue. In this process of healing, larger cuts usually requires for long time to be covered by callus tissue. In a weak bush, the process of healing is extremely slow and hence, are subjected to pathogenic infection Sarma, 1956).

A healthy bush with optimum food (carbohydrate) reserve can resist pathogenic infection and recover completely even after severe pruning. A large percentage of the weight of roots of a vigorous tea bush is starch and hence it was proposed that the starch in the roots may be considered as a criterion of the energy reserves of the tea plant (Tunstall, 1938).

Movement of Photosynthates and Starch Reserves

In a tea bush, apical and axillary buds remain dormant (banjhi) during winter season while maintenance leaves which is the principal site of photosynthesis are active. The photosynthates are distributed in different plant parts in accordance with sink capacity. Metabolic respiration in a tea bush is responsible for loss of around 60 per cent of total photosynthates, while only 8 per cent is partitioned towards growth and development of pluckable shoots (Barua 1971).

In pluckable shoots, active buds alone are the strongest sinks and consume proportionately higher amount of carbohydrates compared to other components like 1 Leaf + Bud and 2 Leaf + Bud. As the buds remain dormant in winter, photosynthates move downward and contribute in starch reserve in roots.

In tea plantation, higher percentage of banjhi (dormant) shoots are observed during the periods of October-February and again in May compared to other months when bushes produce more active shoots. Moreover, a positive correlation has been observed between banjhi shoots on top growth and starch accumulation in root (Manivel, 1981). The effect of banjhi shoots on root starch reserves are noted in Table 1.

Table 1: Effect of banjhi shoots on root starch reserve

Status	Months				
	October	November	December	January	February
Banjhi shoot (%)	32	62	96	98	98
Root starch (%)	10	14	17	17	21

The level of starch content in roots has a significant bearing on both pruning time and recovery of bushes after pruning.

It is evident in Table 1 that, proportion of banjhi shoots increased from October to December and remain in peak during January and February. As there was no top growth, photosynthates produced in leaves were directed towards roots and accumulated as starch. Similar phenomenon was observed in May when proportions of banjhi shoots were more due to inter flush dormancy.

The results of experiment using labelled 14 CO_2 at Tocklai, revealed that photosynthates at maintenance foliage moved upward during the months of March, April, June - September which coincides with active growing season and the periods of maximum harvest. Bidirectional *i.e.*, both upward and downward movement of photosynthates were observed in the months of February, May, October and November while only downward movement to frame and roots were noted in the months of December and January when the growth of tea was very slow with minimum or negligible harvest (Manivel and Hussain, 1982).

It is clear that December and January are the months when accumulation of starch in roots can be considered as optimum compared to other months in a season. Tea bushes, if pruned even severely in these months will recover properly with minimum chances of dieback of pruned sticks and mortality of bushes.

Varietal Difference in Root Starch Reserve

On an average, in plain areas, starch researve in root increases progressievely from November and reach a peak in February (Fig. 1) while tea growing in high altituded areas, the peak in root starch reserve was observed in January (Fig. 2).

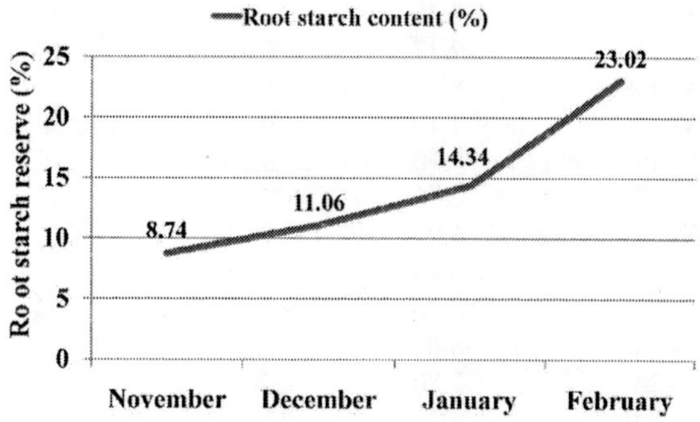

Fig. 1 : Average root starch reserve (%) of thirty (30) TV clones of tea during November – February in plain areas

A distinct clonal variation was observed in respect of starch accumulation in root (Barman *et al.*, 1998) and presented in Table 2. A minimum of 10 per cent starch needs to be present in roots at the time of pruning in plain areas of north east India to optimise recovery and to maximise yield. Although, December is the ideal time for pruning but some of the varieties like TV-

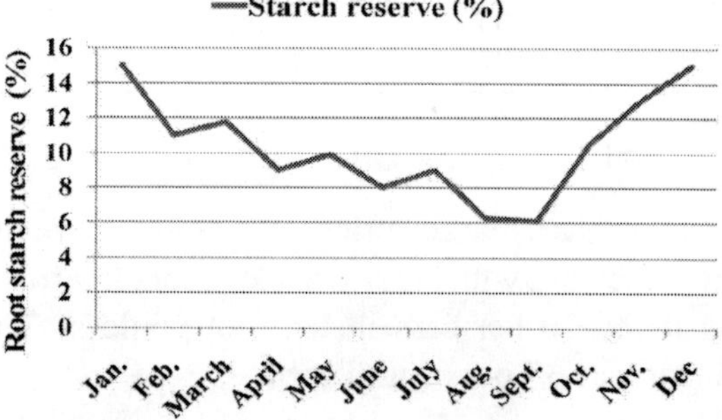

Fig. 2: Average pattern of root starch reserve of Darjeeling clones grown in mid elevation (Saikia *et al.*, 2007)

17, TV-18, TV-19, TV-20, TV-23 should ideally be pruned in January as starch reserve in roots of these clones attain the minimum limit of 10 per cent required for optimum recovery of bushes following pruning.

Table 2: Root starch reserve (%) in TV clones during November – February (Barman *et al.*, 1998)

TV Clones	November	December	January	February	Mean
TV-1	8.68	10.96	13.32	13.34	11.58
TV-2	10.34	13.57	15.56	16.42	13.97
TV-3	8.81	14.28	17.01	16.09	14.05
TV-4	9.64	13.48	15.30	16.01	13.61
TV-5	8.99	11.57	14.40	15.60	12.64
TV-6	11.30	12.17	13.78	15.49	13.18
TV-7	11.73	13.72	18.08	18.72	15.56
TV-8	10.20	12.49	16.66	19.29	14.66
TV- 9	9.11	15.86	16.20	19.63	15.20
TV- 10	10.93	13.92	16.21	18.33	14.85
TV- 11	10.57	10.87	14.42	19.80	13.91
TV- 12	11.73	12.52	16.80	17.54	14.60
TV- 13	8.78	10.91	11.74	12.21	10.91
TV- 14	8.43	10.23	16.01	16.41	12.77
TV- 15	9.37	10.42	12.13	16.28	12.05
TV- 16	8.34	10.44	11.45	16.93	11.79
TV- 17	7.49	9.26	10.60	17.05	11.10
TV- 18	7.10	8.28	12.18	13.18	10.34
TV- 19	7.60	7.59	11.57	15.03	10.45
TV- 20	6.95	7.93	13.65	13.95	10.62
TV- 21	6.89	8.59	12.60	14.93	10.75
TV- 22	8.15	9.04	13.18	15.03	11.35
TV- 23	6.04	8.63	13.43	15.68	10.94
TV- 24	7.80	9.60	12.12	13.57	10.77
TV- 25	9.41	12.84	15.06	18.68	14.00
TV- 26	9.83	10.96	15.76	21.73	14.57
TV- 27	8.01	11.02	13.20	15.36	11.90
TV- 28	8.13	10.16	12.90	13.77	11.24
TV- 29	11.29	16.60	19.42	23.06	17.59
TV- 30	10.30	13.13	14.89	16.47	13.70

Effect of Elevation on Starch Reserve

Elevation has a marked effect on starch reserve in tea. In general, starch reserve in root increase with the rise in elevation. In Srilanka, at sea level, starch content was found 10 per cent in tea root while at 7000 ft above sea level, starch content was 37 per cent (Tubbs, 1934).

Consequently, the phenomenon of die back and mortality of pruned bush was less at higher elevation than tea growing in low elevation.

The effect of elevation on starch accumulation in root was distinctly identified in an experiment with three clones specifically recommended for growing in Darjeeling hills. These clones accumulated higher starch in roots when grown in high altitude tea gardens compared to gardens in low altitude (Table 3).

Table 3: Effect of elevation on root starch reserve (%) of Darjeeling clones

Elevation	Monthwise starch reserve (%) in roots						
	Clone	October	November	December	January	February	March
Low	P312	18.07	17.25	26.28	17.11	9.92	9.54
	B157	13.66	13.06	19.39	18.46	10.58	6.17
	T78	8.22	11.38	19.20	20.41	11.79	6.85
High	P312	18.51	25.29	29.25	22.58	18.49	12.59
	B157	13.95	15.89	25.44	20.24	12.50	10.18
	T78	13.13	14.60	22.58	21.54	18.51	11.43

The effect of elevation on starch accumulation in root was also evident in clones TV-14 and TV-19 (Figure 3) having Assam and Cambod characters, respectively which are mainly grown in the plains.

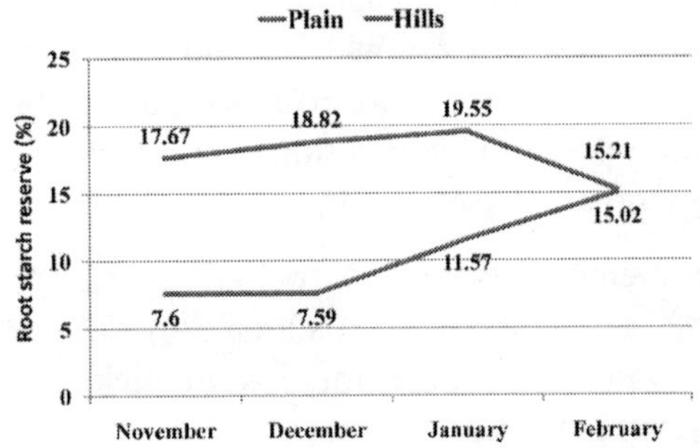

Fig. 3: Effect of elevation on root starch reserve of clone TV-19 during November - February

Similar to clonal variations in the plains, variations in root starch content were observed in clones grown in Darjeeling hills (Barman, 2009) and most of the clones can be pruned in November as root starch reserve exceeds the threshold value of 12 per cent. However, clones like Balasun - 7/1A/76 and Tukdah - 253 preferably be pruned in December for better recovery after pruning.

There are few clones like Tukdah-383, Bunnockburn - 157, Balasun 7/1A/76, Badamtam 15/263 grown in Darjeeling hills accumulate starch less than the threshold value in the month of December. Such clones should preferably be pruned in later part of December – early part of January.

In general, pruning in high altitude areas may best be completed within December to avoid a biotic stress like low temperature and frost which otherwise can hamper post pruning recovery of tea bush.

Starch Accumulation and Age of Tea Bush

Young plants highly succeptible to die back after pruning. Deterioration in health, even mortality of young plants after pruning was observed in commercial plantations in north east India, mainly due to poor starch content in roots at the time of pruning.

It is obvious that young plants are yet to develop a root system which is capable to store enough starch. In the contrary, mature plants with fully developed root system can store more starch to support new growth after pruning (Tunstall, 1938). Therefore, it is important to ensure sufficient starch reserves in roots prior to pruning and the month, January can be considered as the best for pruning in young tea.

Starch Reserve and Thickness (girth) of Roots

There are two types of root in tea, the youngest one is white - creamy in colour and termed as feeder roots while the older one is red in colour hence termed as red root. Feeder roots are responsible for absorption of water and nutrients. Red roots perform the function of conduction of water and food and also act as storage organ of carbohydrate beside anchorage to tea bush.

Experiments conducted by Tocklai revealed that root with a diameter of about 0.7 cm can preserve more starch (%) followed by 0.5 and 0.25 cm thick. It may be due to less respiratory loss in thicker roots as they are relatively older in age (Barman *et al.*, 1998).

Methods to Optimise Starch Reserves before Pruning

As leaf is the principal component for synthesis of carbohydrate, increse in effective leaf area would help to enhance starch accumulation and better partitioning in root particularly during December and January. A maintenance leaf never imports food (assimilate) from other leaves (Manivel, 1978) and retains about 19 per cent of carbohydrate assimilated in each month while rest are distributed in other plant parts (Barman and Saikia, 2005). In commercial cultivation, maintenance leaves in each bush must be productive and should be free from pest and disease to enhance the synthesis of carbohydrate.

Rest to Bushs

Pluckable shoots mainly consist of 2 Leaf + Bud which act as strong 'sink' of photosynthates produced by maintenance foliage. Therefore, it may be assumed that, if 2 Leaf + Bud growth is retained or kept in a physiologically dormant state, the photosynthates will flow to other vegetative parts and the flow will be more towards root during November. Leaving bushes unplucked prior to puning is useful to build up starch in roots (Sarma, 1956, Barua, 1956b, Barua,1966, Rahman, 1978, Rahman,1988) and a period of 4-8 weeks are found optimum for rest.

The length of resting will depend on health of bush and type of pruning to be given. In general, period of rest is directly proportional with the severety in pruning.

Bushes with optimum vigour should not be rested unnecsserily for long period to avoid excessive thickening of the wood and delay in bud break which may reduce the harvest of early season crop. Resting of bush may commence by about the middle of October when the top growth slows down with the onset of the cold weather so that the carbohydrates produced by the leaves are induced to move downwards into the roots for storage (Barua, 1966).

Potash as an Additive is Found to Enhance Starch Accumulation in Roots

Potash known to be beneficial for starch build up in tea. The dose of potash in annual fertiliser application programme should be on the basis of soil test value. In case, available potash in soil is less than 100 ppm, N : K fertilisation should be followed in the ratio of 1: 1.25/1.5 to enhance potash status in soil.

Foliar application of MOP @ 2 per cent at an interval of 21 days prior to pruning enhanced starch accumulation in root to the tune 69 per cent compared to no foliar application (Barman *et al.*, 2003). The benefit of higher starch reserve in root by foliar application of MOP was reflected in increased number of laterals on frame and finally on yield of pruned tea.

Effect of Flower Bud on Starch Content in Young Plants

Young plants are expected to grow vigoursly in a vegetative stage without production of flowers and seeds. In commercial plantation, presence of flower buds in some plants are not uncommon. In general, it is an indication that the plant is in stress either due to biotic or abiotic factors. In young plant, production of flowers and seeds does not have any economic importance rather it consumes a considerable amount of photosynthates assimilated by limited number of leaves meant for growth of root and above ground vegetative organs. Field experiment at Tocklai as presented in Table 4 indicated that flower buds have a deletorious impact on starch accumulation in roots, while its removal can increase starch reserve to an extent of 80 - 90 per cent (Barman *et al.*, 1990).

Table 4: Effect of flower bud, removal of flower bud on root starch in young tea plants

Conditions	Root starch (%)
Without flower buds in November	10.6
With flower buds in November	3.7
Flower buds removed in November	7.0
Without flower buds in December	13.5
With flower buds in December	5.0
Flower bud removed in December	9.0

Flower buds should be removed manually with utmost care without any damage to the axillary bud. It is also important to remove flower buds at very early stage to maximise starch accumulation in root.

Young plants with higher incidence of flower buds should not be considered for pruning, rather should be allowed to recover from stress and improvement in growing condition to avoid die back or mortality.

Effect of Time of Pruning on Total Yield and Distribution of Crop

Effect on yield

Time of pruning has a profound effect on the harvest of total crop in a season. Field observation indicated that December was the best time for pruning for quantity and quality of crop (Wight, 1940) under annual system of pruning which was largely followed till 1960's. Long term experiments carried out in north east India indicated that under annual pruning system, pruning in February – October resulted 5-37 per cent loss in seasonal crop (Table 5) compared to pruning in December (Dutta and Grice, 1966) while, in cases of skiffed teas the loss was less. It was frequently observed that pruning in actively growing months when starch reserve in root was less, result in severe die-back of branches and even mortality of bushes. Untimely pruning may expose bush frame to strong sun leading to severe sun scorch. Lesions caused by sunscorch are the most common entry points of both termite as well as *Poria*. Besides, productive branches are also lost due to die back following damage by sun scorch (Barbora *et al*, 1982).

Pruning in October gives more 2nd flush crop but bushes may suffer from fungal disease induced by sun scorch on frame (Tunstall, 1940). December – January may be considered as optimum period for pruning to harvest the potential crop of the season in north east India.

Climatic parameters play an important role in deciding the time of pruning and skiffing. In general, in drought prone areas, pruning will have to be delayed to reduce the possibility of having too much new tender growth during the droughty period, however, should be completed by middle of January to maximize yield and any further delay in pruning will result loss in total crop to the tune of 10 per cent for every month's delay. In drought prone area, skffing, may be delayed till late January. In non droughty gardens pruning may best be completed within December.

The results of regional trial carried out by Tocklai with the pruning cycle of LP – DS – MS, indicated that pruning when delayed beyond middle of January, there was loss in annual yield to the tune of 17 Per cent in North Bank, 3-6 per cent in Upper Assam, 7-8 per cent in South Bank, 10-11 per cent in Cachar and 3-6 per cent in Dooars depending on weather condition (Chakravartee and Barbora, 1998)

Effect of Time of Pruning on Distribution of Crop

In commercial tea plantation, availability of green leaf /day is important for proper utilisation of manpower required for harvesting, transportation of green leaf, installed capacity of machineries to process green leaf etc. Moreover, price of tea does not remain same throughout the season but usually remain high during 1st and 2nd flush period. Therefore, time of pruning on distribution of crop in a cropping season has a special significance in commercial tea plantations.

Wide variations in spatial crop due to different times of pruning was observed in Assam and Dooars (Table 5, Table 6). Pruning and skiffing done at same time brought about different response in crop in Assam and Dooars which may primarily attributed to prevailing climate. Obviously, time of pruning and skiffing to obtain a preferred crop distribution is location specific.

Table 5: Effect of time of pruning on total crop and periodic crop distribution under annual pruning. Pruning in December taken as the standard of 100%

Pruning months	Monthwise crop distribution (%)		
	Whole season crop	1st and 2nd flush	Main & backend crop
January	104 (+4)	16 (-5)	88 (+9)
February	94 (- 6)	12 (-9)	82 (+3)
March	96 (- 4)	9 (-12)	87 (+8)
April	76 (- 24)	13 (-8)	63 (-16)
May	79 (-21)	20 (-1)	59 (-20)
June	66 (-34)	31 (+10)	35 (-44)
July	63 (-37)	36 (+15)	27 (-52)
August	73 (-27)	39 (+18)	34 (-45)
September	80 (-20)	32 (+11)	48 (-31)
October	77 (-23)	24 (+3)	53 (-26)
November	95 (-5)	23 (+2)	72 (-7)
December	100	21	79

Table 6: Effect of time of pruning and skiffing on crop distribution pattern (%) under different agroclimatic conditions of North East India (Barua *et al.*, 2008).

Type of tea	Early- Crop (March - June)		Mid- Crop (July - September)		End - Crop (October - December)	
	A	D	A	D	A	D
Pruned tea						
Mid October	44	35	56	62	-	3
Mid November	26	26	65	61	9	13
Mid December	18	25	62	56	20	19
Mid January	15	25	61	55	24	20
Early February	12	24	64	54	24	22
Mid February	10	23	64	54	26	23
DS tea						
Mid October	42	34	58	64	-	2
Mid November	33	27	61	61	6	12
Mid December	25	27	59	58	16	15
Mid January	24	24	57	58	19	18
Early February	21	23	60	58	19	19
Mid February	20	22	60	59	20	19
MS tea						
Mid October	46	33	54	66	-	2
Mid November	35	25	57	63	8	12
Mid December	28	25	53	57	19	18
Mid January	24	21	53	58	22	21
Early February	23	23	54	57	23	20
Mid February	22	21	54	57	23	22

A = Assam, D = Dooars

Type of Pruning and Skiffing

Till 1960's in commercial tea plantations, each tea bush was subjected to moderate to severe cut on bush frame keeping either none or minimum wood allowance from previous cut at the end of season and referred to as annual pruning (Rahman, 1978). In a major shift from severe cut in annual pruning, a lighter form of cut on tea bush was introduced and termed as "skiffing". In skiffing, various forms of cut have been in practice primarily to obtain higher annual crop, more early season crop, desired crop distribution in a season, increase the health of bush.

Effect of Pruning and Skiffing on Yield and Quality of Tea

Pruning is associated with loss in crop and the loss is proportional to the severity of pruning. Compared to unprune (UP), the loss in crop could be about 60-70 per cent in the year of medium prune (MP) and about 30-35 per cent in light prune (LP).

Introduction of skiffing in tea in a pruning cycle reduced the loss in crop compared to annual pruning. The possible increase in yield under different skiffing over light prune is noted below:

Deep skiff	:	10- 15 Per cent
Medium skiff	:	15 - 20 Per cent
Light skiff	:	20 - 25 Per cent
Level off skiff and unprune	:	30 - 35 Per cent

Besides effect on total crop, each type of pruning and skiffing, results into a specific pattern of crop distribution.

Pruning and skiffing has a marked effect on quality of tea. With the standard method of manufacture, green leaf harvested from unpruned tea has a lower level of quality than leaf from pruned or deep skiffed tea (Dutta,1959, Rahman and Mitra, 1973). Study conducted at Tocklai (Table 7) revealed that quality of tea increased with the severety of cut in pruning and skiffing in 2nd flush and onwards in a season. However, a reverse trend was noted only in 1st flush (Chakravartee, *et al.*, 1998) .

Table 7: Effect of pruning and skiffing on quality scores by tea taster under different flush of growth in a year.

Type of pruning and skiffing	1st Flush	2nd Flush	Rain crop	Average of full season
Light prune (LP)	8.90	13.64	7.87	10.0
Deep Skiff (DS)	8.75	12.55	7.77	9.89
Medium Skiff (MS)	9.00	11.45	7.66	9.37
Unprune (UP)	9.89	7.75	7.50	8.38

Bio chemical analysis carried out at Tocklai indicated that Theaflavin contents of clones TV-2 and TV-9 were higher in teas made from pruned bushes than those made from unpruned bushes. In addition, both the Tea Taster's at Tocklai and London gave higher valuation to teas made from pruned bushes compared to unpruned bushes (Basu and Dev Choudhury, 1984).

Types of Pruning

Rejuvenation prune

The unproductive wood in frame increases with the increase in age of tea bush. More energy is consumed to support the old and unproductive wood of the bush frame. Rspiratory loss of photosynthates found maximum in stem accounting to about 32 per cent of total respiratory loss in a plant (Barua, 1971). Therefore, removal of excess unproductive wood by lowering down the bush frame through pruning considered essential to reduce respiratory loss and thereby to improve harvest index (Barbora and Hazarika, 1996).

During early years of tea cultivation, collar pruning (very low pruning as close as six inch to ground) was practiced to induce vigour to the old and unproductive bushes in commercial plantations but this practice was associated with higher rate of mortality and crop loss. The per cent death of bush (as high as 25 per cent) and crop loss after collar pruning were directly proportional with the severity of prune (Dutta, 1956). Therfore, the practice of collar pruning was abandoned in commercial plantation (Sarma and Bezbaruah, 1981).

Barua (1971) advocated rejuvenation pruning as an aid to bring up the productivity of old and aging tea areas with following objectives. The method suggested basically involves

1. To remove the unproductive wood in frame infested with diseases and pests.
2. To restore/increase population density by infilling the vacancies.
3. To improve environmental factors like shade, drainage etc.

Practical considerations related with rejuvenation prune are noted below.

(a) Pruning height

i. In plain areas

Cambod/Assam type bushes	:	40 - 45 cm from ground level.

ii. In hills

Assam type bushes	:	25 - 35 cm from ground level.
Chinery type bushes	:	15 - 20 cm from ground level.

(b) Criteria for selection

i. Bushes must be healthy to tolerate shock of severe pruning. In practice, it was observed that recovery of Chinery and hybrid jats has been better than Assam type of bush.

ii. Bushes in the section must not be programmed for uprooting within next 12-15 years.

iii. Bushes should be free from primary root disease.

iv. Incidence of stem disease down to the collar region caused by *Poria hypobrunia*, *Aglospora*, etc should not exceed 20 per cent of the bush.

v. Vacancy in the section should not exceed 25%

(c) Pre pruning field operation

Planning for rejuvenation pruning should be done well in advance and a work plan should be prepared specific to the concerned section after considering all conditions of the field like soil fertility, drainage, health of shade trees, number and type of planting material required for infilling etc.

i. Identify the unwanted and diseased shade trees and ring bark them two year in advance.

ii. Vacant patches should be planted with green crop like Guatemala or Citronella grass, two year in advance to improve soil health.

iii. Plant fast growing temporary shade tree species at least three years in advance.

iv. Ascertain chemical properties of soil. Correct the pH with amendments and improve available potash in soil if required, on the basis of soil test value.

v. Improve drainage as much as possible preferably a pruning cycle in advance.

vi. Raise young plants in nursery well in advance for infilling or interplanting.

(d) Field operation

i. Rest the bushes for a period of 6-8 weeks prior to pruning to enhance health of bushes.

ii. Spray 2 per cent MOP as foliar feeding during the period of rest at an interval of 3 - 4 weeks.

iii. Protect the foliage from leaf defoliating insects during resting.

iv. Apply additional dose of potash and phospatic fertiliser to soil as per soil analysis report.

v. Prune bushes at a height to remove most of the unproductive wood and majority of knots present in frame.

(e) Post prune operation

i. Treat the cut surface either with spray of copper fungicide or with a paste of 20 per cent Trichoderma formulation within 48 hours of pruning.

ii. Remove remaining knots and diseased woods from frame and treat the cut surface with Trichoderma formulation or copper fungicide.

iii. Cover all the cut surface with bituminous paint.

iv. Uproot bushes infected severely by stem disease.

v. Treat the bush frame with alkaline wash.

vi. Level the ground and dig new drains if required.

vii. Infill the vacancies with new plants after the receipt of spring showers or before monsoon. In case of wider spacing, interplanting with a new row of young plants will be a viable option to enhance bush population / unit area.

Medium Prune

There is a rise in frame height and plucking table after completion of each pruning cycle. As the height of plucking table reaches to about 100 cm from ground, plucking becomes difficult and less economic, medium prune is given at a height between 50-60 cm from the ground level, primarily to reduce the frame height. Additional benefits of medium pruning are

i. Reduction of unproductive wood and knots from bush frame.

ii. Infilling of vacancies created due to death of bushes.

iii. Removal of diseased and insect affected branches.

(a) Pre pruning operation

i. Rest the bushes prior to pruning to enhance starch content in roots.

ii. Improve chemical properties of soil on the basis of soil test value and improve drainage of the section.

(b) Field operation

i. In this type of pruning, height is so adjusted that original structure of bush developed by "frame formation prune" are retained while, majority of knots are removed.

ii. To enhance starch accumulation in root, measures like resting of bushes and foliar application of potash prior to pruning should be done as described earlier.

iii. Treat the cut surface either by spray of Copper fungicide or with *Trichoderma* biocide as noted earlier.

iv. Cover the larger cut surface with bituminous paint.

Light Prune

This pruning is commonly practiced after the end of each pruning cycle keeping a wood allowance of 4 - 5 cm above the height of previous pruning mark. The primary objective of light pruning are noted below.

i. Removal of the top hamper of bush
ii. Regulation of the distribution of crop
iii. Reduction in the incidence of pests and diseases
iv. Maintenance of ideal height of plucking table
v. Improvement in quality of tea
vi. Mitigation of the ill effects of drought

In commercial tea plantation, presence of knots (lump) in bush frame is a common occurrence. Knots in bush frame is highly detrimental and associated with reduction in crop, dwarfing of shoot size, die back of plucking points, poor bush health and frequent occurrence of *banjhis* affecting quality of green leaf. Primaries arising from knots go banjhi at lower heights. In addition to loss in crop, these primaries do not provide a good foundation for longer pruning cycles (Barbora and Banerjee, 1988).

Formation of knot is linked with faulty pruning and tipping (Dutta, 1956). Pruning without adequate wood allowance at the time of pruning stimulates the regeneration of multiple primaries from the axil buds of the *janam.* Repeated pruning at same height, wood at the pruned level fuses to form a lump or a compound knot. Movement of nutrients and photosynthates is obstructed to a great extent due to restriction in the development of conducting tissues (Barbora and Banerjee 1988). Experimental findings have confirmed that 4-5 cm of wood allowance should be maintained during light prune operation.

Deep Skiff

With the adoption of 3-4 year pruning cycle, to maintain a balance between yield and quality of crop, deep skiff is imparted in a cycle. This is the severest type of skiffing operation. It is done normally between 12-15 cm above the last height of light prune. At this level, a fork on each of the primary is available thereby ensuring higher crop. The objectives of deep skiff (Dutta, 1964) are

i. To maintain a balance between crop and quality.
ii. Reduction in the ill effect of drought.
iii. Regulation in the distribution of crop.

Medium Skiff

Medium skiff is given to remove the current year's crow's feet. When it is done on light pruned tea tipped at 20 cm or deep skiffed tea tipped at 10 cm, this will be a cut at a height of 15 cm from the pruning mark or 5 cm below the last tipping level. However, when medium skiffing is done after one or more year/years of unprune/light skiff/level off skiff, this skiff is given below the "Crow's feet" formed by the last year's plucking. The objectives of medium skiff are

i. Regulation in the distribution of crop

ii. Reduction in the incidence of excessive banjhi formation

iii. To get more early and total crop (Grice, 1964)

iv. Avoid the problem of excessive creep formed during the previous season

Light Skiff

Light skiff is a cut given at or up to 1 cm above the previous tipping level i.e. just deep enough to remove majority of the plucking points. It would be a cut through the green wood above the crow's feet. The objectives of light skiff are

i. To obtain more total and early crop

ii. To keep the plucking surface flat and levelled

Level of Skiff

Level off skiff is a trimming operation to level the uneven plucking points that protrude above the plucking table at the end of the season. It is a cut which retains more green tips of twigs on the plucking table. It is considered as unprune and is done with same objectives as of light skiff.

Cleaning out

Experiments have shown that the presence of *banjhi* shoots on a pruned bush retard the growth of new shoots, resulting in the loss of quantity and quality in valuable second flush crop (Barua, 1956).

Therefore, *banjhi* shoots along with thin, twiggy and unproductive branches should be removed from pruned frame (Barbora, 1975). Presence of dead wood, diseased wood on frame is detrimental from sanitation point of view and needs to be cut back to healthy wood. The cut surfaces should be polished properly with a sharp knife and to be covered by bituminous paint to restrict the entry of wound parasites during the cleaning out operation following general pruning.

Breathers

In severe pruning like medium pruning, leaving a healthy branch on frame towards South and South – West side has been a practice in north east Indian plantations. Foliage retained as breather contribute photosynthates and mobilise reserve food from roots for development of new shoot after pruning (Rahman 1978, Barbora and Hazarika 1988, Rahman, 1988). Breathers are beneficial towards prevention of sunscorch on frame which otherwise usaully exposed to sun after pruning.

The essentiality of keeping breather is not universal, as bushes which have high in starch reserve at the time of pruning particularly in Darjeeling hills, will not benefit from the breather. However, bushes poor in health, devoid of optimum starch reserve and under severe pruning are to be benifitted from breather. Barua (1972) advocated retention of atleast two breathers on each bush received severe pruning. Breathers are to be removed when majority of dormant bud breaks open.

Bush Sanitation

Alkaline wash

It is often observed that bush frame is covered with lichens, mosses which not only affect the health of bush but also act as a site for pupation of harmful insects like thrips. To eradicate such growth, pruned frame should be treated with Alkaline wash which can be prepared by mixing 6 kg washing soda and 2 kg quick lime in 100 L water and thorough stirring (Barbora and Hazarika, 1996). The mixture can be applied locally or may be sprayed on entire frame. Alkaline wash on frame should also to be applied to sections where plants are infected with fungal disease like Black rot and Thread Blight. In adition to bush sanitation, alkaline wash is useful in softening the bark in old teas and breaking of dormant bud.

Time of use

The use of alkaline wash should be restricted to pruned bush. It is to be applied, after pruning and well before the emergence of new buds from the frame.

Pruning Cycle

Pruning cycle refers to the duration in years between two successive light prunings. In practice, successive light pruning is done after a gap of 3 - 6 years in which bushes are kept in well defined sequence of unprune or skiffed like

LP-UP-UP, LP – UP- DS, LP-DS-UP, LP-UP-DS-UP etc. The system of pruning cycle replaced annual pruning. Pruning cycle provide more flexibility in yield, quality and well distribution of crop in a season.

The concept of pruning cycle were first introduced in the hilly areas of Darjeeling to harvest more total and early crop in the short growing period (Rahman and Mitra, 1973) owing to prevailing cold environment. Thereafter, longer pruning cycles has been a regular practice adopted in other tea growing areas in north east India primarily to harvest more total and early crop.

Each type of pruning and skiffing results in to a different pattern of crop distribution in each month of season and flushing period (Table 8, Table 9).

Table 8: Effect of pruning and skiffing on monthly distribution of crop (%)

Type of prune/skiff	Mar.	Apr.	May	Jun	Jul.	Aug.	Sep.	Oct.	Nov.	Dec.
LP	-	1	5	11	15	17	18	20	10	3
DS	-	3	9	11	17	18	17	15	8	2
MS	-	6	10	13	17	17	16	13	8	0
LS/UP	9	9	10	13	14	14	13	12	6	0

Table 9: Effect of different pruning and skiffing in crop distribution (%)

Type of pruning and skiffing	1st Flush	2nd Flush	Total of 1st 2nd Flushes	Rain crop	Backend	Total
Light Prune	2	20	22	63	15	100
Deep Skiff	3	25	28	63	9	100
Medium Skiff	5	27	32	61	7	100
Light Skiff	10	30	40	55	5	100
Level of Skiff	12	33	45	52	3	100

As regards to average yield/year, three year pruning cycle of LP-UP-UP and LP-DS-UP brought about an increase in yield to an extent of 34 per cent and 26 per cent, respectively compared to light prune year (Chakravartee, 1999).

Choice of Pruning Cycle

One particular type of pruning cycle may not be suitable to all tea growing areas nor even in one particular tea garden. Age and health of bush, kind of planting material (variety), height of plucking surface, soil and climate, type of tea to be manufactured, garden's own preference on quality or quantity, infrastructure of an estate etc. play a deciding role on the selection of pruning cycle.

It has been observed in the plain areas that a 3- year pruning cycle is suitable for young and youngish-mature vigorous tea up to the age of 15 years. Thereafter a 4 - year pruning cycle can be practiced. In case of 3 - year pruning cycle, LP-UP-UP is crop oriented while, LP-UP-DS or LP-DS-UP is quality oriented.

As regard to total crop in a season, 4 – year pruning cycle of LP-UP-DS-UP will have a definite edge over a 3- year pruning cycle of LP-UP-DS, while the former pruning cycle will suffer more than the latter if drought sets in.
Four year cycle is the longest that can be considered for the plains of N. E. India because the wood at the time of next pruning should not be allowed to become thicker than 1.5 cm Rahman and Mitra, 1973). Vigorous bushes under good growing conditions require a shorter cycle of not more than three years to avoid excessive thickening of wood. On the other hand, a four year cycle is considered suitable for average growing conditions.

If the cycle is lengthened beyond this limit, some of the branches become very thick, leading to a substantial reduction in the number of pruning sticks in future. It appears that there is severe competition between branches in the top hamper and the weaker branches which would have survived in a shorter cycle get smothered in a long cycle. This phenomenon is more pronounced in light leaf Assam jats compared to hybrid and Chinary jats.

It has been well established that lighter the skiff, higher will be the early crop and total crop. At the same time, unprune or lighter form of skiff produces shoots inferior in quality compared to light prune (Dutta, 1956) and deep skiff. In extended pruning cycle, lighter form of skiff or unprune teas produce more thin and banjhi shoot affecting quality of made tea. Moreover, unpruned tea normally suffer more from abiotic and biotic stress like drought, diseases like Black Rot, Thread Blight and insect pests like Red Spider mite (Dutta,1956). Hence, where these are prevalent, it is best to have less area under unpruned tea.

Elevation has a marked effect on length of pruning cycle. At higher elevations, low temperature coupled with reduced direct sunlight, in monsoonal period adversely affect the length of active growing period. In general, the rate of growth of tea at high elevation is slow. In Darjelling hills, prior to 1950's two (2) year pruning cycle of LP-UP and three (3) year pruning cycle of LP-UP-UP were widely followed in commercial tea plantations at low and mid elevation, respectively. As price realization was higher in early season crop, the estates gradually adopted extended pruning cycle 4 - 6 years keeping bushes either un prune or skiffed between two light prunes to increase 1st and 2nd flush crop.

Long term experiments in Darjeeling hills conducted by Tocklai brought about following findings (Hazarika *et al*, 1985).

a. Average annual yield (kg/ha/year) showed an increasing trend with increase in the length of pruning cycle to three year or more compared to biennial pruning cycle.

b. In chinary type of bush, LP – UP – LS was found superior than LP –UP - UP in terms of higher average yield/year and harvest of less *banjhi* shoots.

c. A four year pruning cycle of LP – UP – MS - UP brought about higher total yield and higher percentage of early season crop than three year pruning cycle. The four year cycle produced more early season and total crop than six year cycle.

d. In general, lighter the pruning operation higher was the number shoots/kg of plucked green leaf. Light skiffed bushes produced about 51 per cent more shoot per kg of green leaf compared to light pruned bushes. Therefore, deployment of plucker will vary with pruning type.

In addition to elevation, aspects of slopes play an important role in growing condition thereby affecting the length of the pruning cycle. Growth of tea growing in northern aspect of slope is much slower than southern and western aspect due to less availability of sunlight.

Pruning Tools

Traditionally sharp knife made up of mild steel blade and wooden handle fabricated by local blacksmith used for pruning and skiffing in tea plantations.

A knife of 15 cm long blade is considered suitable for pruning in young and youngish - mature tea while 20 cm long blade found convenient for pruning in mature tea. The pruning knife should weigh about 450 gm.

For cleaning out in prune tea, a knife of about 7.5 – 10 cm long blades are to be used. To enhance the perfection in cutting branches during pruning, it is important to use knife of proper specification. Pruning with proper knife would minimize bark damage and splitting of wood. Clean cut on wood is beneficial for covering the cut surface by cambium growth.

Mechanised Pruning

Experiments conducted at Tocklai revealed that shoulder mounted hand operated pruning machine fitted with revolving Chip saw blade of 40 - 100 teeth can be used for pruning (RP, MP and LP) without any detrimental effect on health and yield of tea bush. Use of such machine in tea plantations for pruning is gaining popularity primarily for higher productivity to tide over the shortage of expert manpower.

As regards to effect of mechanized pruning, no adverse effect on health of tea bush were observed (Barbora and Sharma, 1999).

Tipping

In pruning and skiffing, certain amount of foliage is removed from top hamper of bushes. The quantum of removal is directly proportional with the severity of pruning. In case of medium pruning and rejuvenation pruning, bushes are completely devoid of leaves while, removal of leaf is minimal in light skiff.

Following pruning (RP, MP and LP), dormant buds hidden below the bark in frame or in pruned stick become active to grow in to a shoot and finally to a primary branch, adding new leaves capable to perform photosynthesis to meet the food requirements of entire bush. In this process, the primary branches are allowed to grow to a certain height and then decapitated at a predetermined height, parallel to the ground surface primarily to facilitate more lateral growth from axillary buds and a flat surface. The process of decapitation of primaries at a selected height is known as tipping.

Height of Tipping

The height of tipping has a direct bearing on the thickness of primary branches, quantum of leaves, leaf area index, yield in subsequent plucking and quality of harvested green leaf.

Physiological Basis for Deciding Tipping Height

As explained earlier, in pruned bush, no foliage is present for photosynthesis. Therefore, a new layer of productive leaves termed as maintenance foliage are retained on bush before initiation of regular plucking. As a matter of fact, the health and productivity of each bush depends upon the size and efficiency of maintenance leaf.

It is worth to note that 2 leaf and a bud harvested as economic crop is photosynthetically not efficient and largely depends on supply of carbohydrates synthesized by the maintenance foliage. The upward movement of carbohydrates produced by maintenance leaves is maximum during the growing season for the growth of new shoots. The rate of photosynthesis gradually declines from the top layer of maintenance foliage to the lower layer as entry of solar radiation is restricted by self shading by foliage present in top layers (Hadfield, 1963) and the phenomenon is more prominent in Assam type of bushes.

Experiments revealed that the number of primaries on the bush decreased but weight of shoots increased with the increase in tipping height from 5 cm to 35 cm. Similarly diameter of primaries increased with the increase in tipping height. There was positive correlation between the diameter of primaries and the weight of shoots while lower tipping height yielded thinner primaries and higher percentage of *banjhi* shoots (Barua,1959) unsuitable to make good tea.

On an average, five full leaves in each primaries considered optimum for cumulative yield / bush and quality of shoot beside health and vigour of bush. The height in primaries to accommodate five full leaves depend upon the internodal length of leaves originating after pruning which varies with the age and vigour of bush in general and the genetic makeup of varieties in particular. In young vigorous teas and in Assam type of teas the internodal length was found high. Moreover, primaries originating from the centre portion of bush produced leaf with higher intermodal length.

In general, a tipping height of 20 cm to 25 cm has been found to accommodate five full leaves in each primary branch developed after pruning and coincides with the first '*banjhi*' horizon at which majority of primaries go banjhi (dormant) after producing 5 leaves and in no case should be tipped above that level. In case of skiffed bush, tipping height varies between 5 and 10 cm, depending on the severity of skiffing.

The various tipping measures for different prune or skiff imparted to tea bush are noted below

Rejuvenation prune (RP)	:	30 cm to accommodate 5 - 6 new full leaves
Medium prune (MP)	:	30 cm to accommodate 5 - 6 new full leaves
Light prune (LP)	:	20-25 cm to accommodate 5 new full leaves
Deep skiff (DS)	:	10 cm to accommodate 2 new full leaves
Medium skiff (MS)	:	5 cm to accommodate 1 new full leaf
Light skiff (LS)	:	Pluck to janam at same level.

As already discussed, tipping height is linked to health, yield and quality of tea bush and therefore extreme care should be taken in the initial tipping rounds.

In practice, the pluckers are provided with a bamboo stick marked with appropriate tipping measurement, to ensure appropriate tipping height. Measuring sticks are always placed in the centre of bush to initiate tipping and proceed to peripheries at same height.

Summary

Time of pruning should be decided on the basis of starch reserve in roots for better recovery of bush and threshold value of 10 per cent and 12 per cent starch in roots were identified for pruning in plain and in hills, respectively. December appeared to be the best time for pruning in north east India for yield, seasonal distribution and more early crop. Resting of bushes for 4-8 weeks and foliar application of potash Muriate of Potash @ 2 per cent before pruning found beneficial to mobilize more starch reserve in roots. Different type of pruning and skiffing should be accommodated in a 'pruning cycle' considering prevailing weather condition and to strike a balance between yield and quality. Young and vigorous tea bush can be grown in a 3 year pruning cycle while mature teas should be considered for 4 year pruning cycle. Light pruning should be done with an wood allowance of 4-5 cm to avoid knot formation in frame of tea bush. Pruning cuts should be treated with fungicides within 48 hours. Pruning and skiffing should be performed in tea plantations following the standard procedure. Tipping heights for different pruning and skiffing has been standardized to optimize bush health and yield.

References

Barbora A. C. and Sharma J. (1999). Mechanical pruning of tea (*Camellia sinensis*). Two and A Bud 46 (2): 33-35.

Barbora B. C. (1975). The need of bush sanitation in Dooars for increased production. In: Proceedings of the 27th Conference. Tocklai Exp. Stn. pp. 42-43.

Barbora B. C. and Banerjee M. K. (1988). Causes and Prevention of Knot Formation in Tea. Two and A Bud 35: 39-42.

Barbora B. C. and Hazarika R. L. (1996) Rejuvenation of old tea. Two and A Bud 43(1): 10-17.

Barbora B. C., Jain N. K. and Rahman F. (1982). Effects of Different Methods of Bringing up Young Tea on Sunscorch to Bush Frame. Two and A Bud 29 (1): 16-18.

Barman T. S. (2009) Physiological behaviour of tea plant. In Field Management in Tea, Darjeeling pp. 50-72.

Barman T. S. and Saikia J. (2005). Retention and allocation of ^{14}C assimilates by maintenance leaves and harvest index of tea (*Camellia sinensis* L). Photosynthetica 43: 283-287.

Barman T. S., Saikia J. K. and Baruah U. (2003). Foliar nutrition of potash on starch reserve in tea. Two and A Bud 50: 39-41.

Barman T. S., Saikia J. K. and Pathak S. K. (1998). Starch reserve and growth of tea. Two and A Bud 50: 39-41.

Barman T.S., Sarma A.K. and Baruah U. (1990). Sink capacity of flower buds in young tea (*Camellia sinensis* L.). A study of assimilate translocation. Two and A Bud 37: 38-34.

Barua D. N. (1956a). Some physiological aspects of pruning. Two and A Bud 3(4): 10-12.

Barua D. N. (1956b). Some physiological aspects of pruning.In: Proceedings of the 13th Conference. Tocklai Exp. Stn. pp. 48-50.

Barua D. N. (1959). Plucking in relation to pruning. In: Proceedings of the 16th Conference. ITA pp. 64-72.

Barua D. N. (1966). Importance of starch in tea roots. Two and A Bud 13(2): 58-62.

Barua D. N. (1971). Rejuvenation of old tea. Two and A Bud 18(2): 29-33.

Barua D. N. (1972). Rejuvenation of old tea – Part II. Two and A Bud 11(1): 10-13.

Barua S. K., Sarma A. K. and Goswami B. K. (2008) Important considerations in pruning and Skiffing. In : Field management in tea (ed. Goswami B.K.), TRA pp. 91-102.

Basu R. P. and Dev Choudhury M. N. (1984). How plucking and pruning affect quality of plains teas. Two and A Bud 31 (2): 19-21.

Chakravartee J. (1999). Evaluation of agro-technologies in tea plantations of Assam. Two and a Bud 46(1): 9-19.

Chakravartee J. and Barbora A. C. (1998). A few aspects of pruning. In: Field Management in Tea (ed. Chakravartee, J.), Tocklai Exp. Stn. pp. 59-68.

Chakravartee J., Biswas A. K. and Bordoloi P. K. (1998). Effect of pruning cycle for sustained productivity and quality. In: Proceedings of the 32nd Tocklai Conference pp. 22-31.

Das G. M. (1957). Pests in relation to environment. In: Proceedings of the 14th Conference. ITA pp. 24-27.

Dutta S. K. (1956). Results of pruning experiments. In: Proceedings of the 13th Conference. ITA pp. 39-47.

Dutta S. K. (1959). Experiments on pruning and tipping in relation to plucking and yields. In: Proceedings of the 15th Conference. ITA pp.73-92.

Dutta S. K. (1961). Efficient establishment of young tea. In: Proceedings of the 18th Conference. ITA pp. 47-53.

Dutta S. K. (1964). Deep skiffing of tea. In: Proceedings of the 21st Conference. ITA pp. 2-12.

Dutta S. K. (1969) Pruning times and types of crop distribution. In: Proceedings of the 24th Tock. Conference. Tocklai Ex-perimental Station, pp. 39-46.

Dutta S. K. and Grice W. J. (1966). Rains pruning and its effect on total yield and leaf distribution. Two and A Bud 13(2): 70-73.

Dutta, S K. (1957) Cultural practices in relation to environment. In: Proceedings of the 14th Conference. ITA pp. 35-38

Grice W. J. (1964) Light and medium skiffing in tea. In: Proceedings of the 21st Conference . ITA pp. 16-25.

Hadfield W. (1963). Critical studies of the shade problem in tea. In: Proceedings of the 20th Conference. ITA pp. 10-18.

Hazarika U. N., Pradhan S. K. and Sarkar S. K. (1985). Effect of extended pruning cycles of varying durations on yield of tea in Darjeeling. Two and A Bud 32 (1 & 2): 2-6.

Macalpine R. I. (1956). Practical pruning in relation to kind of tea and environment. Two and A Bud 3(4): 17-20.

Manivel L and Hussain S. (1982). Photosynthesis in Tea II. Direction of movement of photosynthates. Two and A Bud 29 (2): 49-52.

Manivel L. (1978). Importance of maintenance foliage in tea. Two and A Bud 25 (2):74 - 75.

Manivel L. (1981). Physiological basis for pruning time in tea. In: Proceedings of the 29th Tocklai Conference. Tocklai Exp. Stn. pp. 44-46.

Rahman F. (1978). Some thoughts on pruning and plucking on mature tea. Two and A Bud 25 (2): 69 – 71.

Rahman F. (1988). Physiology of the tea bush. Two and A Bud 35: 1-14.

Rahman F. and Mitra H. (1973). Factors affecting choice of pruning cycle in mature tea. Two and A Bud 20 (1): 11-13.

Saikia J.K., Rai S., Baruah R.K., Pathak S.K. and Barman T.S. (2007). Physiology of pruning in Darjeeling tea (*Camellia sinensis* L.). Twao and A Bud 54:29-34.

Sarma K.C. (1956). Some pathological considerations related to pruning. In: Proceedings of the 13th Conference. ITA pp. 51-55.

Sarma P. C. and Bezbaruah H. P. (1981). Observations on heavy pruning in relation to kind of tea. Two and A Bud 28(2): 28-31.

Tubbs F. R. (1934). The pruning in tea, Tea Quarterly, TRI Ceylone Vol .VII.

Tunstall A.C. (1938). Diseases affecting the woody portions of the tea bush. In: Proceedings of the 2nd Conference. ITA, Tocklai Exp. Stn. pp. 70-81.

Tunstall A.C. (1940). Branch canker. In: Proceedings of the 4th Conference.ITA,Tocklai Exp. Stn. pp. 85-90.

Wight W. (1940). Branch canker. In: Proceedings of the 4th Conference. ITA, Tocklai Exp. Stn. pp. 90.

Tea: Technological Initiatives, pp. 91-122
New India Publishing Agency, New Delhi, India
Edited by Niladri Bag, Arundhati Bag and L.M.S. Palni

5

Pesticide Resistance in Insect and Mite Pests of Tea in Sub–Himalayan Terai–Dooars Plantations: Status, Detection and Possible Management

Ananda Mukhopadhyay, Somnath Roy, Soma Das, Kumar Basnet

Abstract

*Tea (*Camellia sinensis *L. O. Kuntze), a foliage crop of Sub-Himalayan Terai-Dooars has Tea mosquito bug (*Helopeltis theivora*) (TMB), looper caterpillars [*Hyposidra talaca, Hyposidra infixaria, Biston (=Buzura) suppresseria*] and Red spider mite (*Oligonychus coffeae*) (RSM) as major pests. In conventionally managed tea plantations, synthetic pesticides and acaricides are routinely applied. The pesticide-exposed pest populations show higher tolerance level as evident in their increased LC_{50} values. Augmented activities of detoxifying enzymes viz. general esterases (GE), glutathione S–transferases (GST) and cytochrome P450 mediated monooxygenases (CYP) were recorded in the pesticide-exposed pest populations. Sub–lethal exposure to monocrotophos (36% SL) enhanced the tolerance level in adult TMB by 105 fold through three generations. Increased activity of GE was evident in the pesticide-exposed populations of the tea loopers. Activities of GE and GST in RSM were significantly higher in acaricide-exposed population. Present studies can help developing enzyme-based assay kit for determining level of pesticide-resistance. Integrated resistance management has become an integral part of modern day IPM practices. Adoption of newer chemicals, and cultural practices can minimise synthetic pesticide applications, thereby help reducing resistance development in pests.*

Keywords: *Tea pest, Tea mosquito bug, loopers, Red spider mite, detoxifying enzymes, resistance management*

Introduction

Tea (*Camellia sinensis* L. O. Kuntze) is grown all over the world as monoculture plantation crop in diverse climatic and topographic conditions of the tropical world ranging from mean sea level to 2300m elevation. The beverage is a popular refreshing and invigorating drink with excellent medicinal properties. The prime producers of tea are China, India, Sri Lanka and Kenya. Tea as a plantation crop was introduced in India by the Britishers to the hilly terrains of eastern Himalaya in 1852, but it soon spread to sub–Himalayan foothills of Assam and the adjoining plains of West Bengal and to Nilgiri Hills of South India. Today in India, tea is grown as industrial crop in a vast area of about 563.98 thousand hectares, producing 1208.78 million Kg made tea of which approximately 80% is produced by the tea plantations of North–East (NE) India. In the sub–Himalayan region of West Bengal, tea plantation is spread over in about 1, 22, 620 hectares. In the foothills and plains there are 308 major tea estates and 1, 232 small tea gardens. The region contributes about 25% of the total made tea in India (www.teaboard.gov.in; accessed on 31.12.2014).

Due to favourable climatic and other conditions, the pest activities are very high round the year in tea plantations of Terai–Dooars region in West Bengal (Barbora and Biswas, 1996; Sannigrahi and Talukdar, 2003). The major pest complex of the region includes, Tea mosquito bug (*Helopeltis theivora*), Looper caterpillars [*Hyposidra talaca, H. infixaria, Biston* (=*Buzura*) *suppresseria*], and Red Spider Mite (*Oligonychus coffeae*). Managing pest is becoming a major problem in present day horticulture. In conventionally managed tea plantations several rounds of organosynthetic pesticides are routinely applied throughout the year to keep pest populations under control (Sannigrahi and Talukdar, 2003; Gurusubramanian *et al.*, 2008) which is a burden to planters and a chemical stress to the environment often resulting in resurgence of primary pests (Sivapalan, 1999), secondary pests outbreak (Cranham, 1966), resistance development (Roy *et al.*, 2010a) and environmental contamination including undesirable residues in made tea (Chaudhuri, 1999). Continuous use of synthetic insecticides is known to facilitate the development of higher tolerance or resistance in many insects (Martin *et al.*, 2002; Komagata *et al.*, 2010, Basnet *et al.*, 2014). Over the last two decades, insecticide use for the control of *H. theivora* has doubled, with a concomitant increase in pest control costs (Gurusubramanian and Bora, 2007). About two dozen varieties of chemicals are used in the tea gardens of Terai–Dooars region for managing pests and weeds (Gurusubramanian *et al.*, 2008). In managing pest attack, quinalphos, acephate, chlorpyriphos, monocrotophos, cypermethrin and deltamethrin are being profusely used with an average of 7.5 litres of insecticides applied per hectare per year (Roy *et al.*, 2009a). In recent years consumption of synthetic pyrethroid in tea plantations of the Dooars is growing gradually as for example, the average use of synthetic pyrethroid in the Dooars tea plantation was 10.2% in the year 1998, but in year 2004, it has risen to 40.91% of the total pesticide consumption (Roy *et al.*, 2009a). As a result of such extensive application of pesticides, pests are developing higher tolerance and sometimes resistance. Despite continuous use of pesticides, there are frequent reports of pest control failures (Gurusubramanian *et al.*, 2008; Roy *et al.*, 2009a). Such failures may be caused by a change in the tolerance pattern of the pest species to the applied pesticides.

Pesticide resistance is an evolutionary phenomenon engendered by the challenges from effective pesticides in the environment containing resistant individuals in a pest population. By definition resistance to pesticides is the development of an ability in a population of a pest to tolerate doses of toxicants that would prove lethal to majority of individuals within the same species (Sternersen, 2004).

The terms resistance, less susceptibility, higher tolerance to pesticides are often interchangeably used, although true resistance does not occur unless a heritable genetic change is present (Yu, 2008). In this article, meaning of the term 'resistance' has been used in a broad sense (*sensu lato*) to include both high tolerance and less susceptibility of tea pests to pesticides.

Natural tolerance may apparently show characteristics normally achieved through resistance in a pest. Tolerance in a pest population helps the individuals to withstand the toxic effect of pesticides due to biochemical or physiological attributes that render pesticides ineffective against most individuals. Since 1950s, resistance to pesticides especially to DDT in houseflies has been reported (Sternburg *et al.*, 1954). Insects are known to evolve resistance in about a decade after the introduction of a new pesticide by virtue of their genetic plasticity. The introduced toxicant present in the environment exerts a selection pressure on an insect population and only those individuals that can survive the toxic effect are able to reproduce. Endemic population of a pest usually comprises a variety of biotypes with minute differences from one another. Pesticide selection may favour the fittest in the toxic environment. Selection operates at biochemical, physiological, and behavioural level as well. An individual with allele or gene duplication/amplification that makes it less sensitive to the contaminated toxic environment or in present context, the conventionally managed tea plantations, will have better opportunities to reproduce. Subsequent generations of such pests will therefore have an increasing frequency of these alleles. If the pest organisms cannot be completely eradicated by a pesticide or by other means, resistance/higher tolerance will occur in the pest population in near or distant future.

Major Insect and Mite pests of Tea from Terai–Dooars Plantations of Eastern Himalayas

Insect Pest: Sucking type

Tea Mosquito Bug: Tea mosquito bug (TMB), *Helopeltis theivora* Waterhouse (Hemiptera: Miridae) has emerged as one of the most destructive sucking pests causing considerable economic loss ranging from 25 to 50% (Prasad, 1992; Subramaniam, 1995). It was only a minor pest of tea in yester years (1900–1950) (Das, 1957), for reason (s) little understood, but in course of time it grew to be a major threat to the tea plantations of NE India including the sub–Himalayan region of the Indian states of Assam and West Bengal (Fig. 1a,b). It has been estimated that about 80% of the tea plantations in India suffers from TMB infestation (Muraleedharan, 1992; Gurusubramanian and Bora 2007), hence it has become the pest of national importance. All the developmental stages and adults of TMB primarily feed on the young leaves, buds and tender stems of tea. The bug sucks plant sap and in the process pumps in a mixture of salivary enzymes containing proteases and lipases. A circular spot develops within 15 minutes to 1 hour of sucking. With the passage of time, the spot develops a brown hallow and subsequently dries up, the phenomenon implies action of the enzymes present in the saliva on parenchyma of tea leaves. Along with feeding, the insertion of eggs during oviposition is equally responsible for damaging tea plant. In Terai–Dooars plantation of NE India, the pest is found almost round the year with varying abundance. Synthetic pesticides are being used indiscriminately to curve the menace of this pest since early days (Sannigrahi and Talukdar, 2003; Roy *et al.*, 2009a) leading to rapid change in their susceptibility to pesticides and increase in their tolerance level as well.

Fig. 1: a. Adult tea mosquito bug
b. Tea leaf with puncture marks of tea mosquito bug infestation”

Insect Pest: Chewing Type

Looper Caterpillars: Among the chewing pests *Hyposidra talaca*, *H. infixaria*, *Biston* (=*Buzura*) *suppreseria* and some more species (Lepidoptera: Geometridae) form the looper complex of tea plantations of sub–Himalayan Terai–Dooars region. Of the defoliators of tea, *H. talaca* (Walker) (Fig. 2a) has recently emerged and established as a severe pest largely replacing the earlier known tea looper, *B. suppressaria* (Guene) (Das *et al.*, 2010; Antony *et al.* 2012). Larvae of *H. talaca* are polyphagous in nature and are reported to feed on a variety of trees, shrubs and weeds (Robinson *et al.*, 2010) and can adapt successfully to different alternative host plants showing great phenotypic and physiologic plasticity (Das and Mukhopadhyay, 2014). The species has invaded tea ecosystem from forest possibly due to rapidly changing climate and environment over the last few decades. Moreover, deforestation, diversified cultivation in and around tea plantations and inadequate phytosanitary measures seem to add impetus to their migration (Mukhopadhyay and Roy, 2009; Das and Mukhopadhyay, 2014). Several advantageous features, such as, short life cycle with about 8 generations per year, fast multiplication, no hibernation during winter months and lack of effective natural enemies etc. appear to act in a concerted way in its emergence as a severe tea pest (Das *et al.*, 2010; Anonymous, 2011). In tea, folivorous activity of these pests often leads to heavy damage almost throughout the year in terms of (i) pluckable leaves resulting into crop loss (Fig. 2a) and (ii) maintenance leaves leading to loss of photosynthetic surface of tea bushes (Prasad and Mukhopadhyay, 2013). *H. infixaria* has also emerged as a defoliating pest of tea along with *H. talaca* (Nair *et al.*, 2008a; Das and Mukhopadhyay, 2009; Das *et al.*, 2010). The loopers are voracious feeders and may even cause complete defoliation rendering the tea bushes broom stick like appearance (Fig. 2b). The early stages of these pests are conventionally managed by spraying several rounds of synthetic pesticides viz quinalphos (organophosphate) and deltamethrin (pyrethroid)

Fig. 2: a. Tea leaves injured by *H. talaca* loopers
b. Severe damage of tea plantations by looper caterpillars. *H. talaca* moth shown in inset.

throughout the year. Due to development of higher tolerance as noted recently, managing looper pests by applying synthetic pesticides is becoming increasingly difficult (Mukhopadhyay *et al.*, 2014; Nair *et al.*, 2008b).

Mite Pest

Red Spider Mite

Mites are persistent and the most serious pests of tea almost in all tea producing countries (Cranham, 1966). Among twelve species of mites recorded on tea, the red spider mite, *Oligonychus coffeae* Nietner (Acarina: Tetranychidae) (RSM) is the major one (Banerjee, 1988; 1993). RSM was first discovered on tea in 1868 in Assam, India (Watt and Mann, 1903). Besides India, the mite is fairly available in Bangladesh, Sri Lanka, Taiwan, Burundi, Kenya, Malawai, Uganda, and Zimbabwe (Gotoh and Nagata, 2001). Being polyphagous, RSM is known to attack many tropical plants like cotton, rubber, citrus, mango, jute, oil palm and weeds (Das, 1959). The mite outbreak occurs in NE India including Terai–Dooars of North Bengal region during spring and early summer when the season is dry and temperature is high. The mite attacks the dorsal side of mature tea leaves starting from the mid rib and vein areas and gradually spreads to the entire leaf (Fig. 3a). The infested leaf turns coppery–red and finally drops. During 1950s, the abundance of mites on tea bushes was small with subtle multiplication rate (Das, 1965). With the reduction of rainfall in local sub–Himalayan plantation area and rise in temperature during last half a century, an increase in RSM population especially during winter (January to April) has been recorded (Mukhopadhyay and Roy, 2009). Due to the change in rainfall pattern, RSM infestation continues even up to May–June. Several factors like drought, unpruned tea, sunny days, rough leaf surface and dust accumulation allow easy spread and proliferation of the mite population (Fig. 3b). Severe draught like conditions and lack of adequate irrigation due to depleting ground water has accentuated the RSM infestation in Terai and the Dooars plantations (The Telegraph 14.05.2013). In NE India, attack by this pest may cause a whopping 50%–70% economic loss of tea crop (Subramanian, 1995; Gurusubramanian *et al.*, 2005). High reproductive potential of the mite and very short life cycle, combined with the frequent application of acaricides required to keep mite population below economic threshold level, facilitate the development of resistance in them (Stumpf *et al.*, 2001).

In recent years, local dailies often report huge crop loss in tea sector in NE India including Terai and the Dooars regions of West Bengal resulting from the infestations by TMB, loopers and RSM, and the changed climatic conditions (The Statesman 18.08.2010; The Telegraph 14.07.2011; 23.04.2013; 14.05.2013;

Fig. 3: a. Red spider mite (RSM) infested tea leaves; an adult female RSM shown in inset b.Tea bushes infested heavily by RSM

www.thehindubusinessline.com). Organisations such as Terai Indian Planters' Association, Confederation of Small Tea Growers' Association and advisors of Tea Research Association often claim an increase in production cost of made tea largely due to an excess expenditure on irrigation and pesticides. Huge quantities of pesticides are being applied to control the key pests of tea foliage since last few years. This often has resulted into several hazards including destruction of arthropod natural enemies from the tea ecosystem, environmental pollution and health problems (Das *et al*., 2005). Moreover, pests are becoming more tolerant/resistant to pesticides rendering their management even more difficult (Roy *et al*., 2010a, 2010b).

Nature and Mechanisms of Pesticide Resistance

Development of resistance exemplifies microevolution. Variations within a natural pest population may bless some individuals with traits that make them better adapted to survive insecticide/pesticide exposure. Selection and survival of these pesticide-adapted individuals in the population ensure passing on of the resistance traits to the next generation. Through accumulation of such traits for generations, a population may develop insecticide/pesticide resistance. The mechanisms of developing resistance may be categorised into four kinds. These mechanisms are not mutually exclusive, thus one or more mechanisms may be involved in resistance development in a particular pest population.

i. **Behavioural resistance:** This is the very first level of resistance developed after the initial encounter of an insect/mite with a pesticide resulting development of an altered behavior, which can help them to avoid future contact with the pesticide.

ii. **Resistance due to barrier tissues:** Thickening of the integument of the pest may act as a barrier and result in reduced penetration of the pesticide. Thus the concentration and effect of the pesticide at the target site gets reduced. Excess lipid content or presence of increased fat body cells in the body mass of a pest may provide insulation or cause sequestrations of lipophilic pesticides, thus ameliorating their toxic action at the target tissue.

iii. **Metabolic resistance:** Once the pesticide enters insect/mite system it is rapidly metabolized and detoxified. Detoxification can be divided into phase I (primary) and phase II (secondary) processes. Phase I reactions consist of oxidation, hydrolysis and reduction and involves detoxifying enzymes such as, general esterases (GE) and cytochrome P450 dependent monooxygenases (CYP). The phase I metabolites are sometimes polar enough to be excreted, but are usually further acted upon by phase II enzymes. In phase II reactions, the polar products are conjugated with a variety of endogenous compounds and are subsequently excreted. Glutathione *S*–transferases (GST) play important role in phase II reactions. The most important function of biotransformation is to decrease the lipophilicity of xenobiotics including pesticides, so that ultimately they can be excreted (Yu, 2008). Genetic changes such as, gene up regulation or amplification of enzyme-coding genes may lead to enhanced production of the detoxifying enzymes conferring higher tolerance or resistance. Increased quantities of detoxifying enzymes may also impart resistance by sequestering pesticides.

iv. **Target site insensitivity:** Different classes of pesticides bind to specific target sites after entering the insect/mite system. Alteration at the target site for the pesticide may occur through gene mutation and confer resistance by reduced or no binding of the pesticide at the target site. The target sites involved are generally receptors at sodium ion (Na^+) channels for pyrethroids and DDT, and acetylcholinesterases (AchE) for organophosphates and carbamates, and Gamma aminobutyric acid (GABA) receptor for cyclodienes and gamma HCH.

Status of Resistance/Higher Tolerance to Pesticides

Synthetic pesticides remain the mainstay of pest control program in conventionally managed tea plantations. Repeated and 'no threshold' applications of pesticides have been reported to induce and select more tolerant forms in many insect and mite pests.

Tolerance/Resistance in Tea Mosquito Bug: The pesticide tolerance level is expressed in terms of median lethal concentration (LC_{50}) value for the populations ofTea mosquito bug (TMB) occurring in tea plantations of Darjeeling Hills, Terai and Dooars. Pesticide such as, monocrotophos and endosulfan, representing organophosphate and chlorinated hydrocarbon classes of insecticide respectively have been used for bioassay and the result is presented in Table 1. These insecticides were in use in the recent past or are still being used by some tea growers clandestinely (per comm.).

TMB population of eastern Dooars showed the highest tolerance level to endosulfan, and the second highest to monocrotophos (Basnet and Mukhopadhyay, Unpublished data) which is found similar to the results reported earlier by Roy *et al.* (2009a). From the data presented in Table 1, it is inferred that the insecticide-exposed population are endowed with higher tolerance level as are reflected in their enhanced LC_{50} values.

Table 1: Insecticide tolerance (LC_{50} values) of Tea Mosquito Bug population of North Bengal (Darjeeling Hills, Terai and the Dooars)

Sl. No	Meta-Population	LC_{50} in PPM (Monochrotophos)	LC_{50} in PPM (Endosulfan)	Pest Management Practices
1	Eastern Dooars	16.270	1580.77	Conventional*
2	Central Dooars	17.903	884.950	Conventional
3	Western Dooars	04.592	938.213	Conventional
4	Terai	08.075	——	Conventional
5	Terai	04.580	——	Organic**
6	Darjeeling Hills	00.016	000.090	Organic

*pests are controlled by using synthetic pesticides, **organically managed plantation (with no use of synthetic pesticide)

Source: Basnet and Mukhopadhyay, unpublished data; FTR of UGC – Major research project, 2014

Increased activity of GE, GST and CYP was evident in the insecticide-exposed populations of TMB of both the Dooars and Terai as compared to the unexposed population of organic plantations of Darjeeling (Basnet *et al.*, 2014 and Saha *et al.*, 2012). These findings imply metabolic basis of higher tolerance to be functional in these TMB populations. The detoxification enzyme activity was noted to be higher by 15.4 and 17.6 fold for GE, 1.8 and 1.9 fold for GST and 2.1 and 2.4 fold for CYP in Terai and the Dooars populations respectively that are conventionally managed compared to the organic plantation that was free from inputs of synthetic pesticides (Table 2).

Sub–lethal exposure to monocrotophos (36% SL) enhanced the tolerance level in adult TMB. The LC_{50} value increased about 105 fold when selected by

Table 2: Insecticide susceptibility level and the defence enzymes activities in Tea mosquito bug populations

Population	LC_{50} ppm**	General Esterases Activity (GE)		Monooxygenases Activity (CYP)		Glutathione-*S*-transferases Activity (GST)	
		(μmol α napthol $min^{-1}mg\ protein^{-1}$)	Activity Ratio *	(nmol min^{-} mg $protein^{-1}$)	Activity Ratio*	(μmol $min^{-1}mg\ protein^{-1}$)	Activity Ratio *
Darjeeling	0.016	$0.12{\pm}0.02^{a}$	1	$0.62{\pm}0.11^{a}$	1	$0.24{\pm}0.05^{a}$	1
Terai	08.07	$1.85{\pm}0.17^{b}$	15.4	$1.32{\pm}0.16^{b}$	2.13	$0.42{\pm}0.04^{b}$	8.40
Dooars	16.27	$2.11{\pm}0.12^{b}$	17.6	$1.47{\pm}0.08^{b}$	2.37	$0.45{\pm}0.11^{b}$	9.00

** Toxicity level for Monochrotophos, 26% EC; *Activity ratio=the ratio of the enzyme activity of the conventional field–collected populations to the enzyme activity of the susceptible population (Darjeeling). Values are means±SE. Means followed by the different letter in a column are significantly different ($P = 0.05$) in Tukey's multiple comparison test (HSDa).

Source: Basnet and Mukhopadhyay, Unpublished / FTR UGC – Major research project (2011-2014)

exposure to sub–lethal dose of the insecticide through three generations (Table 3). This observations is corroborated by an earlier work of Roy *et al.*, (2010a), where increase in its resistance ratio in TMB through generations was reported on exposure to sub–lethal dose of the cyclodiene insecticide, endosulfan.

Table 3: Relative tolerance across three generations of Tea mosquito bug when selected against sub-lethal doses of monochrotophos (36 % SL)

Generation	LC_{50} (in ppm)	LFL (inppm)	UFL (in ppm)	LC_{95} (in ppm)	χ^2	Dose vs mortality regression equation
P	08.2[a]	6.60	10.18	25.50	7.27	y=2.9532x-6.5587
F_1	15.48[b]	12.99	18.45	43.84	2.99	y=3.6623x-10.345
F_2	855.38[c]	595.97	1227.74	13623.05	1.07	y=1.3770x-3.169

LC_{95} = the concentration of the pesticide at which 95 % of the tested individual died; UFL = Upper fiducial limit; LFL = Lower fiducial limit
(*Source:* Basnet *et al.*, 2014)

Repeated exposure to sub–lethal doses of insecticide seems to select insecticide–tolerant individuals (Table 3). The tolerance levels were in the order of $P<F_1<F_2$ determined by change in LC_{50} values through generations. Observed and the expected LC values were not significantly different as the calculated value of χ^2 in none of the tested generation was greater than the table value at df = 4 and p = 0.05 level of significance. These observations suggest that continuous application of the same pesticide or pesticides with similar mode of action may select more tolerant forms, and may lead to even development of resistant traits in the pest population.

Table 4: Generation wise activity of defence enzymes in monocrotophos (36% SL) selected population of Tea mosquito bug (mean ± SE; n = 150)

		Defence Enzymes			
Generation	Resistance Ratio	GE Activity (μmol α-napthol min^{-1} mg $protein^{-1}$)	Activity Ratio	CYP Activity (nmol mg $protein^{-1}$)	Activity Ratio
P	0.00	0.50 ± 0.14[a]	0.00	0.83 ± 0.17[a]	0.00
F_1	1.88	2.66 ± 0.82[b]	5.32	1.63 ± 0.15[b]	1.96
F_2	104.32	8.20 ± 0.90[c]	16.40	7.60 ± 1.37[c]	9.50

Resistance Ratio= Ratio of relative tolerance of the progeny population (F_1 and F_2) to the parental (P); GE=General Esterases; CYP = Cytochrome P450 monooxygenases; [a,b,c] mean values with different letters in a column were significantly different at $p<0.05$ level of significance; Activity Ratio = Ratio of defence enzyme activity of progeny population (F_1 and F_2) to that of parental (P).
(*Source:* Basnet *et al.*, 2014)

Activity of GE in F_1 and F_2 generations got enhanced by about 5.32 and 16.4 fold respectively than the parental population (P) (Table 4). Similarly, the activity of CYP also got enhanced by 1.96 and 9.50 fold for F_1 and F_2 generations respectively as compared to P when selected by sub–lethal dose of the insecticide. Increase in activity of GE and CYP by16.4 and 9.50 fold respectively was matched with a rise in resistance ratio in the tune of 104.32 (Table 4), thus emphasizing the involvement of the two detoxifying enzymes in biotransformation of 'monocrotophos'. In insect pests, metabolic detoxification of insecticides is the principal mechanism to survive against xenobiotics/insecticides. Elevated activity of GE has been shown to confer resistance against organophosphate in many insect pests (Hemingway *et al.*, 2004; Wu *et al.*, 2011). Cypermethrin has been shown to enhance the activity of cytochrome P450 (Shono *et al.*, 1979). Exposure to an organophosphate insecticide has also shown the enhancement of CYP, although not in the same proportion as GE. Resistant insects have been known to exhibit greater CYP activity (Feyereisen 1999). As known from their mode of functioning CYP and GE are phase I (primary) detoxifying enzymes that biotransform xenobiotics such as insecticides, by oxidation, reduction and hydrolysis (Yu, 2008).

The relative tolerance and the activity of detoxifying enzymes were found to have high positive correlation coefficients, viz. $r = 0.999$ for GE and 0.994 for CYP (Basnet *et al.*, 2014). Such a positive correlation between relative tolerance to an insecticide and the activities of GE and CYP in different populations of TMB from North Bengal region was also reported by Saha *et al.* (2012). The strong dependence of increasing insecticide tolerance *vis–a–vis* enhanced activity of detoxifying enzymes indicates the possible sequestering role of detoxifying enzyme titres, especially GE and CYP in TMB.

Tolerance/Resistance in Loopers

Bioassay on 3rd instar caterpillars of *H. talaca* and *H. infixaria* collected from pesticide treated conventional tea plantations indicated their increased tolerance to synthetic pyrethroids such as cypermethrin (Das *et al.*, 2010). Among the two looper species, *H. talaca* showed about three time higher LC_{50} value and about two time higher LC_{95} value than *H. infixaria* for the said pyrethroid insecticide. In both the species the LC_{95} value is much higher than the recommended field dose (500ppm) of cypermethrin for tea loopers (Gurusubramanian *et al.,* 2008). Corresponding to the results of the bioassay, the activity of GE was found to be significantly higher in the more tolerant species, *H. talaca* (Table 5) (Mukhopadhyay *et al.*, 2014).

Table 5: Pesticide tolerance level in loopers of *Hyposidra talaca* and *H. infixaria* against Cypermethrin 10% EC and corresponding activity of general esterases

Looper species	LC_{50} (ppm) (LFL - UFL)	LC_{95} (ppm)	General Esterases (GE) activity (μmol α naphthol $min^{-1}mg$ $protein^{-1}$)
H. talaca	330.41 (258.27-422.69)	2195.51	743.4±38.1[a]
H. infixaria	107.767(77.369-150.109)	1393.19	545.1±53.9[b]

[a,b]: mean values with different letters in a column were significantly different at $p<0.05$ level of significance LFL: Lower fudicial limit; UFL: Upper fudicial limit.

(*Source:* Das *et al.*, 2010, Mukhopadhyay *et al.*, 2014)

Bioassays conducted on *H. talaca* population from conventional (pesticide treated) and bio–organic tea plantations (pesticide untreated) using the same kind of pesticide revealed that *H. talaca* population of bio-organic plantations was much susceptible. There was a strong positive correlation ($r>0.9$) between the pesticide tolerance levels and activity of GE in *H. talaca* indicating possible involvement of GE in imparting tolerance to both the looper species (Table 6) (Mukhopadhyay *et al.,* 2014; Das and Mukhopadhyay, unpublished data).

Similar results were obtained in *H. infixaria,* where population reared in the laboratory for three generations without any exposure to pesticides (therefore, may be considered as equivalent to population from bio-organic plantation) were much susceptible to cypermethrin 10% EC whereas the looper specimens which were collected from a conventional tea plantation maintained by periodic pesticide spray were found to be highly tolerant to the same pesticide (Table 7).

Further a survey conducted in Darjeeling Terai region showed, an increased activity of GE in the insecticide-exposed populations of tea loopers (*H. talaca* and *H. infixaria*) as compared to the unexposed populations (Das, S., 2015). Relative toxicity study using some chemical insecticides against *H. infixaria* revealed that the species was highly tolerant to quinalphos and least tolerant to emamectin benzoate (Nair *et al.*, 2008). However, tolerance towards different groups of insecticides such as botanicals, microbials and synthetics (for example, Neem formulation, Btk, Diflubenzuron, and Deltamethrin) increases with progressive advancement of the loopers from early to late instars (Basu Majumder *et al.*, 2012). In addition to enzymatic detoxification, some other mechanism(s) of resistance may be involved with the advancement of larval instars, such as reduced penetration due to thickening of integument and increased fat body insulation.

Table 6: Pesticide tolerance level of *Hyposidra talaca* of organic and conventional tea plantations against Cypermethrin 10% EC and corresponding activity of general esterases

Management Practice	LC_{50} (ppm) (LFL - UFL)	Resistance ratio	LC_{95} (ppm)	General Esterases (GE) activity (μmol α naphthol $min^{-1}mg$ $protein^{-1}$)
Organic Plantation	22.215 (18.30 – 26.97)	1	92.456	470.5 ± 44.6[a]
Conventional plantation	330.41 (258.27-422.69)	14.87	2195.51	743.4±38.1[b]

Resistance Ratio= Ratio of relative tolerance (LC_{50}) of the conventional population to the organic population

[a, b]: mean values with different letters in a column were significantly different at $p<0.05$ level of significance

LFL: Lower fudicial limit; UFL: Upper fudicial limit

(*Source:* Das and Mukhopadhyay, unpublished data)

Table 7: Pesticide tolerance level of *Hyposidra infixaria* population from lab-culture ≈ organic plantation population and conventional (pesticide treated) tea plantations to Cypermethrin 10% EC and corresponding activity of general esterases

H. infixaria	LC_{50} (ppm) (LFL-UFL)	Resistance ratio	LC_{95} (ppm)	General Esterases (GE) activity (μmol α naphthol $min^{-1}mg$ $protein^{-1}$)
Organic plantation	20.26 (14.00-29.30)	1	327.68	343.12 ± 41.37[b]
Conventional plantation	107.77 (77.37-150.11)	5.32	1393.19	545.00 ± 53.00[a]

Resistance Ratio= Ratio of relative tolerance (LC_{50}) of the conventional population to the organic population

[a, b]: mean values with different letters in a column were significantly different at $p<0.05$ level of significance

LFL: Lower fudicial limit; UFL: Upper fudicial limit

(*Source:* Das and Mukhopadhyay, unpublished data)

Tolerance/Resistance in Red spider mites

Red spider mite (RSM) infestation poses a major threat to tea plantations in recent years. Pesticide-exposed populations of RSM of conventionally managed tea plantations of sub-Himalayan Terai and the Dooars regions appears to have developed different degrees of tolerance/resistance to the commonly used acaricides (Roy *et al.*, 2010a). However, relative toxicity of commonly used acaricides of RSM was observed to be much lower in Dooars plantation than those of Terai tea plantations of North Bengal. The mite populations of West Terai and South Terai showed greater acaricide tolerance as compared to eastern Dooars and Western Dooars. Populations of RSM from the Central

Dooars, Western Dooars, and Northern Terai showed intermediate levels of susceptibility to different acaricides. Ethion had the lowest efficacy to kill the pest in all tea-growing regions of North Bengal. Effective field dosages of these acaricides when compared with LC_{95} values showed a significant decrease in the susceptibility of the test population to ethion and dicofal (Roy *et al.*, 2008a, 2008b; 2009b, 2009c, 2009d). However, little tolerance to the newly introduced acaricides like fenazaquin, fenpropathrin, and propargite has been observed, so these were still found effective at doses even lower than the recommended ones (Table 8).

Table 8: Acaricide tolerance level of field-collected red spider mite from different areas of sub Himalayan tea plantation.

Acaricides	LC_{50} (ppm) values from different sub Himalayan regions				
	Eastern Dooars	Central Dooars	Western Dooars	Eastern Terai	Western Terai
Ethion 50 EC	123.01	321.47	366.69	751.02	791.99
Dicofol 18.5 EC	199.58	388.16	201.17	330.58	541.44
Propargite 57 EC	22.94	64.10	62.19	118.48	90.26
Fenpropathrin 30 EC	1.63	3.53	4.73	7.72	12.55
Fenazaquin 10 EC	4.06	6.968	4.35	10.76	16.77
Micro.sulphur 80 W/W	466.67	1188.66	654.85	1110.45	2599.06

(*Source:* Mukhopadhyay *et al.*, 2014)

Higher resistance of RSM to the acaricides in the field populations of south Indian tea plantations (with pesticide exposure) in comparison to laboratory cultured population (without pesticide exposure) has also been observed (Roobakkumar *et al.*, 2012).

Quantitative and qualitative changes of GE depending on pesticide exposure were first reported in RSM from Terai plantation (Sarkar and Mukhopadhyay, 2008). Activity of GE and GST was significantly higher in the whole body homogenates of the female specimens collected from conventional plantation than that from the organic plantation (Table 9) (Das, S., 2015).

Table 9: Variation in the activity of detoxifying enzymes (GE & GST) of red spider mite from different tea plantations of the Darjeeling Terai-Dooars region of West Bengal (Mean±SE, n=20)

Region	TE and management practices	GE (μM/min/mg protein)	GST(μM/min/mg protein)
TERAI	Organic 1	0.181 ± 0.023^{a}	106.63 ± 9.11^{a}
	Organic 2	0.308 ± 0.035^{a}	98.35 ± 8.06^{a}
	Conventional 1	0.704 ± 0.088^{b}	205.62 ± 14.60^{b}
DOOARS	Organic 1	0.330 ± 0.033^{a}	108.08 ± 10.67^{a}
	Conventional 1	0.759 ± 0.076^{b}	202.43 ± 15.41^{b}
	Conventional 2	0.964 ± 0.084^{b}	198.21 ± 15.69^{b}

^{a,b}: mean values with different letters in a column were significantly different at $p<0.05$ level of significance
(*Source:* Das, S. 2015)

Electrophoretic studies further revealed deeply stained esterase bands in female RSM from conventional plantations *vis à vis* faint esterase band with low staining intensity in RSM of organic plantation (Fig. 4). It is generally agreed that increased staining intensities of esterase isozymes are due to increases in the amounts of the enzymes and are associated with resistance (Oakeshott *et al.*, 2005). General esterases show high binding affinity for organophosphates and are known to hydrolyze synthetic pyrethroids (Campbell, 2001; Oakshott *et al.*, 2005; Shan and Ottea, 1998) many of which are used as acaricides in the tea plantations of sub–Himalayan Terai and Dooars.

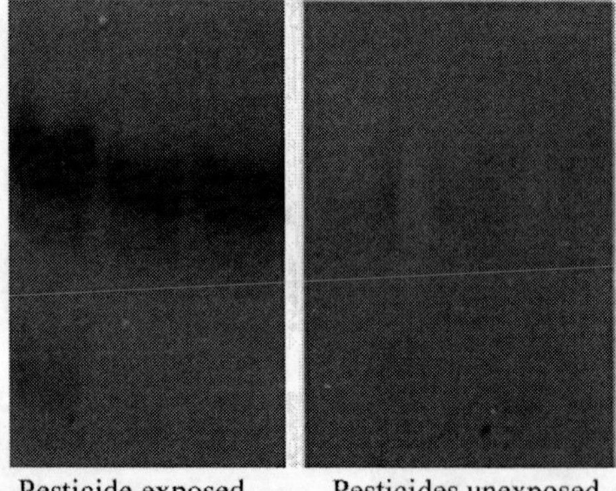

Fig. 4: PAGE profile of esterase isozymes of female RSM (*Oligonychus coffeae*) from pesticide treated and untreated tea plantations. Each lane represents bands of single specimen *Source*: Sarkar and Mukhopadhyay, 2008

Detection of Level of Resistance

Easy detection of level of resistance/tolerance helps in proper monitoring of pest populations that are developing resistance and therefore, also help in formulating management strategies. In order to determine insecticide tolerance or susceptibility level, estimation of LC (Lethal Concentration) or LD (Lethal Dose) values for pesticides are routinely determined. These tests are time consuming, labour intensive and require a lot of live specimen of same cohort.

Alternatively, enhanced detoxifying enzyme activity observed in insecticide-resistant strains of several insect pests (Zhu and Gao, 1998; Maa and Liao, 2000; Zhu and He, 2000; Neus *et al.*, 2008; Perera *et al.*, 2008) can be used as an index in developing isozyme-based detection technique for identifying extremely tolerant insect populations/strains. Generally, the increase in the activity of detoxifying enzymes such as CYP, GE and GST and their isozyme forms observed in arthropod pests on exposure to pesticides is a key biochemical mechanism for bringing about changes in the susceptibility (Soderland and Bloomquist, 1990). The biochemical assays of detoxifying enzymes help in determining the level of tolerance/resistance of a pest population with much lower number of insects to be tested than that of conventional field based bioassays.

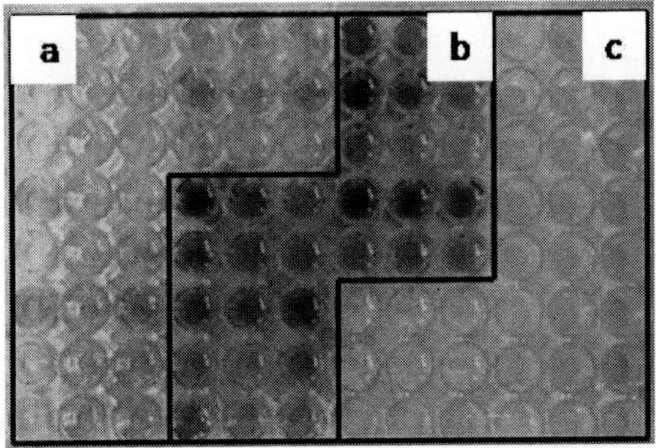

Fig. 5: Microplate showing variable activity of detoxifying enzyme (CYP) evident from the developed colour pattern in: **a.** insecticide-unexposed Tea mosquito bug populations; **b.** insecticide-exposed Tea mosquito bug populations; **c.** Control

Present findings revealed notable variation in activity of GE and GST in the loopers (*H.talaca* and *H. infixaria*) and RSM between pesticide–exposed and unexposed populations. Similarly, in TMB, activities of GE and CYP are significantly different between the pesticide-exposed and unexposed populations. Such enzyme variation gives the opportunity for developing enzyme–based resistance detection kit . Such kits can be even usable at field level as an on–spot tool for quick determination of resistance status. Clear variation of colour intensity in concurrence with the concentration of detoxifying enzymes is evident for GE in case of loopers and RSM, and GE as well as CYP in case of TMB. So these enzymes can be chosen for designing resistance detection kits (Fig. 5). Enzyme-based assay kit for determining pesticide susceptibility may involve microplate, PAGE or dot blot techniques. The dot blot technique will prove handy to tea planters for easy monitoring of the activity and the expression level of detoxifying enzymes, thus reflecting the susceptibility level of a pest population against pesticides. Information on resistance status would guide planters to choose proper pesticide and design pest management strategy, saving both cost of pesticide as well as reducing environmental contamination.

Possible Measures of Resistance Management

Integrated pest management (IPM) is regarded as the most rational alternative to complete dependence on chemical pesticides, especially in permanent agroecosystem such as tea plantations. Development of resistance to synthetic pesticides has become the principal driving force for development of IPM. Strains that are capable of surviving a dose lethal to the majority of individuals in a normal population are termed resistant (french–Constant and Rouse, 1990). As the traditional pesticides are associated with various deleterious effects such as development of resistance, environmental contaminations, public health hazards etc., there is a pressing need for holistic IPM strategies.

Integrated Resistance Management Strategies

Integrated Resistance Management (IRM) strategies have strengthened the pest management systems by identifying appropriate pesticide and pest management tools so as to delay the development of resistance, ensure effective control of target pests and conserve naturally occurring biological controls for enhanced sustainability of ecosystems. IRM strategies is the integration of pragmatic resistant management theories amalgamated with upgraded IPM tactics in an economically viable and eco friendly approach. The detection kit for highly tolerant/resistant pests can help planning proper integrated pest management (IPM) as it is always wise to select and use appropriate concentrations of different pesticides according to the susceptibility of the pest species (Xie *et al.*, 2010). Such biochemical diagnostic kit was developed for identifying insecticide-resistant mosquito species based on GST and hemeperoxidase activity (Brogdon and Braber, 1990, Brogdon and McAllister, 1997). Diagnostic assay based on esterase mediated resistance mechanisms is available for western corn rootworm (Zhou *et al.*, 2002). In India, enzyme-based resistance-monitoring tool for Cotton bollworm has been developed by Kranthi (2005). Holistic approach of IRM in tea would involve many tools for managing the major pest viz. TMB, loopers and RSM (Fig. 6).

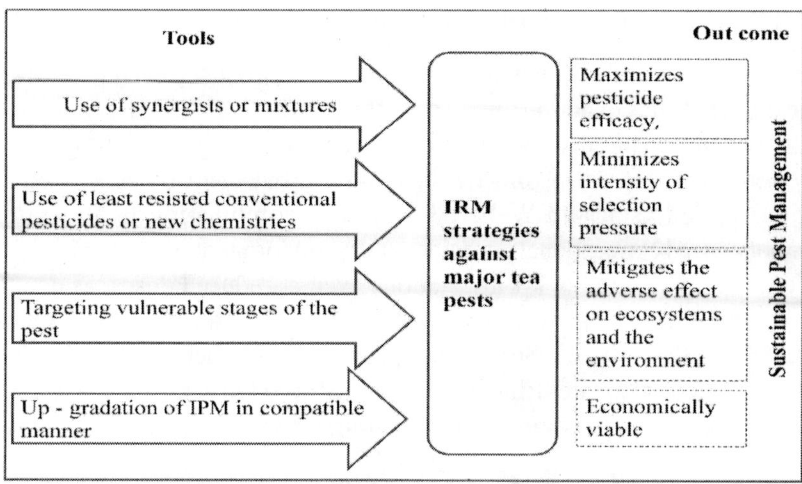

Fig. 6: Integrated Resistance Management strategies

Use of Synergists or Mixtures

Synergists are the natural or synthetic chemicals, which increases the killing power and effectiveness of pesticides, while they themselves being nontoxic. Synergists have been used commercially for about 50 years and have contributed significantly in improving the efficacy of pesticides, particularly after the development of resistance. Use of synergists like Piperonyl butoxide (PBO) (inhibitor of oxidase activity), S.S.S–tributylphosphorotrithioate (inhibitor of GEs activity), Ethacrynic acid, Diethyl maleate and chlorfenethol (inhibitor of GST activity) have been found to increase the lethality of the existing insecticides. Higher efficacy of combinations of pesticide and synergists PBO has been reported in the control of TMB populations from the Dooars plantations which can be used as measures to tackle the insecticide resistant population. The combination of deltamethrin+PBO, quinalphos+PBO and imidacloprid+PBO showed 44.60, 16.01 and 11.14 fold increase of toxicity (synergistic ratio) than the respective insecticide alone (Roy *et al.*, 2009e). Synergistic action of petroleum distilled spray oil (1%), urea (2%), Kalmegh Mother Tincture (*Andrographis paniculata*) (0.5%), Sesame oil (*Sesamum indicum*) (0.5 and 1%) and Jatropha oil (*Jatropha curcas*) (1%) on toxicity of synthetic pyrethroids and organophosphates applied against tea looper pests particularly *H. talaca* could be noted. The oil formulations normally cause the spiracle blockage which generally leads to suffocation and toxic effect on insect integument. Symptoms observed on the integument after topical treatment with 1% of petroleum distilled oil and 2% urea solution are found to disrupt the waxy layer of integument and produce minute holes which may allow penetration of more pesticide inside the insect body (Roy *et al.*, unpublished data).

Joint Toxicity of Pesticide Mixture

Cocktailing of chemical pesticides has joint action when the chemicals involved have different sites of action but do not affect the biological activity of the other in the mixture. This occurs when each component in the mixture acts on a different physiological or biochemical system but contributes to a common response (Anderson and Weber, 1975). The combination treatments with commonly used conventional pesticides and neem formulation showed significant reduction in TMB and RSM infestations even at reduced doses, as compared to sole use of neem formulations or sole pesticide treatments at recommended doses (Roy *et al*., 2010c). Servo agrospray oil (1%) when used as an adjuvant with systemic insecticides (thiomethoxam) reduced the intensity and recurrence of TMB population as compared to the straight application of sole insecticides.

Combating Resistant Pest–Population with Newer Members of Chemical Pesticides

Study on relative susceptibility of tea loopers and TMB to commonly used insecticides suggests the ineffectiveness of synthetic pyrethroids (deltamethrin, lambda–cyhalothrin and bifenthrin) when used against the late instars, therefore use of such insecticide should be discontinued for managing the pest. The neonicotinoid such as Clothianidin 50 WDG, Thiomethoxam 25 WG and Thiacloprid 21.7 SC are found to be very effective in managing TMB population of NE India at recommended dose. Emamectin Benzoate 5% SG and Flubendiamide 39.35% SC show excellent control of tea looper. Ethion and dicofol resistant population of RSM show high level of susceptibility to Etoxazole 10 SC, Fenazaquin 10 EC, Fenpropathrin 30 EC, Fenpyroximate 5 EC/SC, Hexythiazox 5.45 EC, Propargite 57 EC and Spiromesifen 240 SC (22.9 w/v) at recommended dilutions (Table 10). Strategies to minimize selection pressure include either rotation of pesticide groups over space and or time or use of alternative options such as microbial pesticides or ecosystem management or biological control or reduce application frequency.

Targeting Vulnerable Stages of the Pest

Understanding the aspects of biology such as distribution, development and behaviour of pests as well as their relationship with the environment and other organisms are important. This knowledge will often help to identify stages of life cycle of pests and their windows of susceptibility to control measures. One of the pesticide resistant management strategies is the use of chemicals that can effectively control the pest at its vulnerable stage itself. Nymphal stage of TMB, 1st to 3rd instar larvae of loopers and eggs of RSM may be deemed as the most susceptible stages for effective application of pesticide.

Up–gradation of IPM

Since pesticide resistance have emerged as a major threat to pest control programmes, rendering insect pest control ineffective, inefficient and unsustainable, IPM packages need to be reviewed to include IRM (Integrated Resistance Management) as a major component. Development of resistance can be checked if IRM strategies are followed properly. Regular monitoring and analysis of field situations with regard to natural enemies of pests, soil conditions, plant health, influence of weather factors and their relationship for growing a healthy crop need to be undertaken. Such critical analyses of the field situation help in taking appropriate decision on cultivation practices. Field surveillance and scouting need to be done to record the initial development of pests in each area or spot. With the attainment of the economic threshold level (ETL) by pest, plant protection measures should be initiated. Periodically 100 plants or bushes at random in each section need to be surveyed. The percent infestation is to be worked out by analysing 100 shoots from plucker's basket and counting for the number of infested and un–infested shoots. Looper population can be assessed by direct counting of the number of active caterpillars from bushes selected at random from a particular area. To determine the average number of mites per leaf, one hundred leaves need to be sampled from different areas of a particular field and the number of mites on leaves may be counted. Based on field studies on crop loss due to pests, the ETLs for important pests have been worked out. The ETL value for TMB is determined as 5% infestation, and for RSM it has been determined as 4 mites per leaf. However, it must be noted that the ETL value depends not only on the pest population causing quantum of crop loss, but also on the market price of the finished product besides cost of inputs and plant compensatory response to damage. Un-shaded condition is favourable for red spider mite infestation. Therefore planting of shade trees at recommended spacing will reduce mite build up. Detailing of non–chemical based methods that includes cultural practices for the management of major and resistant pests of tea is discussed in following sections in accordance with the suggestions by Roy *et al*., (2013).

a. Tea Mosquito Bug (*Helopeltis theivora*)

- A closer plucking schedule helps to remove inserted eggs and early nymphs of TMB before they cause substantial damage. In severe infestation, level off skiff (LOS) operations should be followed to minimize the infestation by the next generation of TMB.

- Removal of the weeds and alternate hosts of TMB from in and around plantations and the ecotonal region between forest line and tea plantation would result in a good control. The common weeds identified as alternate

Table 10: Pesticides recommended by various agencies for arthropod pests of tea plantations during 2010 to 2014

	TRA (2010)[1]	Govt. of West Bengal (2010)[2]	TRA-CIB (2011)[3]	Tea Board of India (2014)[4]	Planters' Choice (2010)[5]
Insecticides	Acephate 75 WP	Acephate	Azadirachtin 5% EC	Azadirachtin 1 EC	Acephate 75% SP
	Alphamethrin 10 EC	Alphamethrin	Bifenthrin 8% SC	Azadirachtin 5 EC	Acetamiprid 20% SP
	Azadirachtin 5%	Chlorpyriphos	Deltamethrin 10 EC	Bifenthrin 8 SC	Alphamethrin 10% EC
	Bifenthrin 8% SC	Cypermethrin	Deltamethrin 2.8 EC	Clothianidin 50 WDG	Cartaphydrochloride 50% SP
	Clothianidin 50 WDG	Deltamethrin	Diflubenzuron 25 WP	Deltamethrin 2.8 EC	Chlorpyriphos 20% EC
	Cypermethrin 10 EC	Endosulfan	Dimethoate 30 EC	Fenpropathrin 30 EC	Cypermethrin 10% EC
	Diamethoate 30 EC	Fenpropathrin	Endosulfan 35 EC	Phosalone 35 EC	Cypermethrin 25% EC
	Endosulfan 35 EC	Imidacloprid	Fenpropathrin 30 EC	Profenofos 50 EC	Deltamethrin 2.8 EC
	Lambda cyhalothrin 2.5%	Indoxacarb	Fenvalerate 25 EC	Quinalphos 20 AF	Dichlorvos 76% EC
	Monocrotophos 36 SL	Mixure	Flufenoxuron 10 EC	Quinalphos 25 EC	Dimethoate 30% EC
	Phosalone 35 EC	Monochrotophos	Phosalone 35 EC	Thiacloprid 21.7 SC	Endodhan
	Profenophos 50 EC	Quinalphos	Profenofos 50 EC	Thiamethoxam 25 WG	Endosulfan 35% EC
	Thiacloprid 240 SC	Thiacloprid	Quinalphos 20 AF		Ethofenprox 10% EC
	Thiamethoxam 25 WG	Thiamethoxam	Quinalphos 25 EC		Fenpropathrin 10% EC
			Thiomethoxam 25% WG		Fenpropathrin 30% EC
					Imidaclorid 17.8% SL
					Lambda cyhalothrin 2.5% EC
					Lambda cyhalothrin 5% EC
					Malathion 50% EC
					Monocrotophos 36% SL
					Quinalphos 25% EC
					Thiamethoxam 25% WG

(Contd.)

Acaricides	Bifenthrin 8% SC Dicofol 18.5 EC Ethion 50 EC Fenazaquin 10% EC Fenpyroximate 5% EC/SC Hexythiazox 5.45% EC Paraffinic Oil Propergite 57 EC Spiromesifen 22.9% Sulphur 80 WG	Dicofol 18.5 EC Ethion 50 EC Etoxazole 10 SC Fenazaquin 10 EC Fenopyroximate 5 EC/SC Flufenzine 20 SC Flumite 20 SC Hexythiazox 5.45 EC Propargita 57 EC Spiromesifen 22.9 Sulphur 40WP Sulphur 80 WP	Dicofol 18.5% EC Endosulfan 35% EC Ethion 50% EC Mehtyl parathion 50% EC Profenofos 50% EC Propargite 57% EC Sulphur 40% SC Sulphur 80% WGD Sulphur 80% WP Sulphur 80% WW

[1] Bulletin No PP/01/2010, Tea Research Association, Plant Protection Division.

[2] Plant Protection Schedule, Integrated Pest Management, Directorate of Agriculture, Govt. of West Bengal.

[3] Two and a Bud, Volume 56, pp 72-74.

[4] Plant protection code, Ver. 2.0. Tea Board of India, Ministry of Commerce and Industry, Govt. of India.

[5] Personal communication with the managers of tea gardens.

host of TMB are Melastome (*Melastoma* sp), Thoroughwort (*Eupatorium* sp), Fragrant thoroughwort (*Eupatorium odoratum*), Dayflower (*Commelina* spp), Sesbania (*Sesbania cannibina*), Bortengeshi (*Oxalis acetocello*), Mikania (*Mikania macrantha*) and *Duranta repens*.

- TMB prefers moist conditions and mild temperatures. Populations are often higher under heavy shade. Moderate shade of about 60% with improved ventilation is preferable for management.
- During cold weather, pruning/skiffing should be resumed from periphery towards the centre and around 50–60 bushes should be kept untouched for a day or two in the centre to serve as lure and a trap for adults. After thorough spraying of recommended pesticides these trap bushes should be pruned/skiffed.
- Barrier spraying method is recommended for systematic eradication of the pest.
- Applying crude aqueous extracts of plants growing in the areas of tea plantations such as *Clerodendrum viscosum*, *Polygonum hydropiper*, *Cassia alata*, *Xanthium strumarium*, *Vitex negundo* and *Amphineuron* sp. @ 5–8% concentration may check the pest in initial stage of infestation.
- Time of spraying should coincide with feeding/active period of the pest at the bush table i.e., early morning or late afternoon, for more effective and intensive control.

b. Tea Loopers

- Manual removal of caterpillars and moths should be done.
- Collection of chrysalides to be undertaken from the soil around the collar region and cracks and crevices of tea plants in old sections during cold weather.
- Cleaning and bush sanitation with alkaline wash after pruning can help minimizing the looper attack in the following cropping seasons.
- The eggs are laid in clusters inside the cracks and fissure of the bark of the shade trees like *Albizzia odoratissima, A. lebbek, Acacia lenticularis, A.auriculiformis, Derris robusta, Lagerstroemia speciosa, Schima wallichii, Dalbergia sisso, D. assamica* etc. Old trees with lose and cracked barks are preferred. To minimize egg deposition by the looper moths, light scrapping of bark, removing of moss and parasitic and epiphytic plants followed by lime washing of the shade tree trunks up to a height of 6m can check oviposition.

- Eggs laid on shade tree trunk need to be destroyed with mashal (burning torch) or by application of effective insecticide up to around 6m.
- As these caterpillars are polyphagous in nature, cleaning of weed is essential in an around tea plantations.
- Light traps can be arranged to attract and collect moths. For an area of 10 ha one light trap (Actinic BL light or NCIPM, ICAR designed light trap) is advisable. The light traps need to be lighted in the early evening as soon as it becomes dark till 3–4 hours positioned on road and vacant patches inside tea areas at about 0.5m–0.6m above the plucking table, so that it is visible from a distance.
- Locally available fresh weeds need to be collected from tea fields and then burnt along with addition of *dhuna* (carbon di–sulphide) to generate sufficient smoke as an insect repellent against adult stage (moths) of looper and red slug in the tea fields. The smoke generated causes suffocation to the adult moths and they come out from the tea bushes where they usually keep hiding. The emerging adults can thus be, collected with the help of sweeping nets.
- Hand collection of residual population of loopers after insecticide application can be undertaken wherever possible.
- Under severe infestation newly hatched caterpillars disperse from the shade trees with the help of salivary thread in large numbers. In such case spraying of shade tree should be done using foot or power sprayer or other specialised sprayers.

c. Red Spider Mite (*Oligonychus coffeae*)

- Un-shaded condition is favourable for red spider mite (RSM) infestation. Therefore, planting of shade trees at recommended spacing will reduce mite build up.
- The bushes along the motorable roads, which remain covered with dust, are very often found to be severely attacked by RSM. Protect the roadside bushes from dust by growing hedge plants like *Phlogacanthus thyrsiflorus* (titaphool) or applying water on such dusty roads at regular intervals is a good agricultural practice for management of RSM.
- Migration of RSM can be minimised by restricting free movement of the pluckers from un-infested areas to infested areas and also preventing cattle grazing inside the tea plantations area. Removal of alternate hosts (*Borreria hispida*, *Scoparia dulcis*, *Melochia corchorifolia* and

Fussiala suffruticosa) from in and around plantations can give good control.

- The bushes in ill irrigated or water-logged condition experience more damage by RSM than those in well irrigated plantations. Therefore, proper irrigation is needed not only to improve productivity but also to check the build–up of RSM population.
- Red spider mite incidence is high on the bushes receiving heavier doses of nitrogen contrary to potash and phosphorus application which decreases the amount of RSM in tea. Therefore, appropriate fertilization practice is necessary.
- Red spider mite affected fields should get a new tier of maintenance foliage since the infested bushes are very week due to defoliation of maintenance leaves.
- Pesticide spray on both surfaces and foliage is necessary. During full cropping seasons control measures need to be undertaken as spot treatment only.
- RSM are attacked and controlled by several native natural enemies, especially by predatory insects and Phytoseiid mites (Mukhopadhyay and Sarkar 2006). So, conservation of the natural enemies present in the tea ecosystem can be facilitated by minimizing the load of chemicals.
- The crude water extracts of native plants viz. *Clerodendrum viscosum* (leaves and succulent stem), *Melia azadirach* (seed kernel), *Vitex negundo* (leaves and succulent stem), *Terminalia chebula* (dry pericarp of the fruits); *Sapindus saponaria* (dry pericarp of the fruits) and *Nyctanthes arbor–tristis* (leaves) at 5–8% dilutions show effectiveness in controlling RSM population at field level.
- Certain commercial formulations of the entomopathogenic fungi, *Verticillium leccani, Paecilomyces fumosoroseus*, *Hirsutella thompsonii* are found to be effective against RSM.

Discussions

Detection of the status of resistance or high tolerance in the three major arthropod pests of tea from the sub–Himalayan tea plantation opens a new chapter in tea pest management. The mechanisms involved in development of resistance and high tolerance in these pests provide evidence of any one or more of the underlying processes. The mechanisms implied may be:

i. Behavioural avoidance of spray
ii. Reduced penetration of cuticle
iii. Increase in fat body as insulator and storage tissue of pesticide
iv. Enhanced detoxification by detoxifying enzymes
v. Larger or darker forms to withstand higher pesticide doses

Some more mechanisms of resistance may also be operative in these pests but their involvement may only be established with further studies. These hitherto little known mechanisms may include:

i. Lower transport of pesticide to the target site
ii. Reduced bio–activation of organophosphates containing sulphur
iii. Increased excretion of active components of pesticides
iv. Less sensitive receptors
v. Alternative metabolic pathways and some others

Resistant forms/strains of pests cannot be totally eliminated or eradicated as it is of natural occurrence and develop by way of evolutionary process through selection. So, the present strategy harps on delaying the development of resistance. There is no simple method of preventing resistance (Georghiou, 1990). Reviews on pesticide resistance available from Hemingway and Ranson (2000) and Wilson (2001) hint at slowing down the evolutionary process, which implies providing better reproductive possibilities to susceptible individuals.

A prime strategy for delaying the resistance development in pest is to have a refuge crop/area where a population of pest can reproduce without any selection pressure of pesticides. The few resistant individuals created in the pesticide-treated plantation will mate with the susceptible one of the refuge area resulting in heterozygotes that are susceptible to pesticides. In course of time, the resistant gene can get diluted and sunken in the susceptible population. But these strategies need much of fallow land or tea plantation areas that can be kept free of pesticide application. With sounder and reliable non-conventional techniques of pest management in a bioorganic plantation that would co-exist with the conventional ones such a strategy may be contemplated.

Pesticide cocktailing or mixing that have mutually exclusive mode of actions may prove effective in controlling the resistant strains of a pest. Sequential or rotational use of pesticides will not prevent development of resistance if the surviving heterozygote is better than normal susceptible ones. It may only delay resistance development as the individual selection pressure becomes lower (Stenersen, 2004).

As resistance is the result of fundamental biological process leading to evolutionary changes in the traits of a pest population, it will remain a problem in the entire pesticide-era. Though physiological and behavioural mechanisms are implied in development of resistance, target insensitivity to pesticides or their increased detoxification through biochemical and metabolic pathway appear to be of great importance. Another related problem is the development of cross–resistance to a number of related classes of pesticides since their underlying detoxifying mechanism is universally based on the same kind of detoxifying enzymes. Role of GE, GST, and CYP have been found to be active in all the major tea pests considered in the present article for imparting higher tolerance or resistance to specific pesticides. It appears to be a non–ending war, since more and more cases of resistance/tolerance are developing against newer generations of synthetic pesticides that are being sequentially introduced. A better option will be to research more on non-conventional methods of pest management that can be adopted in IPM modules specially meant for the tea plantations of NE India and Darjeeling region, since the tea of this region holds the promise of maximum export.

References

Anderson P.D. and Weber L.J. (1975) Toxic response as a quantitative function of body size. *Toxicology and Applied Pharmacology* 33: 471-483.

Antony B., Sinu P.A. and Rehman A. (2012) Looper caterpillar invasion in North East Indian agroecosystem: change of weather and habitat loss may be possible causes? *Journal of Tea Science Research* 2: 1-4.

Banerjee B. (1988) An Introduction to Agricultural Acarology. Associated Publishing Co., New Delhi.

Banerjee B. (1993) Tea Production and Processing. Oxford and IBH Publishing Co. Pvt. Ltd., New Delhi.

Barbora B.C. and Biswas A.K. (1996) Use pattern of pesticides in tea estates of North East India. *Two and a Bud* 47: 19-21.

Basnet K., Saha D. and Mukhopadhyay A. (2015) Enhancement of Resistance vis-à-vis Defence-Enzyme Activity in Tea Mosquito Bug, *Helopeltis theivora* Waterhouse (Hemiptera: Miridae) Selected Through Exposure to Sub-lethal Dose of Monochrotophos. *Proceedings of the Zoological Society* 68: 184-188.

Basu Majumder A., Pathak S.K. and Hath T.K. (2012) Evaluation of some bio-rational insecticides against the looper complex, Hyposidra spp. in tea plantations of Dooars, West Bengal. *Journal of Biopesticides* 5: 91-95.

Brogdon W.G. and Barber A.M. (1990) Microplate assay of glutathione *S*-transferase activity for resistance detection in single-mosquito triturates. *Comperative Biochemistry and Physiology* B 96: 339-342.

Brogdon W.G., McAllister J.C. and Vulule J. (1997) Heme peroxidase activity measured in single mosquitoes identifies individuals expressing an elevated oxidase for insecticide resistance. *Journal of the American Mosquito Control Association* 13: 233-237.

Campbell B.E. (2001) The role of esterases in pyrethroid resistance in Australian populations of the cotton bollworm, *Helicoverpa armigera* (Hubner) (Lepidoptera: Noctuidae). Ph.D. Thesis: Australian National University, Canberra.

Chaudhury T.C. (1999) Global Advances in Tea Science. Aravali Books, New Delhi.

Cranham J.E. (1966) Tea Pests and Their Control. *Annual Review of Entomology* 11: 491-514. doi:10.1146/annurev.en.11.010166.002423.

Das G.M. (1957) Pests in relation to Environment. *Two and a Bud* 4: 14-14.

Das G.M. (1959) Bionomics of the tea red spider, *Oligonychus coffeae* (Nietner). *Bulletin of Entomologial Research* 50: 265-274.

Das G.M. (1965) Pest of tea in North-East India and their control. Memorandum No. 27. Tocklai Experimental Station, Tea Research Association, Jorhat, Assam, India.

Das S. (2015) A study on some aspects of bio-ecology and variability in defense enzymes of major lepidopteran and mite pests of tea from Darjeeling Terai. University of North Bengal.

Das S. and Mukhopadhyay A. (2009) An insight into the looper complex of tea from Darjeeling terai. In: Ramamurthy V.V. and Subrahmanyam B. (Eds.) National Symposium on IPM strategies to combat emerging pests in the current scenario of climate change; Entomological Society of India, IARI, New Delhi.

Das S. and Mukhopadhyay A. (2014) Host-based life cycle traits and detoxification enzymes of major looper pests (Lepidoptera: Geometridae) of tea from Darjeeling Terai, India. Phytoparasitica 42: 275-283. doi:10.1007/s12600-013-0358-1.

Das S., Mukhopadhyay A., Roy S. and Biswa R. (2010) Emerging looper pests of tea crop from subHimalayan West Bengal, India. Resistant Pest Manage. Newsl. Resistant Pest Management Newsletter 20: 8-13.

Feyereisen R. (1999) Insect P450 Enzymes. *Annual Review of Entomology* 44:507-533. doi:10.1146/annurev.ento.44.1.507.

ffrench–Constant R.H. and Rouse R.T. (1990) Resistance detection and documentation; the relative roles of pesticidal and biochemical assays. In: Roush R.T. and Tabashnik B.E. (Eds.) Pesticide Resistance to Arthropods; Champman and Hall, New York.

Georghiou G.P. (1990) Overview of insecticide resistance. In: Green M.B., LeBaron H.M. and Moberg W.K.) American Chemical Society Symposium Series 421; American Chemical Society, Washington, D.C., pp. 18-41.

Gotoh T. and Nagata T. (2001) Development and reproduction of *Oligonychus coffeae* (Acari: Tetranychidae) on tea. *International Journal of Acarology* 27: 293-298. doi:10.1080/01647950108684269.

Gurusubramanian G. and Bora S. (2007) Relative toxicity of some commonly used insecticides against adults of *Helopeltis theivora* Waterhouse (Miridae: Hemiptera) collected from Jorhat area tea Plantations, South Assam, India. *Resistance Pest Management Newsletter* 17: 8-12.

Gurusubramanian G., Borthakur M., Sarmah M. and Rahman A. (2005) Pesticide selection, precautions, regulatory measures and usage. In: Dutta A., Gurusubramanian G. and Barthakur B.K. (Eds.) Plant Protection in tea: Proceedings of Plant Protection Workshop; Assam Printing Works Private Limited, Jorhat, Assam, India, Tocklai Experimental Station, T.R.A., Jorhat.

Gurusubramanian G., Rahman A., Sarmah M., Ray S. and Bora S. (2008) Pesticide usage pattern in tea ecosystem, their retrospects and alternative measures. *Journal of Environmental Biology* 29: 813-826.

Hemingway J. and Ranson H. (2000) Insecticide resistance in insect vectors of human disease. *Annual Review of Entomology* 45: 371-391.

Hemingway J., Hawkes N.J., McCarroll L. and Ranson H. (2004) The molecular basis of insecticide resistance in mosquitoes. *Insect Biochemistry and Molecular Biology* 34: 653-665.

Komagata O., Kasai S. and Tomita T. (2010) Overexpression of cytochrome P450 genes in pyrethroid-resistant *Culex quinquefasciatus*. *Insect Biochemistry and Molecular Biology* 40: 146-152.

Kranthi K.R. (2005) Insecticide Resistance Monitoring and Management Techniques. Common Fund for Commodities, Amsterdam.

Lopez-Soler N., Cervera A., Moores G.D., Martinez-Pardo R. and Garcera M.D. (2008) Esterase isoenzymes and insecticide resistance in *Frankliniella occidentalis* populations from the south-east region of Spain. *Pest Management Science* 64: 1258-1266.

Maa C.J.W. and Liao S. (2000) Culture dependent variation in Esterase isozymes and malathion susceptibility of diamondback moth, *Plutella xylostella* L. *Zoological Studies* 39: 375-386.

Martin T., Chandre F., Ochou O.G., Vaissayre M. and Fournier D. (2002) Pyrethroid resistance mechanisms in the cotton bollworm *Helicoverpa armigera* (Lepidoptera: Noctuidae) from West Africa. *Pesticide Biochemistry and Physiology* 74: 17-26.

Mukhopadhyay A. and Roy S. (2009) Changing dimensions of IPM in the tea plantations of the North Eastern sub Himalayan region: National Symposium on IPM strategies to combat emerging pests in the current scenario of climate change (ed. by VV Ramamurthy, GP Gupta and SN Puri) Entomology Society of India, New Delhi and Central Agricultural University, Pasighat, A. P. India.

Mukhopadhyay A. and Sarker M. (2007) Natural enemies of some tea pests with special reference to Darjeeling, Terai and the Doors. National Tea Research Foundation, Tea Board, Kolkata.

Mukhopadhyay A., Das S., Roy S. and Saha D. (2014) Stress induced changes in the pests of tea plantations of sub Himalayan Terai-Dooars region: An appraisal. *Two and a Bud* 60: 30-42.

Muraleedharan N. (1992) Pest control in Asia: Tea: cultivation to consumption (ed. by KC Willson and M.N. Clifford, Springer Netherlands, Dordrecht.

Nair N., Sekh K., Debnath M.R., Dhar P.P. and Somchoudhury A.K. (2008a) Biology of *Hyposidra infixaria* Walk. (Lepidoptera: Geometridae), a resurgent looper pest of tea. *Journal of Entomologial Research* 32: 67-70.

Nair N., Somchoudhury A.K. and Dhar P.P. (2008b) Relative toxicity of some chemical insecticides against Hyposidra talaca Walk. (Lepidoptera: Geometridae). *Pestology* 32: 44-46.

Oakshott J.G., Claudianos C., Campbell P.M., Newcomb R.D. and Russell R.J. (2005) Biochemical genetics and genomics of insect esterases. In: Gilbert L.I. and Gill S.S. (Eds.) Insect pharmacology–channels, receptors, toxins and enzymes. Elsevier, London, pp. 229-301.

Perera M.D., Hemingway J. and Karunaratne S.P. (2008) Multiple insecticide resistance mechanisms involving metabolic changes and insensitive target sites selected in anopheline vectors of malaria in Sri Lanka. *Malaria Journal* 7: 168.

Prasad A.K. and Mukhopadhyay A. (2013) A technique to measure the loss in tea crop by the defoliating pest (Hyposidra talaca Walker) on the basis of dry mass and leaf area patrameters. *International Journal of Bio-resource Stress Management* 4: 358-361.

Prasad S. (1992) Infestation of Helopeltis: Tea times (ed. Sandoz (India) Ltd, Bombay.

Robinson G.S., Ackery P.R., Kitching I.J., Beccaloni G.W. and Hernández L.M. (2010) HOSTS-A Database of the World's Lepidopteran Hostplants. Natural History Museum, London, UK.

Roobakkumar A., Babu A., Kumar P. and Muraleedharan N. (2011) Variations in the esterase activity between two different populations of the red spider mite, *Oligonychus coffeae* Neitner (Acarina: Tetranychidae) infesting tea. *Journal of Biosciences Research* 2: 5-9.

Roobakkumar A., Babu A., Rahman V.K.J., Subramaniam M.S.R.S. and Kumar D.V. (2012) Comparartive susceptibility and detoxifying enzyme activities to fenpropathrin in fieldcollected and laboratoryreared *Oligonynchus coffeae* infesting tea. *Two and a Bud* 59.

Roy S., Gurusubramanian G. and Mukhopadhyay A. (2009a) Variation of resistance to endosulfan in tea mosquito bug, *Helopeltis theivora* Waterhouse (Heteroptera: Miridae) in tea plantation of the Sub–Himalayan Dooars, northern West Bengal, India. *Journal of Entomology and Nematology* 1: 29-35.

Roy S., Gurusubramanian G. and Mukhopadhyay A. (2010a) Neem-based integrated approaches for the management of tea mosquito bug, *Helopeltis theivora* Waterhouse (Miridae: Heteroptera) in tea. *Journal of Pest Science* 83: 143-148.

Roy S., Mukhopadhyay A. and Gurusubramanian G. (2008a) Preliminary toxicological study of commonly used acaricides of tea red spider mite (Oligonychus coffee Nietner) of North Bengal, India. *Resistant Management Newslette* 18: 10-15.

Roy S., Mukhopadhyay A. and Gurusubramanian G. (2008b) Susceptibility status of *Oligonychus coffeae* Nietner, (Acarina: Tetranychidae) to commonly applied acaricides in the tea plantation of the North Bengal, India. *SARC Journal of Agriculture* 6: 107-116.

Roy S., Mukhopadhyay A. and Gurusubramanian G. (2009b) Detection of acaricide susceptibility in field populations of tea red spider mite, *Oligonychus coffeae* (Acarina: Tetranychidae) in North Bengal through *in situ* bioassay method. *Current Biotica* 3: 251-254.

Roy S., Mukhopadhyay A. and Gurusubramanian G. (2009c) Monitoring of resistance to commonly used acaricides on tea red spider mite (*Oligonychus coffeae* Nietner) in populations from the Darjeeling plains of North Bengal, India. *Resistant Management Newsletter* 18: 10-14.

Roy S., Mukhopadhyay A. and Gurusubramanian G. (2009d) Status of new acaricides *vis-a-vis* conventional acaricides against the Red spider mite, *Oligonychus coffeae* (Acarina: Tetranychidae) in tea plantations of Darjeeling plains, India. *Romanian Journal of Plant Protection* 2: 1-8.

Roy S., Mukhopadhyay A. and Gurusubramanian G. (2009e) The Synergists Action of Piperonyl Butoxide on Toxicity of Certain Insecticides Applied Against *Helopeltis theivora* Waterhouse (Heteroptera: Miridae) in the Dooars Tea Plantations of North Bengal, India. Journal of Plant Protection Research 49. doi:10.2478/v10045-009-0034-0.

Roy S., Mukhopadhyay A. and Gurusubramanian G. (2010b) Baseline susceptibility of *Oligonychus coffeae* to acaricides in North Bengal tea plantations, India. *International Journal Acarology* 36.

Roy S., Mukhopadhyay A. and Gurusubramanian G. (2010c) Development of resistance to endosulphan in populations of the tea mosquito bug *Helopeltis theivora* (Heteroptera: Miridae) from organic and conventional tea plantations in India. *International Journal of Tropical Insect Science* 30: 61-66.

Roy S., Mukhopadhyay A. and Gurusubramanian G. (2013) Escaping the chemical pesticide trap: Non–chemical managemet of tea pests in north east India. *Two and a Bud* 60: 1-4.

Saha D., Roy S. and Mukhopadhyay A. (2012) Insecticide susceptibility and activity of major detoxifying enzymes in female *Helopeltis theivora* (Heteroptera: Miridae) from sub-Himalayan tea plantations of North Bengal, India. *International Journal of Tropical Insect Science* 32: 85-93.

Sannigrahi S. and Talukdar T. (2003) Pesticide use patterns in Dooars tea industry. *Two and a Bud* 50: 35-38.

Sarker M. and Mukhopadhyay A. (2008) Enzyme Based detection of pesticide resistance in two Major arthropod pest, *Helopeltis theivora* and *Oligonychus coffeae* occurring on tea in the lower elevations and Terai of Darjeeling Hills. N.B.U. *Journal of Animal Sciences* 2: 71-78.

Shan G. and Ottea J.A. (1998) Contributions of monooxygenases and esterases to pyrethroid resistance in the tobacco budworm, *Heliothis virescens*. *Proceedings of National Council of America* 2: 1148-1151.

Shono T., Ohsawa K. and Casida J.E. (1979) Metabolism of trans and cis–permethrin, trans and cis–cypermethrin and deltamethrin by microsomal enzymes. *Journal of Agricultural and Food Chemistry* 27: 316-325.

Sivapalan P. (1999) Pest management in tea: Global Advances in Tea Science (ed. by NK Jain) Aravali Books, New Delhi, pp. 625-646.

Soderland D.M. and Bloomquist J.R. (1990) Molecular mechanism of insecticide resistance. In: Roush R.T. and Tabashnic B.E. (Eds.) Pesticide resistance in arthropods; Chapman and Hall, London, pp. 237-260.

Stenersen J. (2004) Chemical Pesticides Mode of Action and Toxicology. CRC Press, Boca Raton.

Sternburg J., Kearns C.W. and Moorefield H. (1954) Resistance to DDT, DDT-Dehydrochlorinase, an Enzyme Found in DDT-Resistant Flies. *Journal of Agricultural and Food Chemistry* 2: 1125-1130.

Stumpf N., Zebitz C.P.W., Kraus W., Moores G.D. and Nauen R. (2001) Resistance to Organophosphates and Biochemical Genotyping of Acetylcholinesterases in *Tetranychus urticae* (Acari: Tetranychidae). *Pesticide Biochemistry and Physiology* 69: 131-142.

Subramaniam B. (1995) Tea in India. P.I.D. and Wiley Eastern Ltd.

Watt G. and Mann H.H. (1903) The Pests and Blights of the Tea Plant. second edn. The Superintendent, Government Printing Press, Culcutta.

Wilson T.G. (2001) Resistance of *Drosophila* to toxins. *Annual Review of Entomology* 46: 545-571.

Wu S., Yang Y., Yuan G. and *et al.* (2011) Overexpressed esterases in a fenvalerate resistant strain of the cotton bollworm, *Helicoverpa armigera. Insect Biochemistry and Molecular Biology* 41: 14-21.

Xie W., Wang S., Wu Q. and *et al.* (2011) Induction effects of host plants on insecticide susceptibility and detoxification enzymes of Bemisia tabaci (Hemiptera: Aleyrodidae). *Pest Management Science* 67: 87-93.

Yu S.J. (2014) The Toxicology and Biochemistry of Insecticides. CRC Press, Taylor and Francis Group, Boca Ration, FL, USA.

Zhou X., Scharf M.E., Parimi S. and *et al.* (2002) Diagnostic assays based on esterase-mediated resistance mechanisms in western corn rootworms (*Coleoptera*: *Chrysomelidae*). *Journal of Economic Entomology* 95: 1261-1266.

Zhu K.Y. and Gao J.R. (1998) Kinetic Properties and Variability of Esterases in Organophosphate-Susceptible and -Resistant Greenbugs,Schizaphis graminum (*Homoptera*: *Aphididae*). *Pesticide Biochemistry and Physiology* 62: 135-145.

Zhu K.Y. and He F. (2000) Elevated Esterases Exhibiting Arylesterase-like Characteristics in an Organophosphate-Resistant Clone of the Greenbug, *Schizaphis graminum* (*Homoptera*: *Aphididae*). *Pesticide Biochemistry and Physiology* 67: 155-167.

Tea: Technological Initiatives, pp. 123-193
New India Publishing Agency, New Delhi, India
Edited by Niladri Bag, Arundhati Bag and L.M.S. Palni

6

Pests of Tea: Overview and Possibilities of Integrated Pest Management in Indian Tea Scenario

Somnath Roy, Narayanannair Muraleedharan and Gautam Handique

Abstract

Tea, Camellia sinensis (L.) *O. Kuntze, is a perennial crop and grown as a monoculture on large contiguous areas in India. Being a plantation crop, tea provides a relatively stable microclimate and food supply for several notorious pests such as insects, mites, nematodes which cause substantial loss of foliage. However, region-wide variation in faunal diversity exists as a result of autochthonous and heterochthonous recruitment and the influence of climate, altitude and age of the plantation. This publication provides the details on pest information (identification, crop damage, alternate hosts, pest life cycle) and their possible management strategies. On account of complex pest situations, total avoidance of pesticides in tea is not possible; although due to the sensitive nature of the crop pesticide use must be minimal. A tentative IPM strategy for tea cultivation in India has been proposed in this paper which may ensure low – cost, eco-compatible, pest management package with minimal residue problems.*

Kay words*: Tea, pest, lifecycle, damage potential, management, IPM.*

Indian Tea Scenario

Tea, a gift of nature, prepared from the tender leaves and buds of *Camellia sinensis* (L.) O. Kuntze (Theaceae) is the most widely used non-alcoholic beverage all over the world. The tea industry in India is one of the oldest organized industries and our teas are appreciated world over for their unique flavor, aroma, and health benefits. India produces three speciality teas - Darjeeling, Assam and Nilgiri, a logos of Tea Brand India are world famous. In India, tea is grown in a wide amplitude of climatic variables, at latitudes from 8° 12' N in Nagercoil in Tamil Nadu to 32° 13' in Kangra in Himachal Pradesh and at altitudes ranging from near sea level in Assam to 2414 m (7920 feet) above mean sea level (msl) in Korakundha in the Nilgiris in south India (Hazarika and Muraleedharan, 2011). The major tea growing states of India are Assam, West Bengal, Tamil Nadu and Kerela. Other states such as Bihar, Himachal Pradesh, Karnataka, Orissa, Tripura and Uttaranchal also grow tea in small areas. In North East India tea is planted in the Brahmaputra and Barak Valleys of Assam, plains of Dooars and Terai and Darjeeling hills in North Bengal. About 75 per cent of the total tea produced in India is accounted by Assam and West Bengal together. Assam teas are famous for their strong, brisk and full bodied liquor; Nilgiri teas are well known for their delicate flavor, strength and brightness; and the production of the famous Darjeeling tea is aided by the soil and climate of the hills. The tea growing areas of West Bengal are spread over Darjeeling hill slopes and the adjoining plains of Terai and the Dooars. Moreover, Darjeeling District has been declared as an Agri Export zone for producing export-quality tea to the world market (The Statesman, 5 June, 2003) and tea from this region has also gained the 'geographical indicator' status, as many of the tea estates produce 'flavour leaves' bio-rationally or organically. Currently, an area of about 579,000 ha is under tea in India. India is the largest producer and consumer of black tea in the world. Production of tea in 2011-12 was 1095.46 million kg. The country exported 214.35 million kg in 2011-12 (Anon, 2012).

Tea Pest Spectrum

Since the dawn of tea culture, a wide range of pests have been associated with tea plantations. Tea pest and productivity are two antagonistic factors. Pests have largely been responsible for decline in the productivity of tea. Each tea growing area has its own distinctive pests. It is estimated that 1034 species of arthropods and 82 nematode pests infesting tea all over the world (Hazarika *et al.*, 2009). About 300 species of insects and mites have been reported from India (Muraleedharan, 2010). Only a few of them have been regarded as major pests while most of them are minor and localized and caused occasional damage. Most of the pests are polyphagous but, all the pests infest throughout the year and complete their life cycle in tea field. Some insects are often newly added to the list of tea pests. Because once they happen to be on tea and find it acceptable they continue to thrive on the abundant supply of food available to them. On the contrary, the insects, which cause negligible damage to day, may become destructive tomorrow. It is therefore essential that every pest, whether it is a minor or a major one, should not be ignored (Ahmed and Aslam, 2011). Principal arthropod pests infesting tea and their distribution are given in Table 1.

Table 1. Principal arthropod pests of tea in India

Common name	Scientific name	Order	Family	Distribution
A. Pest of Foliage				
Tea mosquito bug	*Helopeltis theivora* Waterhouse**	Hemiptera	Miridae	NEI, SI
Looper caterpillar	*Ascotis* sp.	Lepidoptera	Geometridae	NEI
	Biston (=Buzura) suppressaria Guen*	Lepidoptera	Geometridae	NEI, SI
	Cleora sp.	Lepidoptera	Geometridae	NEI
	Ectropis bhurmitra (Walker)	Lepidoptera	Geometridae	SI
	Ectropis sp.	Lepidoptera	Geometridae	NEI
	Hyposidra infixaria (Walker)*	Lepidoptera	Geometridae	NEI
	Hyposidra talaca (Walker)*	Lepidoptera	Geometridae	NEI
Red spider mite	*Oligonychus coffeae* Nietner**	Acari	Tetranychidae	NEI, SI, HP
White mite	*Acaphylla indiae* Keifer	Acari	Eriophyidae	NEI
Pink or Orange mite	*Acaphyla theae* Watt*	Acari	Eriophyidae	NEI, SI, UTK
	Acaphyllisa parindiae Keifer	Acari	Eriophyidae	SI
Purple tea mite	*Calacarus carinatus* (Green)*	Acari	Eriophyidae	NEI, SI, HP
Yellow mite	*Polyphagotarsonemus latus* (Bank) Ewing*	Acari	Tarsonemidae	NEI, SI
Scarlet mite	*Brevipalpus australis* Baker*	Acari	Tenuipalpidae	SI
	Brevipalpus obovatus Donnadieu*	Acari	Tenuipalpidae	NEI, HP
	Brevipalpus phoenicis (Geijskers)*	Acari	Tenuipalpidae	NEI, HP
Tea thrips	*Scirtothrips dorsalis* Hood**	Thysanoptera	Thripidae	NEI
	Scirtothrips bispinosus (Bagnall)*	Thysanoptera	Thripidae	SI
	Taeniothrips lefroyi (Bagnall)	Thysanoptera	Thripidae	NEI
	Taeniothrips setiventris Bagnall	Thysanoptera	Thripidae	NEI, HP
Jassids/Green fly	*Empoasca flavescens* Fabricius*	Hemiptera	Cicadellidae	NEI, SI, HP
Tea aphid	*Toxoptera aurantii* (Boyer de Fonscolombe)*	Hemiptera	Aphididae	NEI, SI, HP, UTK
Scale insect	*Saissetia coffeae* (Wlk.)*	Hemiptera	Coccidae	NEI, SI
	Saissetia formicarri (Green)*	Hemiptera	Coccidae	NEI, SI
	Eriochiton theae (Green)*	Hemiptera	Coccidae	NEI, SI

Bunch caterpillar	*Andraca bipunctata* Walker*	Lepidoptera	Bombycidae	NEI
Tea leaf roller	*Gracillaria theivora* Walsm.	Lepidoptera	Gracilariidae	NEI
	Caloptilia theivora (Walsingham)*	Lepidoptera	Gracillariidae	SI
Flushworm	*Cydia leucostoma* Meyrick*	Lepidoptera	Tortricidae	SI
Tea tortrix	*Homona coffearia* (Nietner)*	Lepidoptera	Tortricidae	NEI, SI
Red slug catarpillar	*Eterusia magnifica* Butl.*	Lepidoptera	Zygaenidae	NEI
	Eterusia aedea virescens (Butler)*	Lepidoptera	Zygaenidae	SI
Nettle grub	*Thosea* spp.*	Lepidoptera	Limacodidae	NEI, SI
	Parasa pastoralis Butler	Lepidoptera	Limacodidae	NEI
B. Pest of Stem				
Large bark eating borer	*Indarbela quadrino tata* Wlk.	Lepidoptera	Indarbelidae	NEI
Common bark eating borer	*Indarbela theivora* Hamps.	Lepidoptera	Indarbelidae	NEI
Red borer	*Zeuzera coffeae* Nietner	Hemiptera	Coccidae	NEI, SI, UTK
Shot hole borer	*Euwallacea fornicatus* Eichhoff*	Coleoptera	Scolytidae	SI
Scavenging termite	*Odontotermes* sp.	Isoptera	Termitidae	NEI, SI
	Coptotermes heimi (Wasm)	Isoptera	Termitidae	NEI
	Microcerotermes heini (Wasmann)	Isoptera	Termitidae	NEI
	Microcerotermes pakistanicus Ahmed	Isoptera	Termitidae	NEI
Live wood eating termites	*Microcerotermes* sp.*	Isoptera	Termitidae	NEI
	Microtermes sp.	Isoptera	Termitidae	NEI, SI
	Neotermes buxensis Roonwal and Sen Sharma	Isoptera	Termitidae	NEI
	Odontotermes assamensis Holmgren	Isoptera	Termitidae	NEI
	Odontotermes feae (Wasm)	Isoptera	Termitidae	NEI
	Odontotermes parvidus Holmgren	Isoptera	Termitidae	NEI
	Odontotermes redemanni (Wasmann)	Isoptera	Termitidae	NEI
C. Pest of Root				
Cockchafer grub	*Holotrichia* spp.*	Coleoptera	Scarabaeidae	NEI
	Sophrops plagiatula (Brenske)	Coleoptera	Scarabaeidae	NEI

NEI: North East India; SI: South India; HP: Himachal Pradesh; UTK: Uttarakhand
**: Pest of national significance ; *: Pest of regional significance

Crop Loss Caused by Major Insect and Mite Pests

The yield loss due to insect and mite pests may vary from 5 to 55 per cent and in some cases 100 per cent crop loss has been reported (Muraleedharan and Chen, 1997). Damage caused by *Helopeltis theivora* in the tea plantations of South India resulted in 11 to 100 per cent per cent crop loss during the peak season (Rao and Murthy, 1976). The red spider mite, *Oligonychus coffea*, characterized by a high reproductive capacity, resulting in high population level in a short period of time is causing higher economic damage (Das, 1959, 1960). Loss in tea crop due to red spider mite attack in India may be as much as 75 per cent in certain pockets (Roy *et al.*, 2008). It is responsible for 6 to 20 per cent crop reduction in certain parts of North East India (Awasthy and Venkatakrishnan, 1977). The annual crop loss due to red spider mite varies between 17 to 46 per cent (Sarmah *et al.*, 2009). In Munnar, red spider mite caused 14-18 per cent crop loss (Selvasundaram and Muraleedharan, 2003). Rao and Subramanian (1968) obtained 8 to 17 per cent increase in crop by controlling eriophyid mite infeststion. A study showed that the loss due to eriophyids was only around 5 per cent (Muraleedharan and Radhakrishnan, 1989b). The bunch caterpillar, *Andraca bipunctata* causes crop loss up to 20 per cent in severe conditions (Banerjee, 1982). Termite infestation in North East India accounts for 10 to 15 per cent loss in crop (Sen and Chakrabarthy, 1964).

Infestation by pests not only results in crop loss but also adversely affects the quality of processed tea (Muraleedharan and Cheng, 1997). Teas made from flushworm infested shoots have low levels of extractable solids and high crude fibre content. Liquors obtained from such teas are 'flat', the larvae and their excreta present in leaves being responsible for the reduction in quality. If green leaves meant for manufacture contained more than 7 per cent of leaf roller infested shoots, colour of the tea liquor changed to brown and developed an unpleasant taste (Kodomari, 1991). Teas made from thrips infested leaves are of poor quality and contained more stalk and fibre (Murthy and Chandrasekaran, 1979).

Description of Some Insect and Mite Pests Infesting Tea

A. Pests of Foliage

A.1. Insect Pests of Foliage

A.1.1 Tea mosquito bug Helopeltis theivora Waterhouse (Hemiptera: Miridae)

One of the earliest maladies of tea recorded in India was the leaf spotting and dieback caused by *H. theivora*, the most important sucking pest till today, commonly known as tea mosquito bug. This was first noticed in Cachar, Assam in 1865 and at first was thought to be a fungal disease. It was not until around 1873 that it was found to be caused by the feeding of *H. theivora* described by Waterhouse more than a decade later in 1886.

Tea mosquito bug is considered as one of the serious pests of tea in in almost all countries because it attacks only to the young shoots that is the actual crop of tea. In recent years, the unusual weather pattern prevailing for a longer period during the peak season of activity of *H. theivora* makes it further difficult to control this pest.

Marks of Identification

Egg: The egg is elongated, sac like, white and about 0.8-1.0 mm length. The position of egg is marked by of two hairs of unequal lengths projecting from the surface (Fig. 1a & b).

Nymph: The freshly hatched nymph of *Helopeltis* is dirty yellow in colour. The colour of the first and second instar nymph is greenish yellow, but it becomes green in the later instars (Fig. 1c). Nymphs are wingless.

Adult: The adult is a tiny insect with long antennae and wings (Fig. 1d&e). Thorax is yellow in female and black in male. The adult is not a strong flier, the normal distance of flight being from bush to bush between alternate bushes at a time.

a. Egg hairs projecting from the stem surface

b. Eggs of Tea Mosquito Bug

c. Nymph of tea mosquito bug

d. Adult (male)

e. Adult (female)

f. Die back caused due to tea mosquito bug feeding

g. Nature of damage in leaf

Fig. 1 (a-g): Different developmental stages of tea mosquito bug and nature of damage (*Source*: Roy *et al*., 2015)

Host Range

Tea is the most preferred and principal host plant. Beside tea, *H theivora* feeds on a wide range of economic plants including cashew, *Acacia*, cocoa, and camphor and causes considerable damage to pepper. Apart from cultivated plants, they also have non crop host plants (weeds and many jungle plants) which may support their population when the major hosts are scarce (Table 2). This habit may enable them to breed throughout the year, or survive under adverse conditions until the major host is in abundance.

Table 2: Lists of host planta of *H. theivora* present in and around tea plantations of India

Host plant	Family	Economic status	Reference
Acacia mangium Willd.	Fabaceae	Economic plants	Thu *et al.*, 2010
Acalypha indica L.	Euphorbiaceae	Weed	Das, 1965
Anacardium occidentale L.	Anacardiaceae	Economic plants	Ambika and Abraham, 1983; Das, 1984; Sundararaju, 1993
Annona reticulata L.	Annonaceae	Fruit trees	Kalita *et al.*, 2000
Anthocephalus cadamba (Roxb.) J. Bosser	Rubiaceae	Flowering tree	Saha and Mukhopadhyay, 2013
Azadirachta indica A. Juss.	Meliaceae	Medicinal plant	Sundararaju and Sundarababu, 1999
Bidens pilosa L.	Asteraceae	Weed	Saha and Mukhopadhyay, 2013
Bixa orellana L.	Bixaceae	Medicinal plant	Das (1984)
Camellia japonica L.	Theaceae	Flowering Plants	Sudhakaran and Muraleedharan, 2006
Camellia sinensis (L.) Kuntze	Theaceae	Economic plants	Das (1965)
Cannabis sativa L.	Cannabaceae	Cultivated plants	Gogoi *et al.*, 2012
Capsicum spp.	Solanaceae	Fruits plants	Anon., 2007
Ceiba pentandra (L.) Gaertn.	Malvaceae	Economic plants	Das, 1984
Chromolaena odorata (L.) R.M. King & H. Rob	Asteraceae	Weed/ ornamental plan	Srikumar and Bhat, 2013
Cinchona officinalis L.	Rubiaceae	Medicinal plants	Das, 1984
Cinchona sp.	Rubiaceae	Medicinal plants	Das, 1984
Cinnamomum camphora (L.) J. Presl.	Lauraceae	Economic plants	Das, 1984
Duranta repens L.	Verbenaceae	Ornamental plant	Gogoi *et al.*, 2012
Ehretia acuminata R. Brown	Boraginaceae (Ehretiaceae)	Forest tree	Barbora and Singh, 1994
Eugenia jambolana (*Syzygium cumini*)	Myrtaceae	Fruit trees	Kalita *et al.*, 2000
Eurya acuminata D.C.	Theaceae	Medicinal Plants	Das, 1965
Ficus benjamina L.	Moraceae	Tree producing fig	Gogoi *et al.*, 2012

Ficus hispida L.f.	Moraceae	Fig tree	Saha and Mukhopadhyay, 2013
Gardenia jasminoides Ellis	Rubiaceae	Flowering plant	Kalita *et al.*, 2000
Ixora coccinea	Rubiaceae	Flowering plant	Gogoi *et al.*, 2012
Jasminum sandens Vah	Oleaceae	Flowering plant	Das, 1965
Maesa indica (Roxb.) DC.	Myrsinaceae	Weed	Sudhakaran and Muraleedharan, 2006
Maesa ramentacae Wallich	Myrsinaceae	Weed	Das, 1965
Melastoma malabathricum L.	Melastomataceae	Weed	Das, 1965
Mikania micrantha H.B. & K	Asteraceae	Weed	Somchowdhury *et al.*, 1993
Morus alba L.	Moraceae	Economic plants	Barbora and Singh, 1994
Murraya koenigii	Rutaceae	Economic plants	Gogoi *et al.*, 2012
Oxalis acetosella L.	Oxalidaceae	Weed	Barbora and Singh, 1994
Persea bombycina Kost	Lauraceae	Fruits plants	Gogoi *et al.*, 2012
Persicaria (= *Polygonum*) *chinensis* (L.) Nakai	Polygonaceae	Weed	Saha and Mukhopadhyay, 2013
Phlogacanthus pubinervius T. Anderson	Acanthaceae	Medicinal Plants	Somchowdhury *et al.*, 1993
Piper hamiltonii C. DC.	Piperaceae	Medicinal Plants	Gogoi *et al.*, 2012
Piper nigrum L.	Piperaceae	Economic plants	Das, 1984
Plogacanthus thirsyflorus (Roxb.) Nees	Acanthaceae	Medicinal Plants	Gogoi *et al.*, 2012
Premna latifolia Roxb.	Verbenaceae	Medicinal Plants	Barbora and Singh, 1994
Psidium guajava L.	Myrtaceae	Fruits plants	Saha *et al.*, 2012
Pteridium aquilinum (L.) Kuhn	Dennstaedtiaceae	Eagle fern	Saha and Mukhopadhyay, 2013
Sida cordifolia L.	Malvaceae	Weed	Gogoi *et al.*, 2012
Smilax herbacea L.	Smilacaceae	Weed	Somchowdhury *et al.*, 1993
Theobroma cacao L.	Malvaceae	Economic plants	Anon., 2007
Clidemia hirta (L.) Don	Melastomataceae	Economic plants	Ragesh, 2013

Nature of Damage

The nymphs and adults of *H. theivora* suck the sap of the young leaves, buds and tender stems and while doing so, they injects toxic saliva which causes the breakdown of tissues around the site of feeding. Within 2-3 hours of sucking a circular spot is formed around the feeding point and in 24 hours it becomes translucent, light browning (Fig. 1f). Within a few days the spots appear as dark brown sunken spots which subsequently dry up. The badly affected leaves become deformed and even curl-up. In addition, due to oviposition, the tender stems develop cracks and over-callusing which lead to blockage of vascular bundle thereby affecting the physiology causing stunted growth and sometimes die-back of the stems (Fig. 1g) (Rahman *et al.*, 2005).

A single fully grown nymph (5th instar) is the most voracious feeder among the life stages, producing the most and largest feeding lesions (Bhuyan and Bhattacharyya, 2006). An adult could make 150 feeding spots in a day (Hainsworth, 1952). A single female can produce lesions over an area of 412.43mm^2/day (Kalita *et al.*, 1995).

Seasonal Incidence

Adults and nymphs of *Helopeltis* could be seen on tea bushes almost throughout the year but peak in the incidence in North East India is noticed during June and July, often extending up to September when the number of rainy days is large. In sub Himalayan Dooars tea plantations the population of *H. theivora* is abundant throughout the year. However, the lowest population of this pest is noticed during the winter months of December to February. The population usually began to build up in the months of May/June reaching a peak during September to November (Roy *et al.*, 2009a).

Life History

Egg: The female starts egg laying within a few hours after mating. The eggs are laid singly into the green tissues of growing shoots or young stems, buds and midribs of young leaves. Often, eggs are laid in clusters. A single female may lay about 220 eggs in 36 days during March-April. About 4-10 eggs are laid per day. The incubation period varies in different seasons (Table 3).

Nymphs: There are five nymphal instars. Nymphs are wingless. From the second instar onwards, they bear a scutellar horn. Wing pads develop in the third instar. Duration of nymphs varies with the seasons (Table 3).

Adults: The longevity of male varies between 26 and 55 days while that of female is between 32 and 36 days during November to March.

Table 3: Duration of life stages of *H. theivora* in the climatic conditions of Assam and Dooars

Life stages	Duration in days			
	Nov.-Jan	.Mar-April	June-July	Aug.-Sept.
Assam condition				
Incubation period	16-20	7-8	4-6	5-6
Nymphal period	19-20	16-20	9-14	8-10
Total duration (Egg to adult)	35-40	23-28	13-20	13-16
Dooars condition				
Incubation period	6-18	7-15	6-8	4-8
Nymphal period	9-20	8-13	8-10	8-10
Total duration (egg to adult)	15-32	16-24	14-17	13-16

(*Source:* Anon., 2010 and Roy *et al.*, 2009b)

Varietal/clonal Preferences

Chinery and hybrid tea jats were invariably found more susceptibility to *H. theivora* than 'Assam' varieties (Anstead and Ballare, 1992; Rau, 1940). However in Assam, all tea clones were found infested and among them TV-1 was the most susceptible (Das, 1984). The clones TV11, TV17, TV21, TV25 and TV26 had been reported considerably tolerant to this pest (Somchowdhury *et al.,* 1993). In an extensive study by Roy *et al.* (2009a) screened 28 tea cultivars which are common in North East India and among them TV1, TV12, TV23, TS653 and TV16 were the most susceptible to *H. theivora* infestation. TV4, TV11, TV28, TV29 and ST449 were less susceptible and TV2, TV9, TV17, TV18, TV20, TV25, TV26, TV30, Teenali 17, TS652, TS491, P126, TV7, TV10, TV14, TV19, TV22, and TS426 were moderately susceptible. No clone was immune to infestation by *H. theivora.* The South Indian tea clones UPASI- 2, 3, 7, 9, 22 and AKK-1 were the most susceptible to *H. theivora* infestation. UPASI- 10, 12, 14, 15 and 17 were less susceptible and UPASI- 1, 11, 13 and ATK- 1 were moderately susceptible (Sudhakaran, 2000).

Survival Potential

The ability to survive under inimical environment is called survival potential. The survival potential of a pest helps to overcome the adversities of life process through some contrivances such as protective devices, mimicry, food habit, shelter, migration and laggard stage.

Table 4: Response of *Helopeltis* to different factors

Factor	Response
Temperature	Positively thermotropic
Humidity	Positively hydrotropic
Light	Negatively phototropic
Cloud	Positive
Hosts specificity	Polyphagous
Protection	Eggs embedded in plant tissues
Dispersion	Slow flier

Factors Influencing the Population Build-up

- Several alternate weed species as host
- Extended pruning cycle
- Same chemicals being repeatedly used in the same fields for longer period
- Density of shade

- Microclimate
- Lack of proper understanding of the bio-ecology and monitoring of the pest
- Spraying interval

A.1.2. Looper Complex

In recent times besides the common looper (*Buzura suppressaria*), other species of looper caterpillars *viz*., *Hyposidra talaca, Hyposidra infixaria, Ectropis* sp., *Ascotis* sp. and *Cleora* sp. are also found to occur in tea plantations of Assam and North Bengal. The outbreak of these loopers are considered to be a culminating effect of factors like climatic changes, deforestation, large scale use of inorganic insecticides earlier against other pests of tea etc. As these insect pests have number of advantages like very short life cycle with about 8 generations in a year, fast multiplication, no hibernation during winter months and lack of effective natural enemies at present etc. *H. talaca* also known as 'black inch worm' is found to be the most predominant species among the looper complex infesting tea (Chutia *et al*., 2012).

Nature of Damage

The damage symptoms of different species are more or less similar. The newly hatched caterpillars make small holes along the margins of young leaves and then eat from the margins. Young ones are very often seen on growing shoots and buds scrapping on the leaf surface. The grown up caterpillars prefer semi mature to mature leaves. From late third instars, they become voracious feeders. In severe conditions, bushes are completely stripped of foliage. Even bark of small branches is also eaten away in absence of foliage (Fig. 4c).

Common looper, *Biston* (=Buzura) *suppressaria* (Lepidoptera: Geometridae)

This is the earliest known species of looper caterpillar infesting tea. The moth is grayish white, finely speckled with black (Fig. 2a). The males are generally smaller than the females and are easily distinguished by the bipectinate (feathered) antennae. The female moth lays eggs in heaps, each containing 200-600 eggs covered with a buff coloured hairs. The most common site for the deposition of eggs is the trunk of shade trees or any other tall trees in the vicinity of tea fields. The young caterpillar is dark brown with greenish white lines along the back and side (Fig. 2b). The body colour soon turns to light green and with age, it acquires a brownish grey colour similar to that of a mature twig of tea. There are five larval instars. The full grown caterpillar moves down to ground for pupation at a depth of 2.5-5 cm in the soil under tea bushes. The duration of pupal stage is about three weeks in summer and more than three months during cold weather. The life cycle is completed in about 72 days during March-May and in about 60 days in June-July.

Table 5: Duration of different stages of *B. suppresaria*

Stages	Duration in days	
	March-May	June July
Egg	14.5	13.0
Caterpillar	38.5	29.0
Pupa	19.0	18.0
Total duration (Egg to adult)	72.0	60.0

Black looper, *Hyposidra* spp.

Two species of black looper *viz., H. talaca and H. infixaria* are now predominant in tea plantations of Assam and North Bengal. They also known as 'black inch worm'. The caterpillars of *H. talaca* and *H. infixaria* are polyphagous in nature and reported to feed on a number of forest plants and weeds from India; Malaysia and Thiland (Browne, 1968; Mathew *et al.*, 2005; Winotai *et al.*, 2005; Das and Mukhopadhyay, 2008; www.mothsofborneo.com).

Hyposidra talaca (Lepidoptera: Geometridae)

Early instars caterpillars are brown to black in colour with transverse six white bands on the dorsal side of the body (Fig. 2i). In mature larva the bands are disappeared and change its colour to brownish grey with white spots on the body. The caterpillar passes through 5-6 larval instars. Pupation generally takes place in soil around the collar region of tea bushes and shade trees. In old tea bushes, it prefers the cracks and crevices of the frame to pupate. The full grown caterpillar is measured about 45-50 mm in length. Adult moths are greyish to light brown in colour and the females are much larger than the male with a bulging abdomen. (Fig. 2j). Adult longevity is about 4-8 days in summer and 4-6 days in winter.

Hyposidra infixaria (Lepidoptera: Geometridae)

Early instars caterpillars are brown to black in colour with transverse six white bands on the dorsal side of the body. The colour of mature larva is brownish grey with two transverse lines at the thoracic region of the body (Fig. 2c). The adults are pale or greyish in colour (Fig. 2d). The distinguishing character of this species is that the male moths bear a tuft of hairs at the abdominal tip.

a. Adult of *B. suppressaria* **b.** Larva of *B. suppressaria*

c. Adult of *H. infixaria* **d.** Larva of *H. infixaria*

e. Adult of *Ectropis* sp. **f.** Larva of *Ectropis* sp.

g. Adult of *Ascotis* sp. **h.** Larva of *Ascotis* sp.

i. Adult *of H. talaca* **j.** Larva of *H. talaca*

k. Adult of *Cleora* sp. **l.** Larva of *Cleora* sp.

Fig. 2 (a-l): Different developmental stages of looper complex in tea

Table 6: Duration of developmental stages of *H. talaca* and *H.infixaria*

Stages	*H. talaca*	*H. infixaria*
Egg	6.0±0	8.0 ±0
First instar larva	4.64±0.125	5.44±0.114
Second instar larva	3.25±0.116	3.80±0.092
Third instar larva	3.80±0.141	3.96±0.080
Fourth instar larva	4.41±0.150	4.14±0.105
Fifth instar larva	11.58±0.335	10.38±0.165
Pupa	18.94±0.300	11.14±0.203
Total development period	55.00±0.280	47.82±0.163

(*Source:* Das *et al.*, 2010)

Ectropis sp. (Lepidoptera: Geometridae)

The early instar caterpillars are brownish in colour with black stripes on the dorsal side of the body. The full grown caterpillar is about 30- 35 mm in length and look like a dried up twig of tea bush (Fig. 2e). The adults are pale yellow in colour with black wavy lines. The males are smaller than the female with bipectinate antennae (Fig. 2e).

Ascotis sp. (Lepidoptera: Geometridae)

The early instar caterpillar is light green in colour (Fig. 2h). A matured caterpillar changes its colour to greenish grey with a pair of tubercle located dorsally on the abdominal region of the body. The adults are greyish in colour with brown spots on each wing (Fig. 2g).

Cleora sp. (Lepidoptera: Geometridae)

The wing base of moth is white with creamy to black or even greyish black (Fig. 2k). Unlike *Ascotis* sp. the forewings are somewhat rounded. Fully grown caterpillars are robust with orange head and thoracic legs. There is a pair of reddish brown tubercles dorsally and a smaller pair dorso-laterally in anterior abdominal segment of full grown caterpillar with a narrow transverse greyish bar anterior to them (Fig. 2l).

A.1.3. Tea jassid/Green fly, *Empoasca flavescens* Fabricius (Homoptera: Cicadellidae)

The tea jassid *E.* (=*Amrasca*) *flavescens* Fabricius (Homoptera:Cicadellidae) is an important sucking insect pest of tea, generally during the first and second flushes (March to June) in tea plantations of North-East India. It is commonly known as 'tea green fly', 'tea jassid' or 'tea leafhopper'. The pest remains active at various levels of intensity throughout the season.

Marks of Identification

Egg: The egg is elongate, somewhat narrower at one end, slightly curved with a smooth surface and measures about 0.5 mm long and 0.25 mm width. The colour of the egg is almost white when freshly laid but the later the colour changes to pale yellow.

Nymph: The newly hatched nymph is small colourless with pink eyes but soon after feeding it becomes slightly yellowish green in colour (Fig. 3a).

Adult: The adult is a small yellowish green insect, the forewings being pale yellow. The female measures about 2.75 mm long while the male is about 2.50 mm long.

Nature of Damage

The insect sucks the sap of growing shoots and leaves. Nymphs and adults of *E. flavescens* are the most important stages that damage the tea plant (Zhu *et al.*, 1993; Zeiss and Braber, 2001) and are mainly phloem feeders. Sucking of the sap of young leaves and tender shoots causes the main damage. The nymphs are responsible for greater damage than the adults. During feeding, the rostrum is inclined downwards and the stylets are inserted into the plant tissue. They feed on the contents of the phloem vessels and reach this tissue through cortical cells. Once penetration has started, saliva is injected into the plant. The injury to the plant is probably the combined effect of feeding in the vascular tissue and the action of certain enzymes present in the saliva. The plant sap is drawn into the alimentary canal by the activity of the cibarial sucking pump that is provided with strong dialator muscles. The attacked leaves become dry, uneven and usually curl downwards, getting inverted and boat-shaped. The margins turn brown and subsequently dry up. This characteristic symptom is known as 'rim blight' or 'hopperburn'(Fig. 3b) (Das, 1965). The midrib and veins of the affected leaf also show somewhat brownish discoloration. Leaf structure and texture of clones determine the tolerance level against *Empoasca* spp. Because *Empoasca* spp. suck the sap from the phloem, their stylets must traverse the cuticle, upper epidermis, palisade tissue and spongy mesophyll. Therefore, the thickness of the palisade tissues, subepidermis and collenchymas under the main vein has a significant negative correlation with leafhopper density in tea (Hazarika *et al.*, 2009). Thick and spongy cells physically prevent the probing of the leaf tissues by *Empoasca* spp. and feeding may induce the salicyclic acid and reactive oxygen species pathways, and two jasmonic acid/ethylene-dependent defence signaling pathways (Jin and Baoyu, 2007). The 'hopperburn' and other symptoms are caused mainly by interference with the translocation of food materials and water due to the physical plugging of the xylem and the destruction of the phloem cell (DeLong, 1971).

Life History

The eggs are embedded singly in the soft tissues of tea bushes such as mid-rib, veins of young leaves, petioles and tender stems. During the cold weather the eggs may occasionally be found in the older leaves. The site of oviposition can be noted by the swelling of the tissues. The incubation period is 10-13 days in March, 9-11 days in April and 6-8 days in May, June and July under the climatic conditions of North East India. The young nymph on hatching pushes its way out of the tissues leaving a minute hole and the tissues around the opening become discoloured. There are five nymphal instars and the total duration of nymphal instars vary from 12-15 days in March, 10-13 days in April and 8-10 days in May-June under climatic conditions of North East India.

Seasonal Incidence

Populations of the insect occur on tea bushes throughout the year. During cold weather very few insects are found on pruned and skiffed tea bushes but a large number of them may be found in unpruned, young and nursery tea plants. The pest is more active during March to July. With the rise in ambient temperature from March the insect multiplies rapidly to assume serious proportions in May-June; the attack continuing up to July. From August, the population suddenly declines to a negligible number though there may be a slight increase in November. The attack, however commences later in Darjeeling where June and July are regarded as the "green fly" season.

Varietal Preference

Tocklai released clones like TV1, TV7, TV6, TV9 are more susceptible to jassid attack. The Assam kinds of tea are generally more susceptible than China hybrids.

A.1.4. Tea Thrips, *Scirtothrips dorsalis* Hood (Thysanoptera: Thripidae)

Once tea thrips was considered as a minor or an occasionally serious pest in localized areas of tea plantations (Das, 1965), but are now established as serious and regular pests in tea plantations of Northeast India (Mukhopadhyaya and Roy, 2009; Saha *et al.*, 2012).

Marks of Identification

Egg: The egg is bean-shaped, slightly narrower at one end and is almost colourless when freshly laid.

Nymph: The newly hatched nymph is almost white but soon after sucking of plant sap, the colour gradually changes to pale yellow. The second instar nymph is orange yellow (Fig. 3c).

Pre-pupa: The pre-pupa can be recognized by the free antennae directed forward while in the pupa; the antennae are reflected over the head to reach the middle of the pro-thorax.

a. Nymph of jassid/greenfly

b. Hopperburn symptom caused by jassid

c. Damage symptoms caused by thrips (*S.dorsalis)*

d. Adult thrips (*S. dorsalis)*

Fig. 3 (a-d): Different developmental stages of tea jassid and thrips and their nature of damage

Adult: The adult insect is pale yellow in colour, the abdomen being paler. The female measures 1.05 mm long and 0.19 mm width. The male measures 0.71 mm in length and 0.14 mm in width. (Fig. 3d).

Host Range

Besides tea, it is also reported to attack various vegetable crops, cotton, citrus and other fruit and ornamental crops Southern Asia (Ananthakrishnan, 1993). The pest has been reported to attack more than 150 hosts (Mound and Palmer, 1981; Seal *et al.*, 2010).

Nature of Damage

S. dorsalis feeds on the meristems of host plants' terminals and on other tender above ground parts, creating feeding scars, distortion of leaves and discoloration of buds, flowers and young fruit. It does not feed on mature tissue. It causes damage by sucking the contents out of individual epidermal cells, which leads to necrosis of the tissue. During feeding on a young leaf, it makes small slits by inserting the stylets and sucking the sap oozing through the wound. The sucking marks are made one after one, forming thin pale lines on the underside of leaves parallel to the main vein. Initially, such tissue has a silvery sheen, but soon, the damaged areas turn brown or black (Fig. 3d). In heavily infested tea plantations, it causes 'silvered' or 'sand papery lines' (Fig. 3c). Damaged leaves become thicker and harder than the normal ones, duller with darker green colour, and often puckered or deformed (Das, 1965).

Life History

The eggs are laid singly in the tissues of leaf buds and young leaves. Eggs remains completely embedded in the tissue. A female may lay 2 to 3 eggs in a day. The developmental period of egg stage lasts for 7-8 days in March, 6-8 days in April and 6-7 days in May and June. Pupation takes place in lichens and mosses growing on tea bushes but sometimes in cracks and crevices on the stems. The total duration of nymphal stages (first and second) is 6.0, 5.3-5.7 and 5.0 and 4.3 days in March-April, May and June respectively. The life cycle is completed in 17.6, 16.7, 15.4-15.6 and 13.4-13.6 days in March, April, May and June respectively.

Varietal Preferences

Some of the Tocklai released clones like TV1 and TV2 are highly susceptible to thrips attack.

A.1.5. Red slug caterpillar, *Eterusia magnifica* Butl. (Lepidoptera: Zygaenidae)

Egg: Egg is oval, about 1.0 mm long, pale yellow when freshly laid, gradually turning to greenish colour. But a day or two days before hatching, it assumes slightly reddish tinge.

Caterpillar: Freshly hatched caterpillar has a dirty white colour with two sub-dorsal strips running almost parallel from the third thoracic to the eighth abdominal segment. There are three rows of small round hair-bearing tubercles on each side of the body (Fig. 4a). The full grown larva is slug like, about 25 mm long, brownish red to brick red with a brown head which is retractile. In addition to three pairs of thoracic legs, it bears five pairs of prolegs, the anal pair being the largest. There are three rows of prominent tubercules on each side of the body, each tubercle with two or three hairs and one or two pores.

Pupa: The pupa is pale yellow to yellowish brown, the head and thorax being slightly darker.

Moth: The moths are brilliantly coloured insects with a wing expanse of 55-65 cm (Fig. 4b). The head, thorax and two basal segments of the abdomen are black, suffused with blue and the remainder of the abdomen is pale yellow in female but tipped with black in the male. The antennae of the males have long pectinations and those of females are less pectinate except at the tip (Fig. 4b).

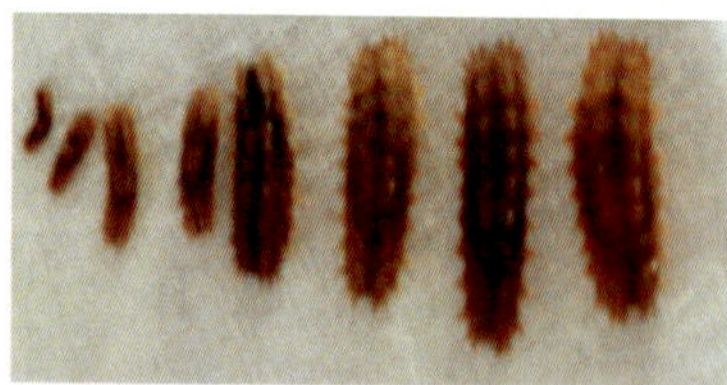

a. Different instars of red slug caterpillar

b. Mated pair of red slug moth in field

c. Damage caused by red slug caterpillar in tea bushes

Fig. 4 (a-c): Different larval instars and adults of red slug caterpillar and nature of damage

Nature of Damage

The caterpillars are usually active during the early and later parts of the day till night. During the hottest hours of the day they are found occasionally feeding in shaded places, but usually descend to the ground and remain concealed under dried leaves, prunings, clods of earth or any convenient shelter. Some are found to climb up the shade trees in the forenoon coming down late in the afternoon. The caterpillar prefers mature leaves. If these are not sufficiently available like in pruned tea, the caterpillar might attack the bark of one or two year old stems of tea bushes. The early instar caterpillar eats out holes and bites off edge of leaves and when they grow in size the whole of the leaf is damaged (Fig. 4c). In severe infestation, the bushes may may be completely stripped of their leaves.

Life History

Egg: The eggs are laid in groups on the undersurface of leaves, and on branches or stems of tea bushes. A female may lay up to 1200 eggs. The incubation period during May is 8-12 days.

Caterpillar: There are five larval instars. The caterpillars become full-grown in four to five weeks in May and June but the larval period is longer during cold weather.

Pupa: Pupation takes place in closely woven, roughly oval, pinkish cocoon usually formed in the folds of leaves, but occasionally in the forks of branches. Cocoon may be formed in dried fallen leaves and prunings on the ground, particularly during a severe attack. The pupal stage lasts for 18-21 days in May and June.

Adults: The moth usually emerges during night. During strong sunshine hours, they hide in shaded places but during cloudy days they are sometimes found to fly and take rest on shade trees. In the evening they may be found swarming round the shade trees or any tall tree, and are attracted to light during dusk.

There are four broods in a year. The approximate time of occurrence of moths and caterpillars of different broods are given below:

Moths	**Caterpillars**
December to March	February to April
May to June	End of June and July
End of July and beginning of August	End of August and throughout September
End of September and October	November to January

(*Source:* Mann and Antram, 1906)

A.1.6. Bunch Caterpillar, *Andraca bipunctata* Wlk. (Lepidoptera: Endromidae)

Egg: The eggs are yellow and slightly oval (Fig. 5a).

Caterpillars: The freshly hatched caterpillar is light yellow with the head and first thoracic segment blackish brown (Fig. 5b). The body is covered with long, stiff, white hairs which are replaced by fine short hairs in the following instar. The full grown caterpillar is about 65 mm long. The head and prothoracic segments are brownish black and the body is densely covered with fine hairs. It is tawny yellow with a reddish tinge and broad, blackish brown, transverse bands running on each body segment crossed by a number of yellowish longitudinal lines. In addition to the three thoracic pairs of legs, there are four pairs of prolegs and a pair of claspers at the posterior end of the body.

Pupa: Pupa is yellowish brown in colour (Fig. 5c).

Adult: Male moth, dark rusty in colour has reddish undersurface and dark brown antennae. Female moth is pale reddish brown with white antennae (Fig. 5d, e).

Nature of Damage

Young caterpillar remains on branches below the foliage and feeds on epidermal tissues, margins and the entire leaf (Fig. 5). Leaf appears translucent due to feeding of epidermal tissues. First instar caterpillar generally feeds only on epidermal tissue, second instar larva feeds marginal area of the leaf and third and four instar caterpillar's feeds on entire leaf.

Life History

Egg: The female starts laying eggs the following night after mating and the process continues for 2-5 days. The eggs are deposited in clusters on the undersurface of the leaves and occasionally on the stems and each cluster consist of a series of eggs arranged in rows. A cluster may contain about 120 eggs and total number of eggs laid by a female may be 500 or more

Caterpillars: There are five larval instars. Total larval duration is 3-4 weeks.

Pupa: Pupation takes place on the ground and form cocoons amongst dried leaves and pruning litter. Pupal stage lasts for about 16-20 days in April-June, 46 days in July-September, and 68-120 days in October-February.

a. Egg **b.** Larvae

c. Pupa

d. Male **e.** Female

Fig. 5: Caterpillars and adults of *Andraca bipunctata*

A.1.7. Tea Leaf Roller, *Gracilaria theivora* Walsm. (Lepidoptera: Gracillariidae)

Marks of Identification

Egg: Egg is minute, oval, colourless and almost transparent.

Larva: The full-grown larva is 8-10 mm long (Fig. 6). The head is black and the body is pale yellow in colour. In general appearance the caterpillar is similar to that of the flushworm but differs from the latter in the absence of "Check spot' and fourth pair of prologs on the 6th abdominal segment.

Pupa: The pupa is very slender, 5-6 mm long, brown above, and yellowish underneath.

Adult: The moth has narrow elongated fringed wing with an expanse of 10-12 mm. The forewing is purplish brown with iridescent tints, crossed by triangular yellow spot near the middle. The hindwing is grayish brown.

Life History

Egg: The eggs are laid singly, usually one egg on a leaf, but two to three eggs can occasionally be found on a leaf.

Larva: The larval stage lasts for 10-14 days

Pupa: The pupal period is 9-13 days during the main season.

Fig. 6: Tea leaf roller, *Gracilaria theivora*

A.1.8. Nettle grub, *Parasa pastoralis* (Lepidoptera: Limacodidae)

They resemble a slug in the form of the body and their gliding motion, and are brilliantly coloured, usually green, with multi-coloured markings (Fig. 7). They have a short, thick, fleshy body, a small retractile head, minute thoracic legs, each ending in a claw, and secondary sucker discs on the abdominal segments. The nettle grubs are provided with tufts of hairs or series of branching spines which are poisonous and painful, and their presence makes it difficult to work in an infested area.

Fig. 7: Nettle grub, *Parasa pastoralis*

Nature of damage

The caterpillars usually attack older leaves and the damage thus caused to mature bushes in mild attacks may not be appreciable but in severe outbreaks both the young and older leaves are equally attacked. The young plant may be completely stripped of their leaves.

Life History

Egg: Egg is yellow, flat and overlapping like scales, are laid in clusters on the undersurface of the leaves near the top of the bush. They hatch in about 4 days in July.

Larva: The first instar larvae remain gregarious and feed on tissues in patches on the undersurface of the leaves, leaving the upper epidermal membrane intact. After the first moult, they disperse to other parts of the bush or to neighbouring bushes and feed on the edge, then devour parts or whole of the leaf, especially at the lower part of the bush. The larva becomes full grown in about three weeks when it forms an oval cocoon on the stem or in the fork of a branch and pupates therein.

Pupa: The pupal period is about four weeks in July and three to four months during the cold weather. There are three broods in a year and the second brood is the most destructive.

A.1.9. Flushworm, *Cydia leucostoma* (Meyrick) (Lepidoptera: Gracillariidae)

Egg: The egg is minute, slightly oval and yellowish in colour.

Larva: The full grown larva is about 9-10 mm in length and is greenish or brownish. The head is yellowish or brownish and is characterizcd by the presence of a black "Check spot" behind the simple eyes on each side. There are four pairs of prolegs on the 3-6 abdominal segments and a pair of claspers.

Pupa: The pupa is about 5-6 mm long and 1.2-1.5 mm broad; greenish yellow when freshly formed, but eventually turns brown.

Adult: The moth is a tiny insect with a wing expanse of 10-13 mm. The head and thorax are blackish and the abdomen is grayish brown. The forewing is a mixture of dark brown and grey tinged with violet in the middle and streaked with yellow and white. The hindwing is dark brown.

Nature of Damage

The larva normally attacks the young shoots. The just hatched larvae tie up the margins of two or three tender leaves and form a case enclosing the bud inside. The young larva feeds by scraping off the tissues of the upper surface, occasionally eating away the apical portion of the bud. As a result, the leaf becomes rough, thick and somewhat brittle in texture, with a brownish discolouration of the damaged surface. The affected leaf also presents a crinkled appearance. The affected shoots cannot reach full development and the internodes are greatly shortened and the tender stem bends over as a result of binding of two or more leaves together.

Life History

Egg: The eggs are laid singly on young leaves usually on the underside of the second or third leaf.

Larva: The duration of the larval stage is about 3-4 weeks.

Pupa: The caterpillar comes down the shoot to an old leaf, makes two inclusions, 12-15 mm apart, 4-5 mm deep, on the margin and turns the flap over to pupate inside. The flap is usually folded on the upper side, occasionally on the underside of the leaf. Pupation also very often takes place on the 'fish leaf'. The pupal period lasts for 11-12 days in March.

A.1.10. Tea tortrix, *Homona coffearia* (Lepidoptera: Totricidae)

Egg: The eggs are somewhat oval in shape and pale yellow in colour.

Larva: The full grown caterpillar is about 23 mm long and green in colour. The head and the prothoracic (Fig. 8) are shining black and this character alone is sufficient to distinguish it from the flushworm with which it is often confused. The body is sparsely covered with thin hairs. There are four pairs of prolegs on 3rd-6th abdominal segments, and a pair of claspers on the anal segment.

Pupa: The pupa is tawny yellow to dark brown in colour. The dorsal surface of the abdomen is furnished with two rows of teeth in each segment. The anal spine (cremaster) is stout and has eight incurved hooks. The male pupa is smaller than that of female.

Adult: The female moth differs considerably from the male in colour and size and other characters (Fig. 8). The female is usually of brownish yellow colour. The fore wing of the female has a medial brown band running obliquely from the costal margin to the inner margin, a small darker patch near the coastal margin and a larger one near the inner margin and the apex pointed with a darker patch on it. The whole wing is traversed by a number of dark brown wavy lines, which are often indistinct. The hind wing is orange yellow. The male is smaller than the female and is of dark grey colour. The thorax has tufts of grey scales. The oblique brown medial band on the forewing is more is more conspicuous, a black spot on the band near the costal margin and a few black scales on the inner margin. The shoulder Fig. of the wing, a characteristic of the male moth, arise from the costal margin based to the black spot, and runs incurved towards the base a sort of platform. The hind wing is grey with golden glare. Wing expansion of male and female are 21-23 and 25-27 mm respectively.

Fig. 8: Adult Tea tortrix, *Homona coffearia*

Nature of Damage

The caterpillar's spins a web tying up the two halves of a young leaf, usually the first leaf, enclosing the bud in it for shelter and feeds inside and damaging the tip and the sides of the leaf (Das, 1957). It frequently changes its shelter and attack fresh leaves, obviously, a number of leaves on different shoots may be damaged until the caterpillar becomes full grown. Sometimes it fastens two leaves, one above the other, and being stationed in between, feeds upon the edges.

Life History

Egg: The eggs are laid in clusters on the upper side of leaves, each cluster containing as many as 117 eggs overlapping like scales of fish. The incubation period is 6-8 days in March.

Larva: The caterpillars become full grown in 3-4 weeks (usually 26 or 27 days in March)

Pupa: Full grown caterpillar pupates on the leaf where it has been lately feeding, or moves to another leaf, ties up the two halves, or folds the margin to pupate therein. It does not form any cocoon, but spines a cob-web around it before pupation. The pupal period is six to eight days in March and April.

A.1.11 Scale Insects and Mealy Bugs

Scale insects (Fig. 9) and mealy bugs (Superfamily: Coccoidea) are of great economic importance and they infest the vast majority of cultivated plants.

The females of coccoids are wingless, degenerated in structure with vestigial legs and antennae and short rostrum. Males are winged, with only the first pair well developed. They reproduce either sexually or parthenogenetically. Some are oviparous while others are ovoviviparous or viviparous. The first instar nymphs crawl around till they settle down with their rostrum attached to the host plants and gradually assume their scale like appearance. Scale insects and mealy bugs excrete honeydew and are attended by ants.

Fig. 9: Scale infestation

Mealy bugs (Pseudocccidae) are covered by a powdery coating or waxy filaments and they have a clearly segmented body and well developed legs.

Many species of coccoids are known to infest tea. They attack leaves, tender stems and occasionally roots.

Three species of scales *viz.*, *Saissetia formicarii*, *Eriochiton theae* and *Saissetia coffeae* occur on tea in India. In Darjeeling district, these species are responsible for considerable damage to young and mature tea bushes by their persistent attack.

Saissetia coffeae (Wlk.) (Homoptera: Coccidae)

Distribution: It has been found to occur on tea in most of the tea growing areas of India.

Habit: Congregated on young stems, a few are found scattered on leaves, particularly on the under surface. Form of coccid varies somewhat in relation to the position. Adult female almost hemispherical on a flat surface, but may be somewhat elongated on thin stems. Derm polished, pale brown to chestnut brown in colour depending on age, and speckled with minute translucent cells. Margin usually flattened.

Life history: Eggs are seen under the scale and the crawlers disperse immediately after hatching and attach themselves to the tender plant parts. The male forms a cocoon after the second moult and become a winged adult. Females are sedentary.

Damage: The attack is not widespread, and is usually confined to isolated bushes or patches of tea. On mature tea bushes, only a few shoots are usually found to be heavily infested. Young plants are sometimes severely attacked, resulting in die-back while the seedlings may be killed outright. The pest is usually kept in check by parasitic fungi.

Saissetia formicarii (Green) (Homoptera: Coccidae)

Distribution: It is reported throughout the plains of North East India

Habit: It occurs on stems of young and mature tea bushes and also on tea seed trees, invariably enclosed in the nests of ants. Adult female broadly oval to almost circular in outline. Median dorsal area highly convex with four longitudinal series of pale yellow spots. Margin usually flattened. Derm thin and soft. Anal operculum situated on a raised area. Colour varies from grayish fulvous to chestnut brown.

Damage: The affected branches become unproductive, where the attack persists at the same place for some years, irregular swellings are formed as a result of death of the bark and subsequent callus growth underneath. In case of severe damage, the young branches may die.

Eriochiton theae Green (Homoptera: Coccidae)

Distribution: It is common on young and mature tea in many tea estates of Darjeeling where it is considered to be one of the most serious pests of tea (Das, 1961). In the plains it has never been found on plucking tea but occurs abundantly on tea seed trees (Das, 1961).

Habit: Occurs mostly on stem and undersurface of leaves; young stems of one or two years old plants are preferred. The infested branches can be easily recognized by the presence of white felted covering of waxy-secretion and by black sooty fungi which grow profusely on honey dew excretion falling on leaves and stems of infested and adjoining bushes and also on the vegetation underneath. On tea seed trees in the plains, the coccid is invariably attended by the ant, *Oecophylla smaragdina* on exposed stems, occasionally enclosing it in its nests. However, in the Darjeeling district, it is occasionally attended by a small number of the ant, *Ceratopheidole bhavanae* (Bingham), which has its nests in the ground (Das, 1959). Adult female completely enclosed in a nest made up of white fluffy matter, oblong oval in shape, broader behind, rather strongly convex with a small circular aperture at the hind end and a very distinct median longitudinal ridge with numerous transverse ridges and furrows on each side. Male puparium oval, glassy, with a dorsal median ridge bearing long curled, white filaments and found congregated, particularly on the undersurface of the leaves.

Damage: The affected leaves turns yellow and ultimately fall off. The infested branches become weak and unproductive. In a severe and persistent attack, the bark dies and callus growth takes place underneath. Subsequently, irregular swellings are formed on the stems. Severely affected branches die-back sooner or later. If the attack persists year after year, there will be gradual loss of branches and the tea bushes may be killed ultimately.

A.2. Mite Pests of Foliage

Tea plant is commonly attacked by four species of mites namely, red spider, scarlet, pink and purple mites. Mites are sucking pests *i.e.*, they feed on the leaf sap with their sucking type of mouth parts.

A 2.1. The Red Spider Mite, *Oligonychus coffeae* Nietner (Acari: Tetranychidae)

Among mite pests, the most important one is the red spider mite, *Oligonychus coffeae*, which occurs in almost all the tea growing countries of south-east Asia. *O. coffeae* is the largest of all tea mites and can be easily seen by the naked eyes.

Marks of Identification

Egg: Eggs are reddish, spherical (0.10 mm) and are provided with a small filament. Before hatching, eggs assume a light orange colour (Fig. 9a)

Larva: The freshly hatched larva is yellowish orange in colour, six legged, round in shape and measures 0.15 mm in length.

Nymphs: The protonymph, 0.20 mm long, has four pairs of legs and an oval body. Anterior part of the body is pale crimson while the abdomen is deep reddish brown. The deutonymph is similar to protonymph except for the size and shape. Sexes can be differentiated at this stage. The female deutonymph is slightly bigger, resembles the adult and measures 0.25 mm in length (Fig. 10b).

Adult: The adult female is elliptical in shape, bright crimson anteriorly and dark purplish brown posteriorly. It measures 0.35-0.45 mm in length. Males can be distinguished by their small size and tapering abdomen. Males are 0.25 -0.35 mm long (Fig. 10b.).

a. Eggs of red spider mite

b. Nymphs and adults of red spider mite

c. Typical symptom of red spider mite damage in tea leaf

d. Damage symptom in field

Fig. 10 (a-d): Different developmental stages and nature of damage caused by red spider mite

Host Range

Red spider mite was first recorded on coffee and later on tea in Sri Lanka. Tea is the most preferred and principal host plant of the red spider mite. In India, red spider mites pose a serious threat to the crop (Hazarika *et al.*, 2009).

Table 7: Alternate hosts of red spider mites

Alternate host	Nature of plants	References
Ageratum conyzoides	Weed	Das 1965
Albizzia falcate	Shade tree	Das 1965
Aristolochia sp.	Creeper, Economic plants	Das 1965
Bidens sp	Weed	Das 1965
Borreria hispida	Weed	Das 1965
Convovulus sp	Weed	Rao, 1969
Crassocephalum sp	Weed	Das 1965
Dadaps (Erythrina *lithosperma*)	Tree	Selvasundaram and Muraleedharan 2003
Derris robusta	Shade tree	Andrews, 1918
Drymaria cordata	Weed	Rao, 1969
Eugenia cumini	Tree	Light, 1927
Eugenia sp	Tree	Das 1959
Grevillea banksii	Ornamental plant	Rao, 1969
Grevillea robusta	Ornamental plant	Rao, 1969
Grevilliea	Ornamental plant	Selvasundaram and Muraleedharan 2003
Hibiscus abelmoschus	Vegetable plants	Das 1965
Hibiscus ficulneus	Flowering plants	Das 1965
Hibiscus panduraformis	shrub-like perennial herb	Das 1965
Indigo sp.	Economic plants for blue color dye preparation	Selvasundaram and Muraleedharan 2003
Ipomaea sp	Weed	Rao, 1969
Litsea lancifolia	Weed	Das 1965
Litsea polyantha	Weed	Das 1965
Melastoma malabrathicum	Weed	Das 1965
Melochia corchorifolia	Weed	Das 1965
Ornamental Camellia	*Ornamental plants*	Selvasundaram and Muraleedharan 2003
Oxalis corniculata	Weed	Rao, 1969
Scoparia dulcis	Weed	Das 1965
Tephrosia candida	Weed	Das 1965
Tephrosia candida	Weed	Andrews, 1918
Triumfetta neglecta	Weed	Das 1965
Urena lohata	Weed	Das 1965

Nature of Damage

Infestation by red spider mites starts along midrib and veins and gradually spreads to the entire upper surface of leaves. (Fig. 10c) As a result of feeding, the maintenance foliage turns ruddy bronze, making red spider infested fields distinct even from a distance (Fig. 10d). Severe infestation by this mite ultimately leads to defoliation. They normally infest the upper surface of mature leaves, though in extreme cases they frequent the lower surface of older leaves as well as young leaves. White specks of cast skins are noticed on upper surface of severely damaged leaves. The mites spin a web of silken threads on the leaf and protect themselves from adverse weather conditions

Life History

Egg: The female lays eggs 24 hours after emergence. Eggs are laid singly along midrib and veins. Usually 4-6 eggs are laid per day. A female lays 80-130 eggs during her life time. Incubation period is 4-6 days.

Immature: There are three developmental stages, the six legged larva, the protonymph and the deutonymph. The larva, after an average feeding period of 2 days enters the first quiescent stage. The activity phase of the protonymph is two days and thereafter it enters the second quiescent stage. It enters the third quiescent stage after 3-4 days. The time required for development from egg to adult is 10-14 days (Table 8).

Adults: Both males and females are sexually mature on emergence. Mating takes place immediately after the emergence of females. The maximum longevity of female is 25-30 days. The male, female ratio is 1:2.5. There are several generations in a year. The duration of the life cycle is shorter (9-12 days) in summer months and in winter the life cycle is prolonged up to 25 days. It is possible that further variations in the duration of the life cycle may occur under fluctuating weather conditions.

Table 8: Life history of *O. coffeae* on tea at three different temperatures

	Duration in days (Mean ± SE)		
Life stages	20 ± 2 ºC	25 ± 2 ºC	30 ± 2 ºC
Pre-ovipositional period	1.50 ± 0.24	1.00 ± 0.58	0.90 ±1.64
Ovipositional period	17.80 ± 2.20	18.60 ± 0.97	25.60 ± 2.42
Incubation period	8.70 ± 0.52	6.40 ± 0.42	5.80 ± 0.46
Larval period	1.80 ± 0.40	1.30 ± 0.36	1.10 ± 0.41
Protonymphal period	1.70 ±0.76	1.40 ±0.41	1.20 ± 0.32
Deutonymphal period	1.80 ± 0.56	1.70 ± 0.95	1.50 ± 0.25
Total nymphal period	5.30 ± 0.90	4.40 ± 1.10	3.80 ± 0.60
Total developmental time	14.00 ±1.80	10.80 ± 1.45	9.60 ± 1.20
Longevity of male	8.60 ± 0.75	12.10 ±1.10	14.50 ± 0.95
Longevity of female	19.20 ± 1.10	21.70 ± 1.16	26.90 ± 1.90
fecundity (No. eggs)	80.2 ± 10.75	107.4 ± 12.83	128.8 ± 14.40

Source: (Selvasundaram and Muraleedharan, 2003)

Seasonal Incidence

The red spider is active and breeds on tea throughout the year in North East India (Das, 1959a, Choudhury *et al.*, 2006, Mukhopadhyay and Roy, 2009). Populations begin to build up in early March and reach their greatest density during late March and early April. Injury becomes most severe during May and June or until the monsoon rains wash off and kill all active forms on the leaves. In South India the incidence of red spider mite is high during January to May and low during June to December. The population buildup begins in February or March and reach peak in April/May. Population density was low in wet, rainy months of July to October. However, tea bushes in a few pockets located mostly in the unshaded and rocky areas always harboured the mites (Selvasundaram and Muraleedharan, 2003).

Factors Influencing the Occurrence of Red Spider Mite

Drought: Prolonged drought causes stunting of growth of the bushes which renders the bushes more liable to red spider mite attack.

Pruning: Mites generally persist on old leaves during the cold weather which are responsible for attack in the following growing season. Therefore, the type of pruning which removes more old leaves from the bushes during the cold weather also reduce the possibility of attack.

Rain: Heavy rains wash away the mite population mobile stages to a great extent. However, light rain with intermittent sunny days may increase the population density.

Planting material: Assam hybrids are normally less severely attacked than China hybrids.

Road side bushes: Road side bushes are more prone to spider mite attack because mites prefer dusty leaf surface for oviposition.

Weeds: It has been observed that areas which are periodically cleaned of weeds and grasses, particularly during the early part of the season, suffer less from red spider mite attack than uncleaned areas because large number of weeds have been recorded to harbor red spider mites of all stages.

The red spider mite is always less serious on bushes under shade trees. The degree of attack varies with density of shade; the heavier the shade the lesser is the attack.

Manure: Nitrogenous fertilizers induce growth and thereby reduce the susceptibility of bushes to red spider mite attack. Phosphate and potash both alone or in combination have no observable effect on red spider mite attack both in shaded and unshaded areas. Organic manures such as cattle manure and compost increase the incidence of this pest.

Drainage: Areas having poor drainage system and temporarily waterlogged suffer more from red spider attack than well-drained areas.

Soil acidity: The tea plant grows better in fairly acidic soils having pH between 4 and 5.5. The bushes grown in soils with sub-optimal pH are generally week, and are, therefore, more liable to red spider mite attack.

A.2.2. The Scarlet Mite, *Brevipalpus* spp.

Brevipalpus spp. is a well known pest of tea. In north east India *B. phoenicis* is the widely prevalent species while in south India, *B. australis* is the common scarlet mite.

Nature of Damage

Scarlet mite attacks the undersurface of leaves, chiefly along the mid-rib and at the base of petioles and young stems are also attacked. The damage is characterized by general yellowing of leaves, brown scurfy discolouration along the mid-rib on the undersurface, which with the increased damage extends on both sides of the mid-rib, and also on the basal part which subsequently dries up. Occasionally in case of severe infestation, the underside of the affected petiole sp;its and dries up causing premature leaf fall and the bushes might be completely defoliated. Axillary buds are also attacked which cease to grow. The infested portion of bark also split, dry up and become swollen due to callus growth underneath. The effect of damage is chronic and insidious. The infested bushes produce spindly shoots and suffer from debilitation and consequently they survive only as unproductive bushes. Under droughty conditions, scarlet mites may be often responsible for the death of the young plants.

Brevipalpus phoenicis (Geijskes) (Acari: Tenuipalpidae)

Marks of Identification

Egg: The eggs are oval and bright red in colour, measure 0.09 mm in length and 0.06 mm in width with one end slightly broader.

Larva: The freshly hatched larva is dull red and has three pairs of legs.

Nymphs: Protonymphs and deutonymphs have four pairs of legs. Deutonymphs are similar to protonymph in appearance but larger in size.

Adult: The adult female is flat, somewhat elongated oval, scarlet in colour, with black markings on the dorsal aspect.

Seasonal Incidence

The mite persists on tea bushes in varying densities throughout the year. Peak population is reached during May-August in Assam.

Life History

The female usually starts egg laying after 3 to 5 days of emergence during March to August and in cold weather, it varies between 7 and 10 days. Eggs are laid on the leaf stalks, at the base of the leaf hairs, cracks and cravices on stems, leaves etc. Each female normally lays one egg per day but occasionally 2 to 3 eggs are also laid. Egg laying is not continuous, often it is interrupted for a day or two during summer but during cold weather the cessation may even last for a fortnight or so. The highest number of eggs laid by a female was found to be 47 during an oviposition period of 40 days in March-April and minimum number was 8 in August. The incubation period was found to be 11-13 days in March, 8-9 days in April, 6-7 days in July and 17-25 days during December-January. After the first moult it becomes protonymph and the second moult it becomes a deutonymph and after the final moult the adult emerges. Each moult is preceded by a short quiescent period.The larval stage is 4-6 days in March, 2-3 days in April, 1-2 days in July and 7-10 days in December-January. The protonymphal period is 2-4, 2-3, 2 and 2-5 and 6-8 days in March, April, July, November and December respectively. The deutonymph stage lasts for 4-5 days in March, 2-3 days in April and July, 4-7 days in November and 7-9 days in December. The life cycle is completed in 24-28 days in March, 21-25 days in April, 18-20 days in July, 27-35 days in November-December and 55-60 days in December-January.

Brevipalpus australis Baker (Acari: Tenuipalpidae)

Marks of identification

Egg: Eggs are bright red, elliptical and laid in clusters of 5 to 7 on the undersurface of the leaves.

Nymphs: The nymphal stages are eight legged and resemble the adult.

Adult: Adult mite is scarlet red in colour and obovate in shape. They are slightly bigger than the eriophyid mites and can be seen with the naked eye.

Life History

Incubation period ranges from5 to 7 days and the larva emerging from the egg is also scarlet in colour. The three legged larva, after 4-5 days enters a quiescent stage, the protochrysalis, which in turn moults into the protonymph after 3-4 days. The protonymph is active for 5-6 days. The quiescent stage that follows the protonymph is called the deutochrysalis. This stage moults into the deutonymph in about 3 days and the latter feeds on the plant sap for nearly 5 days. Before moulting into the adult, the deutonymph again enters a dormant stage, the teleiochrysalis, for a period of 3 days. The total developmental period from egg to adult varied from 30-36 days.

A.2.3. The Pink or Orange mite, *Acaphylla theae* (Watt) Keifer (Acari: Erriophyidae)

This eriophyid mite has been reported to cause considerable damage, particularly to young tea. This is also widely spread in all the tea growing districts of South India and North East India. The pink mite is so small that it is very difficult to see them without the help of a magnifying glass. Very often, its damage is attributed to soil conditions, drought, or some unknown cause.

Marks of Identification

Egg: Eggs are globular in shape (0.035 mm in diameter) and almost colourless but with a shining surface.

Nymph: The nymph is elliptical in shape, white at first, gradually becoming orange. It has two pairs of legs directed forward.

Adult: The adult female is carrot shaped (0.14 to 0.19 mm long) and is orange in colour.

Nature of Damage

It attacks both the surface of the leaf, but is more prevalent on the undersurface. They may be present on mature leaves but they colonies on young tender leaves and the top shoots. Petioles and tender stems are also not spread. The affected leaves become pale coloured, often leathery. The veins and the margins of the leaves particularly on the undersurface show a brownish discoloration which is the characteristic of its damage. The affected leaves may show crinkled appearance and the bushes present a sticky appearance with stunted growth.

Seasonal Incidence

The pest can be found on tea bushes almost throughout the year. But, under favourable weather conditions, with fewer rains, the population increases rapidly during the early part of the flushing season from March reaching a peak in June-July. From August the population of mite starts declining, reaching a low point in December-January.

Life History

The females lay eggs singly on the surface of the leaves. The incubation period is 6-7 days in December and January and 3-4 days in March-April and about 3 days in July and August. About 2-4 eggs are laid per day by a female.Nymphs undergoes two moult to become an adult. The nymphal period is 6-7 days in January, 5 days in March and 3-4 days in July and August. The life cycle is completed in about 8 days in March, 6 days in July and August and 13 days in Decemebr and January.

A.2.4. The Purple Mite, *Calacarus carinatus* (Green) (Acari: Eriophyidae)

This pest cause extensive damage to both young and mature teas throughout the tea growing district of North East and South India.

Marks of Identification

Egg: The egg is sub-hemispherical, colourless and almost transparent and measures about 0.07 mm in diameter.

Nymph: The young nymph is pear shaped tapering posteriorly and is cream coloured gradually turning purple. It has two pairs of legs placed anteriorly. The five longitudinal white waxy bands start appearing in the late first instar of nymphs. The colour of the body gradually becomes deeper in subsequent instars.

Adult: The adult female is deep purple, 0.15-0.20 mm long spindle shaped with five prominent white waxy ridges on the dorsal aspect of the abdomen.

Nature of Damage

It attacks both the surface of leaves but is more prevalent on the upper surface. Older leaves are preferred but during heavy build up young leaves are also equally infested. Petioles and tender stems are also occasionally attacked. The characteristic damage symptom of purple mite is copperish bronze discolouration of leaf surface which is more marked at the margins, particularly on the upper surface. The affected leaves eventually dry up and fall off. The badly infested leaves when carefully examined with hand lens are found to be dusted with white particles which are nothing but the cast-off skins of young mites. Serious damage is usually caused during the early part of the season, particularly when the drought is prolonged. Young tea is also occasionally attacked during the later part of the season and the attack often persists throughout the cold weather.

Seasonal Incidence

This mite is found on the tea bushes throughout the year in greater or lesser numbers. The population of the mite tends to increase from October and a serious attack may develop before the cold weather. The rate of multiplication of the mite increases from late February or early March and consequently there is heavy build-up of mites, the young and mature tea bushes are badly attacked during that period. Once the rain starts, the mite population persists on the lower surface.

Life History

The eggs are laid singly on the leaves. The female lays 1-4 eggs in a day with a maximum of 30 eggs during its life span. The oviposition period varies from 10 to 13 days. The incubation period is 6-8 days in January, 4-5 days in March and 3-4 days in July and August. Nymph moults three times before becoming an adult. Each moult is preceded by a short quiescent stage. The duration of the nymphal stage is 5-6 days in January, 4-5 days in March and April and 3-4 days in July and August.

B. Pests of Stems

B.1.The shot-hole borer, *Euwallacea fornicatus* Eichhoff (Coeoptera: Scolytidae)

The shot-hole borer is the most important pest of tea in south India.

Nature of Damage

Females of this ambrosia beetle construct galleries in tea stems (Fig. 11), leading to branch breakage and consequent crop loss. The galleries are mostly seen in the primary branches, formed after pruning. However, in many places attack is also found on the old wood, even at the collar region. Pruning removes a large part of the affected stems and also the beetle populations. During the year of pruning, there would be practically no new attack, mainly due to the physiological changes taking place in the bushes immediately after pruning. Fresh infestation vigorously starts from the end of second year.

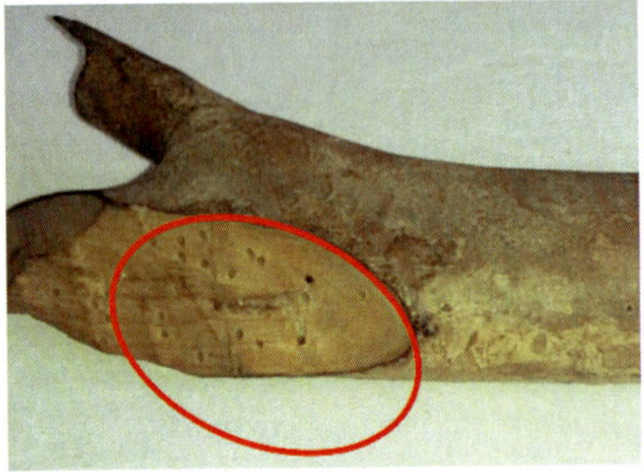

Fig. 11: Damage symptom of shot-hole borer (*Euwallacea fornicatus*) in tea

The holes made by the beetle are less than 2 mm in diameter and the galleries made in the stems may be circular (spiral), longitudinal (vertical) or mixed type.

Life History

Female beetles are about 2 mm long and black in colour. Males are almost half the size of females, devoid of wings and incapable of making galleries. Males are very few and the male, female ratio is generally 1:8. Pre-ovipositional period is 8-10 days. Eggs of shot-hole borer are white, elongate oval and are laid inside the galleries. In laboratory conditions, ovipositional period extend up to 20 days. Eggs hatch in 4-6 days and the grubs feed on the fungus, *Fusarium bugnicourtii* Brayford growing in the galleries. The spores of this fungus are carried by the females and once they bore into the stems, the galleries are lined by this fungus. There are three larval instars and the duration of larval stage is 16-18 days depending on weather conditions. Pupae are whitsh and this stage lasts for 7-9 days. Total developmental period is 27-33 days under South Indian weather conditions. Unfertilized eggs develop into males. Immediately after emergence, adults are pale white with a brownish tinge and they assume black in 9-10 days.

B.2. The red borer *Zeuzera coffeae* Nietner (Hemiptera: Coccidae)

The common red (coffee) borer attacks many forest trees and plantation crops like tea, coffee and cocoa. The stem boring caterpillar is mostly a problem in young tea but fortunately widespread damage does not occur.

Nature of damage

Usually, young stems are bored by the caterpillar. As the larva grows, the tunnel is also extended and holes are made at intervals to eject the excreta and wood particles. In young plants, the tunnel may run through the main stem, even up to root level. Such badly affected plants cannot be saved.

Fig. 12: Adult of red borer *Zeuzera coffeae*

Life history

Adult moths have white wings with many black spots (Fig. 12) and the wing expanse is 3 cm. Eggs are laid in a string and appear like beads on a thread. These strings of eggs are deposited on stems, often in crevices and the young caterpillars emerge in 10 days. They suspend themselves by silken threads and get dispersed with the aid of wing. The larvae landing on tea bushes bore in to young stems. Usually the larvae tunnel downwards, devouring the woody parts, especially the pith. In course of time, the tunnels are extended to thicker branches. The mature larva is about 3.5 cm long and purplish brown or reddish brown in colour. Larval development is completed in 4-5 months. Pupation takes place in a special chamber, with the head of pupa pointing towards the future exit hole. Pupal period lasts for a month.

B.3. Termites (Isoptera: Termitidae)

Many termite species caused considerable damage to tea bushes and shade trees. All termite colonies are governed by a caste system. Queen termites (Fig. 13e) are integral to the founding and growth of termite colonies (Fig. 13d). Young and mature tea as well as cuttings in vegetatively propagated nursery is often damaged by the termites. Among the termite castes, workers are very dangerous to tea. They are responsible for major damage to young and mature teas plants especially in Cachar and North Bank areas of Assam and Dooars in North Bengal. Das (1962) reported that at least 15 per cent of the total crop is annually lost due to the attack of termites. Termites that are responsible for damage to tea bushes may be classified into two groups:

B.3.1. Live wood eating termites

They form innumerable oblong oval nests of about 3 cm diameter in the soil up to a depth of 40 cm (Fig. 13a). These attack living tissues of tea bushes and are considered to be primary pests of tea. They excavate galleries within the healthy wood of tea bushes (Fig. 13b). Tea plants in India are infested primarily by three species of live wood eating termites, namely *Microtermes* sp., *Microcerotermes pakistanicus* Ahmed and *Microcerotermes* spp. Live wood eating termites are smaller than the scavenging termites.

B.3.2. Scavenging termites

They have their nests in mounds or underground. The height of the mound is about one meter and diameter is 30-50 cm. These generally attack dead and dying tissues and are regarded as secondary pests of tea. There are six species of scavenger termites, namely *Odontotermes assamensis* Holm, *O. parvidens* K & N Holm, *O. feae* (Wasm), *Coptotermes heimi* (Wasm), *Neotermes buxensis* Roonwal and Sen Sharma, and *Capritermes* sp.

Nature of damage

Nature of damage caused by both scavenger and live wood eating termites is similar, but the injury caused by the former is slower and confined to a few unhealthy mature bushes, while the injury caused by the live wood termites is rapid and occurs in more or less severe form in young as well as in mature sections. The scavengers first attack the dead wood resulting from fungal or borer attacks under cover of earth runs, and also extend the damage to the adjacent healthy tissues by removing the dead wood (Fig. 13c). The live wood-eating termites attack the tissues of the bark of healthy young bushes. Once they get entry, they excavate fine tunnels into the hearthy wood and either hollow out the stem, leaving only a thin layer of wood and bark or replacing the whole woody part with earth materials leaving the bark outside (Fig. 13b, c).

a. Earth rum made by live wood cutting caterpillar

b. Live wood cutting termite damage

c. Damage in tea bush caused by scavenging termite

d. Nest of live wood cutting termite

e. Termite queen

Fig. 13 (a-e): Termite nest, queen and nature of damage caused by termites

Seasonal History

The initial termite activity starts from August to October and peak period of their activity is seen from November to February. Therefore, termite prone areas should be kept under vigilance from the end of August.

Life History

The developmental stages of the termite include the egg, the immature or nymphal forms and the adults which are separated into several different castes such as workers, soldiers, king and queen. The workers and soldiers are always wingless and consist of both males and females but they never or rarely produce any young. The winged true males and females are formed within the colony and come out from the nest in swarms for nuptial flight during April-May with the onset of rains. After mating, they shed their wings and establish a new colony.

Why Termite is a Problem Today?

- More attention is generally paid to defoliator pests, which are often responsible for immediate damage to tea but termite damage went unnoticed for years.
- Awareness of the planters and scientists was less than today.
- No systemic chemical was used in tea against termites.
- Cultural and mechanical controls were not highlighted.
- Persistent chlorinated hydrocarbons were only effective against scavenging termites.
- Concealed/ subterranean habit
- High fecundity
- Availability of alternate host plants
- Damage is comparatively slow.
- Intensive cropping, especially monocropping reduces fertility and structure of soil making it more conducive to termite attack.

C. Pests of Root

C.1. Cockchafer Grubs (Coleoptera: Scarabaeidae)

Cockchafer grubs or white grubs (Fig. 14a) often cause considerable damage to young tea plants and seedlings in nurseries. Several species of cockchafer *viz*., *Holotrichia impressa* Burm., *Sophrops plagiatula* (Branske), *S. iridipennis* (Branske), *Anomala bilobata* Arrow and *Adoretus versutus* Her. are found in tea plantations of India. *H. impressa* has been found to be mainly responsible for causing damage to young tea in the Dooars and Darjeeling. In Assam particularly in the North Bank of the Brahmaputra, the grubs of *S. plagiatula* were found to be mainly responsible for damage.

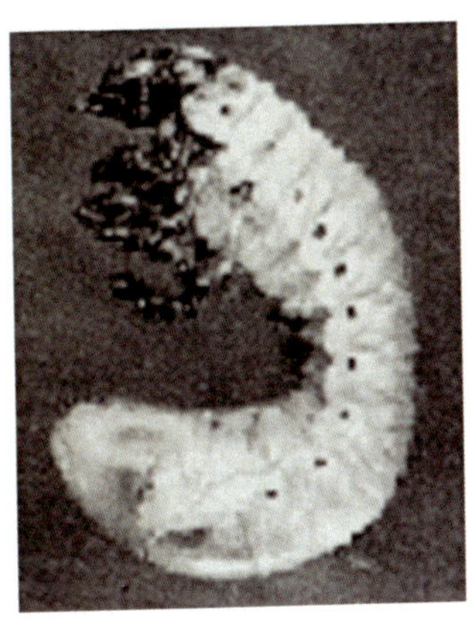

a. Grub

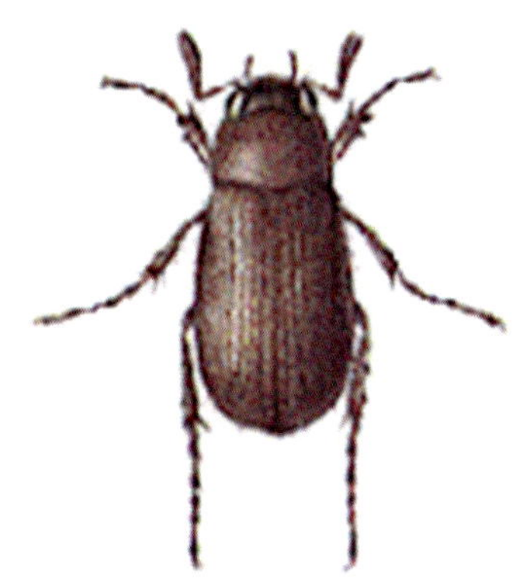

b. Adult

c. Cockchafer damage

Fig. 14 (a-c): Life stage and damage symptom of Cockchafer bettle

Egg: Eggs are oblong and glistening white.

Grub: Grubs whitish in colour and full grown larvae with dark brown mandibles.

Pupa: Pupa yellowish brown in colour measures 20-22 mm in length.

Adults: Colour of the adults varies according to species. Generally most of these beetles are dark brown in color and possess lamellate antennae (Fig. 14b).

Nature of Damage

The white grubs attacks the roots of young tea (below 3 years) but the serious damage is done more to clonal than to seedling plants (Das, 1965). The grub eats away the bark in the collar region of the plant just below the soil surface, either in a ring or in patches, resulting in death of the plant (Fig. 14c). The damage is healed up by callus growth if debarking is less. The extensive callus just above the damaged part may lead to the death of the plant when debarking is done in a girdle. The damage done by cockchafer is very similar to that of manure damage, but in the latter case the bark remains intact. The adult beetles feed on tea plants and shade trees.

Life history

The adults are nocturnal and start emerging from the soil during March, and the emergence continues till August. The peak activity of the beetles is seen during April to June. The adults emerge from the soil between 6 and 6.30 pm. The beetles produce a buzzing sound when emerge and settle on the trunk of shade trees and other structures (fencing, light posts etc.) where mating takes place. Eggs are laid in the soil at a depth of 4-15 cm. There are three larval instars. The early instar grubs feed on the roots of weeds, grasses and young plants; semi decomposed and decomposed vegetable matter. They also attack the young roots of shade plants and green crops. Pupation took place in soil inside earthen cocoon. The biology of *H. impressa* and *S. plagiatula* are depicted below:

Table 9: Duration of different life stages of *H. impressa*

Life stages	Duration (days)
Pre-ovipositional period	2-6
Ovipositional period	1-2
Incubation period	10-11
Grub period	104-142
Pupa	12-15
Total developmental period (egg to adult)	126-168

Table 10: Duration of different life stages of *S. plagiatula*

Life stages	Duration (days)
Egg period	12-14
First instar grub	28-35
Secondt instar grub	112-119
Third instar grub	140-147
Pupa	12-14
Total developmental period (egg to adult)	304-329

Integrated Management

During the last several decades, the control of pests in tea field is predominated by the use of synthetic chemicals. Though broad spectrum pesticides offer powerful incentives in the form of excellent control, increased yield and high economic returns, they have serious drawbacks such as development of resistance to pesticides, resurgence of pests, outbreak of secondary pests, harmful effects on human health and environment and presence of undesirable residues. For the last few years, there are welcome efforts to adopt non-chemical control methods and to evolve an integrated pest management (IPM) programme against the pests infesting tea.

IPM is a sustainable approach for managing pests by combining biological, cultural, physical and chemical tools in a way that minimizes economic, health and environmental risks. The IPM concept is not based on the complete elimination of pests; rather it is based on the principles of keeping pest population within tolerable limits with the help of agronomic adjustments, bush sanitation and rational use of safer agrochemicals. Practicing IPM does not preclude the use of pesticides but reduces it to a minimum level by means of their judicious, need based application in integration with other available methods of control like cultural, mechanical, physical, biological and chemical methods.

Components of IPM

A. Pest Monitoring

Early detection of pests is critical to achieve successful control of damage caused by pests. Therefore, a regular system of monitoring should be in place in the cultivation of tea even though monitoring devices like light traps, yellow pan water traps, yellow sticky traps and pheromone traps may be useful, regular visual monitoring is the fool-proof method to detect pests early for initiating pest management programmes (Rabindra, 2012).

Sampling Procedure for Assessing the Incidence of Pests

The density of the pests in each area has to be assessed regularly during the pest season to take up appropriate and timely control measures. Methods of assessment for important pests are discussed below:

Tea Mosquito Bug

The percentage of infestation has to be assessed by collecting 100 shoots from pluckers' basket and counting the number of infested shoots.

Red Spider Mite

One hundred leaves may be sampled from different areas of the particular field and the number of infested leaves may be counted to determine the percentage of infestation.

Pink and Purple Mites

Pink and purple mite populations have to be assessed at periodical interval by collecting 100 leaves from 100 bushes selected at random from each area. From each leaf, pink and purple mites may be counted with the help of hand lens. Or the method followed for red spider mites can also be followed.

Caterpillars

Flushworm, leaf roller and tea tortrix population can be assessed by counting the number of infested shoots from bushes selected at random from the particular area.

Shot Hole Borer (SHB)

To assess the extent of SHB infestation, the field has to be divided into 2 ha blocks and from each block two hundred stem cuttings are to be taken at random. Care to be taken to collect stems of 1-1.5 cm diameter and 20 cm long.

Economic Threshold Level (ETL)

The ETL for important pests has been worked out based on extensive studies on crop loss to pests. (*Source*: Muraleedharan, 2010).

Pests	Economic Threshold Level
Tea mosquito	5 per cent infestation
Thrips	3 thrips per shoot
Shot hole borer	15 per cent infestation
Flushworms, Leaf rollers and tea tortrix	5 infested rolls per bush
Red spider mites	50 per cent infested leaves
Pink and purple mites	5 mites per leaf

B. Host Plant Resistance

Planting of pest resistance cultivars/clones is one of the acknowledged components of integrated pest management. Even low levels of resistance are important since the need for other control methods can be reduced. Chinery varieties are more susceptible to the attack of *O. coffeae* and *Brevipalpus* spp. Age specific survivorship and fecundity rate of *O. coffeae* were found to be higher when reared on China jats, rather than on Assam varieties. The pink mite, *Acaphylla theae* prefers Assam cultivars (Loganathan, 1992). Incidence of *Helopeltis theivora* is more on China jats and the clone TV-1 (Tocklai Vegetative-1), extensively planted in North East India, is highly susceptible to its attack. It is also on record that dark leaved varieties are more prone to damage by tea mosquito bug than the light leaved ones (Anand Rau, 1940). The flushworm, *Cydia leucostoma* preferred the clone, UPASI-17, considered 'Cambod' in nature, to many other clones (Muraleedharan, 1991b). Soft wooded tea plants are easily damaged by termites. Similarly, clones with high content alpha spinasterol are susceptible to attack by the shot-hole borer, *Euwallacea fornicatus*. Clones with high alpha spinasterol content are susceptible to damage by shot hole borer. Certain Sri Lankan selections like TRI 2024 and TRI 2025 which are popular in South India should be avoided in shot hole borer prone areas.

C. Cultural Practice

Cultural control apparently is the most economical and widely applicable method of pest control. This involves the intelligent manipulation of all aspects of crop husbandry. In tea culture, certain routine cultural practices such as plucking rounds, adjustment of pruning cycles, the modification of shade trees and timely weed control may be effectively employed as pre-emptive measures of pest control (Sasidhar & Sanjay, 2000). This approach of pest control is cheap, risk free and often effective for long period without adverse effect on the environment.

C.1. Plucking

Populations of leaf folding caterpillars such as flushworms, leaf rollers and tea tortrix can be suppressed to a considerable extent by their manual removal during plucking. "Black plucking" in combination with insecticide application is employed for effective control of tea mosquito bug (Chowdhury, 1993). Harvesting of shoots at closer interval will result in the removal of eggs of tea mosquito bug and deny suitable material for feeding and oviposition. Cultural approaches like black plucking (BP), and level off skiff (LOS) along with chemical spraying significantly decreased the infestation level of *Helopeltis* (9-50 fold) and increased the crop yield (2-3 times) in comparison with spraying only chemicals without cultural operations (Rahman *et al.*, 2005) (Table. 11)

Table 11: Effect of cultural practices on the infeststion of *H. theivora*

Cultural practices	Treatment	Pre-treatment		Post-treatment	
		Per cent infestation/plot	Mean crop yield (kg)/plot/ week/kg	Per cent infestation/ plot	Mean crop yield (kg) /plot/week/kg
BP	ILE	76.5	1.80	1.3	21.0
BP	LEB	81.6	1.45	1.2	26.0
LOS	ILE	79.5	1.50	0.2	27.8
LOS	LEB	80.0	1.80	0.2	29.0
Normal	ILE	87.0	1.70	9.8	10.0
Normal	LEB	80.0	2.20	10.2	10.5
Control	-	80.0	2.50	82.0	3.2
CD (p=0.05)	-			6.45	8.52
CV(%)	-			4.88	7.63

ILE: Imidacloprid, lamda cyhalothrin and etofenprox

LEB: Lamda cyhalothrin, etofenprox and beta cyfluthrin

BP: Black plucking (Removal of infested leaves) with chemical spraying

LOS: Level of skiff (Removal of infested leaves along with tender stems below the table) with chemical spraying

Normal: Normal plucking with chemical spraying

Control: No cultural operation and pesticide spraying

C.2. Pruning and Skiffing

The operation of pruning and length of pruning cycle are important factors in pest ecology. Pruning removes a large part of the foliage and stems along with the pest. However, the newly emerging foliage is nutritionally more attractive to certain insects such as aphids, thrips, flushworms and leaf rollers. In North East India, in a pruning cycle of three years, the tea bushes in the first year of skiffing were less infested by red spider mite (*O. coffeae*) than those skiffed for two consecutive years (Das, 1965). Introduction of extended pruning cycle is supposed to have increased the activities of tea mosquito bug in North East India since the pest finds an undisturbed place for hibernation during winter and gets a supply of tender shoots earlier in the season. If the tea fields located near forests and prone to *Helopeltis* attack are kept under normal pruning, the activities of this pest can be reduced. Light pruning of the infested areas will reduce the intensity of attack in the following year as most of the eggs embedded in the shoots and leaves will be destroyed during pruning (Das, 1965). Minimizing new access points by removal of dead woods, cankers and snags will lessen fresh attack by termites. Proximity of termite affected areas should be taken in to consideration before selecting an area for new planting. Damage due to shot-hole borer in South India is more severe in the third and fourth years of the pruning cycle, and there exists an exponential relation between the percentage of borer attack and the age of the field since pruning.

C.3. Shade Regulation

The planting of shade trees and many ancillary crops in the tea ecosystem is considered to be a necessary practice. In tea, shade regulation plays a predominant role in pest suppression. Infestation by mites and thrips is seen more in tea fields devoid of shade. Densely shaded areas are prone to the attack of *Helopeltis*. Certain shade trees like *Indigofera* and *Albizzia* are the alternate hosts of several caterpillar pests. So, the adoption of recommendations on shade management will help to prevent the excessive build up of thrips, mites and *Helopeltis*. Moderate shade status (60 per cent) coupled with cultural operations like black plucking and level off skiffing protected the crop from *Helopeltis* (Rahman *et al*., 2005) (Table.12).

Table 12: Effect of shade status on *H. theivora* infeststion

Site	Shade intensity	Per cent infeststion		Treatment (June-August)	Mean crop/month (kg) (July- October)
		Pre-treatment	Post-treatment		
I Unshaded	Full sunshine	59.9	10.0	BP,three rounds of pesticides*	18.2 (12 harvest)
II Moderate shade	60% shade	35.4	3.5	One round of pesticide	23.0 (12 harvest)
III Unshaded	89% shade	90.0	3.0	BP, LOS, two rounds of pesticides	20.2 (11 harvest)
CD (p=0.05)	-	-	2.89	-	-
CV(%)	-	-	1.42	-	-

C.4. Field Sanitation

Field sanitation assumes significance in the management of several pests. Weeds offer excellent hiding places and serve as alternate hosts for *Helopeltis* and red spider mites. Weeds like *Mikania cordata, Bidens biternata, Emilia* sp., *Polygonum chinese, Oxalis acetosella, Malastoma malabethricum* and *Lantana camara* offer excellent hiding places and serve as alternate hosts for the Tea mosquito bug. Weeds such as *Malastoma malabethricum* and *Urena lobata* act as alternate hosts of red spider mite. Weed free cultivation and preventing trespassing of cattle, goat, and other animals from red spider mite-infested fields reduce its spread. *Ageratum conizoides, Borreria hispida Commelina benghalensis, Pouzolzia indica* and *Oxalis corymbosa* are alternate hosts of root knot nematode. So, growth of host plants in and around tea fields should be controlled and this will help to reduce the growth of pest population.

C.5. Fertilizer Application

Application of higher levels of potassium fertilizers is known to reduce the incidence of pests in several crops. Application of high rate of muriate of potash to the soil in the first year of the pruning cycle significantly reduced the infestation by shot-hole borer.

C.6.Trap Crop

Studies related to the use of trap crops in tea are scarce. A trap crop also manipulates the habitat in an agroecosystem, which can be included under the ecological engineering approaches for the purpose of IPM (Gurr *et al.*, 2004). Border plantings of *Adhatoda vesica* serve as a barrier for red spider mite,

Oligonychus coffeae (Watt & Mann, 1903). As such, susceptible tea clones such as Tocklai vegetative clone TV1 to *H. theivora* may be utilized as the trap crop (Hazarika *et al.*, 2009).

D. Physical Control

Physical controls aim to reduce pest populations by using devices which affect them physically or alter their physical environment.

D.1. Manual Removal of Larvae and Pupae

Populations of foliage feeding caterpillars such as looper, bunch caterpillar, faggot worms, flushworms, leaf roller and tea tortrix can be reduced to a great extent by manual removal of larvae and pupae. Tea mosquito bugs lay large number of eggs on the broken ends of plucked shoots and intensive manual removal of stalks during plucking will help to reduce the eggs and thereby the incidence of tea mosquito bug.

D.2. Soil Stirring or Forking

Soil stirring or forking around the collar region during December/January, ground leveling, filling up the collar depressions and alkaline (lime) wash of bush frames leads to the reduction in population of many pests like termites, bunch caterpillar, looper, nettle grubs,.red slug, psychids, common bark eating borer *etc.* besides improving the bush health, enabling them to withstand the attack by pests resulting in higher productivity.

D.3. Heat Treatment and Soil Solarisation

Soil used in the nursery may be heated to 60-65°C for killing the infective juveniles of soil nematodes.

D.4. Use of Light Traps

Fluorescent light traps are useful in attracting the moths of caterpillars and other insects. These traps can be set up during the seasons of moths/ beetle emergence and the attracted moths /beetles can be killed mechanically or by using insecticides. These traps are useful for monitoring the activity of the pests and also as a tool for their suppression.

E. Mechanical Control

Mechanical methods are manual devices utilized for pest suppression. There have only been a few attempts to utilize this method for tea pest management. However, few methods have been developed and practiced in tea plantation in India and other countries for the control of termites.

E.1. Mound Digging Process

Termitaria are architecturally designed dome shaped close system of earthen mounds that provide natural protection from adverse environment. '*Queen*' lives inside the mound and reproduce infinitesimal progenies to build up the population. The mechanical control method to destroy the termitaria seems to be a plausible solution for termite control.

F. Biological Control using Multicellular Organism

Insect parasitoids and predators play a vital role in the natural regulation of many tea pests.More than one hundred and seventy species of predatory and parasitic insect and mites have been reported from tea ecosystems of India (Muraleedharan and Selvasundaram, 2002). Efforts towards the conservation and augmentation of natural enemies in the tea ecosystem, could offer significant advances in biological control in tea in future.

F.1. Mites

Several species of predatory mites have been reported , mostly belonging to Phytoseiidae, Stigmaedae and Tydeidae. *Amblyseius herbicolus* and *Euseius ovalis* are the two common phytoseiids feeding on the eriophyids, *Acaphylla theae* and *Calacarus carinatus*. The stigmaeid, *Agistemus fleschneri* is an important predator of eggs and nymphs of *Oligonychus coffeae* in North East India. In South Indian tea fields, *A. herbicolus* is the most common predator of the mite pest, *Acaphylla theae*. Populations of this predator reached the peak during December and February and a single predator could consume more than 400 pink mites during its life span. The coccinellids, *Cryptogonus bimaculatus*, *Jauravia quadrinotata*, *J. soror*, *J. opaca*, *Monochilus sexmaculatus* and *Stethorus gilvifrons* are also predators of phytophagous mites.

F.2. Thrips

Anthocorids belonging to *Anthocoris* and *Orius* and the predatory thrips, *Aeolothrips intermedius* and *Mymarothrips garuda* are important natural enemies of thrips reported from India. In North East India, a neuropteran predator, *Chrysoperla carnea* was found effective against thrips and tea mosquito bug when applied @ 500-1000 adults/ha.

F.3. Scale Insects and Mealy Bugs

A few parasitoids of scale insects and mealy bugs have been reported from India. *Coccophagus cowperi* and *Encyrtus infelix* heavily parasitise *Saissetia coffeae*. The black scale *Chrysomphalus ficus* is attacked by six parasitoids and predators.

F.4. Caterpillar Pests

Leaf folding caterpillars, *Cydia leucostoma* is parasitized by nine species of braconids, two ichneumonids and one encyrtids in addition to a pupal parasitoids belonging to *Ascogaster*. Among the larval parasitoids *Apanteles aristaeus* is the most common species. The leaf roller, *Caloptilia theivora* is heavily parasitized by the eulophid, *Sympiesis dolichogaster*. The incidence of parasitism is between 20 and 83 per cent (Muraleedharan and Selvasundharam, 2002). The tea tortrix *Homona cofferia* is attacked by one egg and nine larval parasitoids. The ichneumonids larval parasitoid *Phytodietus spinipes* plays a significant role in the population regulation of this tortricid. *Apanteles fabiae* and *A. taprobanae* parasitise the looper caterpillar, *Buzura suppressaria*. Another braconid, *Cotesia ruficrus* also attacks looper. Tachinid, *Cylindromyia* sp. is the chief larval parasitoid of *Andraca bipunctata*, the bunch caterpillar in Assam.

F.5. Tea Mosquito Bug

The eggs of tea mosquito bug are parasitized by a mymarid, *Erythmelus helopeltidis* Ghan. The incidence of parasitism in the field varied between 52 and 83 per cent. The neuropteran predator *Chrysoperla carnea,* the spider *Oxyopes* sp. praying mantids, mermethid nematodes (*Hexamermis* sp.), reduviid bug act as potential natural enemies of tea mosquito bug.

G. Biological Control using Microbial Agents

Several microbes have been reported to be pathogenic to insect pests. Bacterial insecticide, *Bacillus thuringiensis* has been effectively used for the control of looper caterpillars (Borthakur and Raghunathan, 1987) and other lepidopteran pests (Selvasundaram and Muraleedharan, 2000). White muscardine fungus, *Beauveria bassiana* (Selvasundaram and Muraleedharan, 2000, Rahman *et al.*, 2006 b), *Cephalosporium* sp. (Hazarika *et al*., 1994); Agnihothrudu, 1999); *Verticilium lecanii* (Barua, 1983); *Paecilomyces fumoroseus* (Barua, 1983), *P. carneus* (Hazarika *et al.*, 1994); *Hirsutella thompsonii* (Debnath, 2004a); *Metarhizium anisopliae* (Debnath, 2004 b) and NPV (Antony *et al.*, 2011) are effective and have been used widely especially in organic gardens of Darjeeling against tea mosquito bug, mites, thrips, jassids, shot hole borer, termites, aphids, scale insects and lepidopteran pests.

H. Pheromonal Control

Pheromone traps are generally used for monitoring pest population and mass trapping of adult pest species. Pheromone traps of *Spodoptera litura* (cutworms) infesting tea are commercially available and can be used for controlling the populations of the moths. The pheromone of tea leaf roller has been field tested. Similarly, the female sex attractants of the flushworm (*Cydia leucostoma*) have also been identified recently. Studies on the sex pheromones of tea mosquito bug are in progress.

I. Plant Based Products for Pest control

Certain wild and weed plants available in and around tea gardens having pesticidal properties could be utilized for tea pest control. The pesticidal properties of different native plants are summarized in Table 13. Leaves, succulent stems and rhizome of available wild/ weed plants are collected locally from tea estates during the stalk/off season, chopped, and dried under shade. The dried material is kept in a 100 kg capacity polythene bag. Two or four kg of dried plant material is taken into a 200 litre capacity drum, 200 litres of water is added to it and the material is soaked for 48 hours. After 48 hours, the extract filtered using the muslin cloth. The filtrate is used for direct spray @ 200 litres per ha of tea area. Fresh plant material (4 kg/ha) can also be used instead of dried one (2kg/ha) (Gurusubramanian *et al*., 2008a).

J. Use of Inorganic Compounds/ Hydrocarbon Oils

Formulations of sulphur and lime sulphur are effective against mites. Spray oil from paraffinic base has been found effective against eriophyid and red spider mites. Since this oil does not leave any residue in tea, it could be incorporated in to the mite control programme in tea.

K. Chemical Control

Over the years, the pattern of pesticide usage on tea in India has followed the world trend. In recent years there has been a greater dependence on the use of pesticides. The average quantity used of chemical pesticides used was estimated to be 11.5 kg/ha in the Assam valley and Cachar, 16.75 kg/ha in Dooars and Terai and 7.35 kg/ha in Darjeeling (Barbora and Biswas, 1996). In a survey, synthetic pesticides constituted 85 per cent of the total pesticide used, while, 15 per cent were of organic and inorganic origin in the tea gardens of Dooars. In this, acaricides accounted for 25 per cent (3.60 litre/ha) and insecticides 60 per cent (8.46 litre/ha). Within the synthetic insecticides, organophosphate compounds (64 per cent- 5 rounds per year) were the most preferred followed by organochlorine (26 per cent – 2 rounds per year) and synthetic pyrethroids (9 per cent- 7 rounds per year) (Sannnigrahi and Talukdar, 2003). At present it is a global concern to minimize the pesticide residue in though even, a number of synthetic chemicals are available for pest control in tea, the choice is limited for their widespread application owing to the Maximum

Table 13: Botanicals and their pesticidal properties against different tea pests

Target pests	Developmental stages	Plants and family	Part used	Pesticidal properties	Reference
Tea mosquito bug *Helopeltis theivora*	Nymphs and adults	*Clerodendron infortunatum* stem (Verbenaceae)	Leaves and succulent	O, A,I	Gurusubramanian *et al.* (2008a), Roy *et al.*, 2008a
		Clerodendrum viscosum Ventenat (Verbenaceae)	Leaves and succulent stem	O,A, I	Gurusubramanian *et al.* (2008a), Roy *et al.* 2010a
		Clerodendron inerme (Linn.) Gaertn. (Verbenaceae)	Leaves and succulent stem	O,A, GR	Gurusubramanian *et al.* (2008a)
		Pongamia glabra Vent. (Fabaceae)	Leaves	A	Gurusubramanian *et al.* (2008a)
		Pongamia pinnata L. (Fabaceae)	Leaves, seed Kernel	A, Repellent effects	Gurusubramanian *et al.*, 2008a; Deka *et al.*, 1998; Mamun and Ahmed, 2011
		Swietenia mahagoni (L.) Jacq. (Meliaceae)	Seed kernels	I	Mamun and Ahmed, 2011
		Xanthium strumarium L. (Asteraceae)	Leaves and succulent stems	O, A, GR	Sarmah and Bhola, 2008; Sarmah and Bhola, 2011
		Pogostemon parviflorus Benth (Labiatae)	Leaves and succulent stems	Growth inhibitory action, decreased fecundity	Gurusubramanian *et al.*, 2008a; Rahman *et al.*, 2007; Rahman and Nath, 2012
		Annona squamosa L. (Annonaceae)	Leaves and succulent stems	A	Gurusubramanian *et al.* (2008a)
		Lantana camara L. (Verbenaceae)	Leaves and succulent stems	A, I, GR	Gurusubramanian *et al.* (2008a)

Target pests	Developmental stages	Plants and family	Part used	Pesticidal properties	Reference
		Adhatoda vasica Nees (Acanthaceae)	Leaves and succulent stem	O, A	Gurusubramanian *et al*., 2008
		Polygonum orientale	Leaves	O, A	Gurusubramanian *et al*. (2008a)
		Melia azedarach L. (Meliaceae)	Seeds, leaves	A,O	Gurusubramanian *et al*. (2008a), Mamun and Ahmed, 2011
		Acorus calamus L. *(Acoraceae)*	Leaves	O, A	Sarmah and Bhola (2011)
		Azadirachta indica A.Juss. (Meliaceae)	Leaves and succulent stem	A	Sarmah and Bhola, 2008
		Chrysanthemum sp. (Asteraceae)	Dried flowers	A	Mamun and Ahmed, 2011
		Datura metel L. (Solanaceae)	Leaves and succulent stem	I,A,O	Mamun and Ahmed, 2011
		Polygonum hydropiper (L.) Delabre (Polygonaceae)	Leaves and succulent stem	A	Mamun and Ahmed, 2011 Rahman and Nath, 2012; Sarmah and Bhola, 2011
		Polygonum orientale L. (Polygonaceae)	Leaves and succulent stem	GR, A	Gurusubramanian *et al*., 2008
Red spider mite *Oligonychus coffeae*	Adult	*Clerodendron infortunatum* L. (Verbenaceae)	Leaves and succulent stem	O,A	Gurusubramanian *et al*. (2008a)
		Xanthium strumarium L. (Asteraceae)	Whole plant	O, A	Gurusubramanian *et al*. (2008a)

Target pests	Developmental stages	Plants and family	Part used	Pesticidal properties	Reference
		Clerodendron infortunatum L. (Verbenaceae)	Leaves and succulent stem	O, AC	Gurusubramanian *et al.* (2008a)
		Acorus calamus L. (Acoraceae)	Rhizome	O, AC	Gurusubramanian *et al.* (2008a)
		Polygonum hydropiper(L.) Delabre (Polygonaceae)	Whole plant	AC, A, O	Gurusubramanian *et al.* (2008a)
		Pongamia pinnata L. (Fabaceae)	Leaves and succulent stem	AC	Gurusubramanian *et al.* (2008a)
		Azadirachta indica A.Juss. (Meliaceae)	Kernel	O, AC	Gurusubramanian *et al.* (2008a), Roobakkumar *et al.* (2010)
		Lantana camara L. (Verbenaceae)	Leaves	O, AC	Gurusubramanian *et al.* (2008a), Sarmah *et al.* (1999)
		Terminalia chebula Retz. (Combretaceae)	Dry pericarp of the fruit		Roy *et al.* (2014)
Bunch caterpillar *Andraca bipunctata*	Third instar larva	*Clerodendron infortunatum* L.(Verbenaceae)	Leaves and succulent stem	A	Gurusubramanian *et al.* (2008a)
		Polygonum hydropiper(L.) Delabre (Polygonaceae)	Whole plant	A	Gurusubramanian *et al.* (2008a)
		*Azadirachta indica*A.Juss. (Meliaceae)	Kernel	O, I	Gurusubramanian *et al.* (2008a)
		Eupatorium glandulosum (Asteraceae)	Leaves and stems	A	Gurusubramanian *et al.* (2008a)
		Urtica dioca L.(Urticaceae)	Leaves and stems	A	Gurusubramanian *et al.* (2008a)

Target pests	Developmental stages	Plants and family	Part used	Pesticidal properties	Reference
		P. runcinatum Buch.-Ham. ex D. Don(Polygonaceae)	Leaves and stems	A	Gurusubramanian *et al.* (2008a)
		Artimisia vulgaris L. (Asteraceae)	Leaves and stems	A	Gurusubramanian *et al.* (2008a)
Red slug caterpillar *Eterusia magnifica*	Third instar larva	*Azadirachta indica* A.Juss. (Meliaceae)	Kernel	I	Gurusubramanian *et al.* (2008a)
		Melia azedarach L. (Meliaceae)	Kernel	I	Gurusubramanian *et al.* (2008a)
Clania cramerii	Third instar larva	*Azadirachta indica* A.Juss. (Meliaceae)	Kernel	A	Gurusubramanian *et al.* (2008a)
Termite *Microcerotermes* sp.	Workers	*Azadirachta indica* A.Juss. (Meliaceae)	Kernel	A	Gurusubramanian *et al.* (2008a)
		Carica papaya L. (Caricaceae)	Unripe fruit	I	Gurusubramanian *et al.* (2008a)
		Tagetes erecta L. (Asteraceae)	Whole plant	I	Gurusubramanian *et al.* (2008a)
Tea thrips & Jassids *Scirtothrips dorsalis* &*Empoasca flavescens*	Nymphs and adults	*A. conyzoides* L. (Asteraceae)	Whole plant	I	Gurusubramanian *et al.* (2008a)
		Artemesis vulgaris L. (Asteraceae)	Leaves	I	Gurusubramanian *et al.* (2008a)
		Melia azedarach L. (Meliaceae)	Seeds &leaves	I	Gurusubramanian *et al.* (2008a)
		Azadirachta indica A.Juss. (Meliaceae)	Kernel	I	Gurusubramanian *et al.* (2008a)

O: Ovicidal, A: Antifeedant, I: Insecticidal, AC: Acaricidal, GR: Growth regulatory

Residue Limits (MRLs) declared by European Union (EU) and Codex Alimentarius Commission and EPA of USA, Food and Agricultural Organization (FAO), World Health Organization (WHO), German Laws (GL), European Economic Commission (EEC/EC) etc. The list of CIB-TRA approved agrochemicals along with MRL is given in Table 14. Safety harvest intervals were also established in tea based on the field data generated during last few years. This information will help the industry to decide the harvesting interval after the application of the chemicals such as ethion (7 days), dicofol (6 days), endosulfan (3 days), quinalphos (4 days), chlorpyrifos (12 days) and deltamethrin (6 days) and to keep their residue levels below the MRL prescribed by European Union.

Need based, judicious and safe application of pesticides is the most vital aspect of chemical control measures under IPM strategy. It involves developing IPM skills to play safe with environment by proper crop health monitoring, observing ETL and conserving the natural bio-control potential before deciding in favour of use of chemical pesticides as a last resort. The following suggestions have important implications on the success of control measures in the context of IPM strategy:

1. Pesticides should be applied only when it is absolutely essential.
2. Use only the pesticides which are recommended by Tea Research Institutes (UPASI TRF/TRA).
3. Follow the recommendations on the dosage and dilution rates stipulated for each pesticide and follow all safety measures while applying pesticides.
4. Spray the approved insecticides/acaricides only after plucking, so that the minimum safe period is always maintained.
5. Use pesticides on a rotation basis.
6. Use neem based/microbial formulations.
7. Do not use pesticides which have crossed shelf life (Date expired chemicals).
8. Use only conventional sprayers, bot knapsack and motorized with proper functional nozzles.
9. Sample of tea may be analyzed for the residues of the commonly used pesticides.

Table 14. List of CIB-TRA approved Agro-chemicals for use in Tea by the member gardens of TRA (as on 1st July, 2014)

Name of the Chemicals	MRL(ppm) India	MRL(ppm) EU	MRL(ppm) Japan	Chemical Nature (Classification) /	Mode of action	Dilution	Active ingredient hectare	Dosage/ ha/ml or g/ha
Bifenthrin 8 SC	-	5	25	4th generation pyrethroid	Effective by contact or ingestion affects the central and peripheral nervous system by interfering with sodium channel gating	1:1600 (62.5 ml in 100 L)	20 g ai/ ha	250 ml
Dicofol 18.5 EC	5	20	3	Organochloride	Contact: nerve poison	1:400 (250 ml in 100 L)	185 g ai/ ha	1000 ml
Ethion 50 EC	5	3	0.3	Organophosphate	Contact: inhibitor of acetylcholinesterase	1:400 (250 ml in 100 L)	500 g ai/ ha	1000 ml
Fenazaquin 10 EC	3	10	[UL 0.01]	Quinazoline	Contact: Mitochondrial complex I electron transport inhibitors; Ovicides	1:400 (250 ml in 100 L)	100 g ai/ ha	1000 ml
Fenpropathrin 30 EC	-	2	25	3rd generation pyrethroid	Contact and stomach action : nerve poison	1:1600 (62.5 ml in 100 L)	75 g ai/ ha	250 ml
Fenpyroximate 5 EC/SC	-	0.1	10	METI	Contact: Mitochondrial complex I electron transport inhibitors	1: 2000 (50 ml in 100 L)	10 g ai/ ha	200 ml
Hexythiazox 5.45 EC	-	4	-	Thiazolidionone	Compounds of unknown or non-specific mode of action	1: 2500 (40 ml in 100 L)	8.72 g ai/ ha	160 ml
Propargite 57 EC	10	5	5	Organosulfur	Contact on juveniles and adults by inhibition of ATP synthesis	1:400 (250 ml in 100 L)	570 g ai/ ha	1000ml

Name of the Chemicals	MRL(ppm) India	EU	Japan	Chemical Nature (Classification) /	Mode of action	Dilution	Active ingredient hectare	Dosage/ ha/ml or g/ha
Sulphur 80 WG	-	-	-	Sulphur	Contact & fumigant: Respiration inhibitors	1:200 (500 g in 100 L)	1600 g ai/ ha	2000 g
Wettable Sulphur 40 WP	-	-	-	Sulphur	Contact & fumigant: Respiration inhibitors	1:200	1600 g ai/ ha	2000 g
Spiromesifen 240 SC (22.9 w/v)	-	50	30	Ketoenols	Contact: lipid biosynthesis inhibitor; ovicide	1:1000 (100 ml in 100 L)	91.6 g ai/ ha	400 ml
Azadirachtin 5 EC	-	0.01	-	Neem based	Contact and Stomach poison: antifeedant and growth disruptor	1: 1500 (66.6 ml in 100 L)	13.33 g ai/ ha	266.4 ml
Deltamethrin 2.8 EC	-	5	10	2nd generation pyrethroid	Contact: nerve poison	1:2000 (50 ml in 100 L)	5.6 g. ai/ ha	200 ml
Phosalone 35 EC	-	0.1	2	Organophosphate	Contact and influential mode of action	1:400 (250 ml in 100 L)	350 g. ai/ ha	1000 ml
Profenofos 50 EC	-	0.1	1	Organophosphate	Contact and stomach action: Acetylcholine esterase inhibitors	1:1000 (100 ml in 100 L)	200 g. ai/ ha	400 ml
Quinalphos 25 EC	0.01	0.1	0.1	Organophosphate	Contact action: Acetylcholine esterase inhibitors	1:400 (250 ml in 100 L)	250 g. ai/ ha	1000 ml
Quinalphos 20 AF	0.01	0.1	0.1	Organophosphate	Contact action: Acetylcholine esterase inhibitors	1:400 (250 ml in 100 L)	200 g. ai/ ha	1000 ml
Clothianidin 50 WDG	-	0.7	-	Neonicotinoids	Systemic action: affect on central nervous system	1:4500 (22.22 g in 100 L)	44.44 g. ai/ ha	88.88 g
Thiacloprid 21.7 SC	-	10	30	Neonicotinoids	Systemic action: affect on central nervous system	1:10000 (100 ml in 100 L)	86.8 g. ai/ ha	400 ml
Thiamethoxam 25 WG	-	20	20	Neonicotinoids	Systemic action: affect on central nervous system	1:4000 (25 g in 100 L)	25 g. ai/ ha	100 g

Requirement of Spray Fluid

The volume of spray fluid requires for pesticide applications depend on the age of the tea crop, terrain and type of spraying equipment employed. A maximum of 350 litres of water is sufficient to cover one hectare while using low volume sprayers (power sprayers). In case of high volume sprayers (hand operated knapsack sprayers) a maximum of 450 litres of spray fluid will be required to cover one hectare. A higher spray volume will be required in special situations like post prune insecticide application for the control of shot hole borer or termite.

The knowledge of the target site for different pests and method of spraying is very much essential for efficient control of tea pests. The target site of spray fluid for different pests is depicted in the Table 15.

Table 15: Target site of spray fluid for different tea pests

Pest	Target of spraying	Remark
Tea mosquito bug	Whole canopy; weeds and soil surface	Adopt barrier spray
Red spider and purple mite	Both surface of leaves of whole canopy	Thorough coverage
Pink mite	Both leaf surface of top and mid hamper	Thorough canopy
Scarlet mite	Lower surface of leaf and stem of whole canopy	Spray from old to young leaves
Thrips	Tender shoots, lichens and mosses, crack and crevices of frames and soil surface	Thorough spray
Jassids	Under surface of leaves of top and mid hamper	Thorough spray
Looper caterpillar	Whole bush canopy and shade tree trunk	Spray at early stage and thorough coverage
Bunch caterpillar	Mid and bottom hamper stage and thorough coverage	Spray at early
Red slug	Mature leaves, whole canopy, young stems and trunk of shade trees	Thorough spraying of whole bush
Termite	Whole frame, collar and soil around collar	Thorough drenching
Shot hole borer	Stems	Thorough drenching of stems

Nematode Pests

Nematodes or eelworms are soil-inhabiting organism. A large majority of the soil-inhabiting nematode live saprophytically, while the plant-parasitic species constitute a smaller percentage, feeding on the roots of plants, either as ecto-parasites or as endo-parasites. Several genera of plant parasitic nematodes *viz.*, *Pratylenchus*, *Helicotylenchus*, *Tylenchorrhynchus*, *Tylenchu*,

Hoplolaimus, *Rotylenchus*, *Xiphinema*, *Paratylenchus* and *Aphelenchoides* have been recovered from rhizophere soil in Northeast India (Das, 1965). Among the pathogenic eelworm attacking tea roots, *Meloidogyne* species (*M. incognita*, *M. javanica* and *M. hapla*) are the prevalent and prominent ones in Northeast India, causing severe damage to tea seedlings and young clonal plants by attacking the roots and stunting of tea seedlings in nurseries up to three years (Roy and Rahman, 2013). The root lesion nematode, *Pratylenchus* sp. is also reported and few occasions to cause severe damage to the roots of tea seedlings (Das, 1965).

Muraleedharan (1991) reported about the infestation of roots of mature tea bushes by *M. brevicauda* and young tea by *M. javanica*, *Pratylenchus* sp. and *Helicotylenchus* sp. in South India.

Symptoms of Eelworm Attack in Tea

Usually tea seedlings infested with these parasitic nematodes are not always killed outright and symptoms of attack can be mistaken for drought, mineral deficiencies, lack of organic soil constituents, water logging and so on (Das, 1965). In the field, patches of plants are found stunted, wilted, chlortic, or generally unhealthy. These are caused by atrophied root systems due to rotting, galling or by killing the fine tender roots and lateral rootlets. The roots of tea when infested by *M. incognita* tend to form very small galls that look like beads on a string. Direct damage by imparing root spread, and thus the plants' ability to take up water and nutrients weakens growth and health of seedlings is greatly debilitated.

Crop Loss

The amount of damage by nematodes causes an overall loss in production of healthy tea seedlings of 5-10 per cent or as much as 50 per cent in heavy infested areas (Mukherjee, 1960).

Alternate Host Plants

There are 53 weed species, 9 species of shade trees and rehabilitation crops are reported to serve as alternate host for root knot nematode, *M. incognita* and *M. Javanica* (Basu, 1974, 1975, Basu and Banerjee, 1967 and Basu and Gope, 1971).

Management of Eelworm

The following management practices should be followed to minimize the infestation of eelworm.

Replanting

Useful step should be taken to incorporate nematicide during the time of new palnting of tea. Nursery soil can be treated with systemic granular nematicide such as carbofuran 3G or phorate 10 G @1g/sleeve/kg of nursery soil.

Rehabilitation

Recent research findings showed the nematode population in soils could be contained below critical level (7.00/10g soil) in an ecofriendly method by planting and lopping the green crops like Guatemala and Citronella grass. Nematode population in Guatemala and Citronella plots were 2.98 and 4.56 respectively which were below the critical level. So, Guatemala and Citronella can be planted before planting tea for improving soil properties as well as for suppressing the nematode population (Mamun *et al*., 2011).

Nursery

- Soil from the nursery site should be tested for eelworm population and acidity status. If the population of eelworm is found to be 6 or above per 10 g of soil tested, it is considered to be unsuitable for use.
- Cultivation work (ploughing and harrowing) for preparing the nursery bed should be done to expose and dry the undecomposed weeds and roots of the plants. All sorts of mulching materials should be kept away from the seed nursery to avoid nematode infestation.
- Plant parasitic nematodes can be killed by uniform heating (after sieving) of the soil up to 60-70 °C for 4-5 minutes on plain tin sheets. The soil can be used after heat treatment.
- Undecomposed cow dung is a threat for tea nursery.

References

Agnihothrudu V. (1999) Potential of using biological control agents in tea. In: Jain NK (ed) Global Advances in Tea Science, Aravali Books, New Delhi, pp. 675-92.

Ahmed M. and Aslam A.F.M. (2011) Recent concepts and strategies for tea pest management in Bangladesh. *Tea J. Bangladesh* 40: 18-36.

Ambika B., Abraham C.C. (1983) New record of *Helopeltis theivora* Waterhouse and an undetermined species of *Helopeltis* (Miridae: Hemipetra) as potential pests of cashew, *Anacardium occidentale* Linn. *Indian J. Entomol*. 45(2), 183-184.

Ananda Rau S. (1940) *Helopeltis* : A summary of information collected through replies to the questionnaire with comments there on. *UPASI Tea Sci. Dep. Misc. Pap.* 2:1-8.

Andrews E.A. (1918) An Experiment on the treatment of red spider by insecticides. *Quarterly Journal of the Indian Tea Association.* pp. 46-49.

Anonymous (2010). Special Bulletin on Tea mosquito bug, *Helopeltis theivora* Waterhouse (Bulletin No. PP/01/2010). published by Plant Protection Division,Tocklai Tea Research Institute, Tea Research Association, Jorhat, India.

Anonymous (2012). Production and Export of tea, Annual Report, 2011-12, Govt. of India, Ministry of Commerce, Department of Commerce and Industry.

Anonymous (2007). Overview of forest pests Indonesia. In: Forest Health & Biosecurity Working Papers. Forestry Department, Food and Agriculture Organization of the United Nations, FAO, Rome, Italy, pp. 7-8.

Anstead R.D., Ballare E. (1922) Mosquito blight of tea. Planter's Chron. UPASI, Coimbatore, 17: 443-455.

Antony B., Sinu, A.P. and Das S. (2011) New record of nucleopolyhedrosis viruses in tea looper caterpillar in India, J. Invertebr. *Pathol*. 108: 63-67.

Awasthy R. C. and Venkatakrishnan N. S. (1977). Benefit evaluation of Tocklai recommendations III. Control of red spider. *Two a Bud* 24 (2): 37-38.

Banerjee B. (1982). A strategy for the control of *Andraca bipunctata* Walker on tea. Crop Protection 1 (1): 115-119.

Barbora B.C., Singh K. (1994). All about *Helopeltis theivora*: a serious pest of tea. *Two a Bud*., 41: 19-21.

Barua G.C.S. (1983). Fungi in biological control of tea pests and disease in north-east India. *Two a Bud* 30: 5-7.

Basu S. D. and Banerjee B. (1978). Effect of infestation of *Meloidogyne incognita* Chitwood on some ancillary plants grown with tea in N.E. India. *Two a Bud* 25(1): 28-29.

Basu S.D. (1975). Weed hosts of root-knot nematode. *Two a Bud* 22 (2): 91.

Basu S.D.(1974). Weed hosts of root knot nematodes in tea estates around Jorhat. *Two a Bud* 21 (2): 50.

Bhuyan M., Bhattacharyya P.R. (2006) Feeding and oviposition preference of *Helopeltis theivora* (Hemiptera: Miridae) on tea in Northeast India. *Insect Science* 13: 485-488.

Borthakur M.C. and Raghunathan A.N. (1987) Biological control of tea looper with *Bacillus thuringiensis*. *J. Coffee Res*. 17: 120-121.

Browne F.G. (1968) Pests and diseases of forest plantation tree: an annotated list of the principal species occurring in the British Commonwealth. Oxford: Clarendon Press.

Chowdhury P.K. (1993) Package for eradication of tea mosquito bug, *Helopeltis theivora*. *Two a Bud*. 40: 6-8.

Chutia B.C., Rahman A., Sarmah M., Barthakur B.K. and Borthakur M. (2012) *Hyposidra talaca* (Walker): A major defoliating pest of tea in North East India. *Two a Bud*. 59: 17-20.

Das G. M. (1965) Pests of tea in North East India and their control. Tea Research Association, Tocklai Experimental Station, Assam. p. 115.

Das G.M. (1959) Bionomics of the tea red spider *Oligonychus coffeae* (Nietner). *Bull. Ent. Res*. 50 (2): 265-275.

Das G.M. (1960) Occurrence of the red spider *Oligonychus coffeae* (Nietner) on tea in North East India in relation to pruning and defoliation. *Bull. Ent. Res*. 51(3):415-426.

Das G.M. (1961) Coccoids on tea in north east India. *Indian J. Entomol*. XXIII (IV): 245-256.

Das S. and Mukhopadhyay A. (2008) Host based variation in life cucle traits and general esterase level of the tea looper *Hyposidra talaca* (Walker) (Lepidoptera: Geometridae). *J. Plantation Crops* 36 (3):457-459.

Das S., Mukhopadhyay A. and Roy, S. (2010) Morphological diversity, developmental traits and seasonal occurrence of looper pests (Lepidoptera: Geometridae) of tea crop. Journal of Biopesticides, 3(1 Special Issue): 16-19.

Das S.C. (1984) Resurgence of tea mosquito bug *Helopeltis theivora* Waterhouse, a serious pest of tea. *Two a Bud* 31, 36 38.

Das S.C. (1984) Resurgence of tea mosquito bug, *Helopeltis theivora* Waterh, a serious pest of tea. *Two a Bud* 31: 36-39.

Debnath S. (2004a) Natural occurrence of entomopathogenic fungus, *Hirsutella thompsonii* on red spider mite, *Oligonychus coffeae* (Nietner) infesting tea plants, *Camellia sinensis* L. (O) Kuntze in North East India. In: Proceedings of the international conference of tea culture and science, ICOS-2004, November 4-6, 2004, Shizuoka, Japan. pp. 1204-1206.

Debnath S. (2004) Prospects of native entoimogenous fungus *Metarhizium anisopliae* var *major* for integrated control of termite pests of tea in North East India. In: Proceedings of the 10th meeting of the IOBC/WPRS working group on breeding for resistance to pests and diseases, 15-19, September, 2004, Bialowieza, Poland, pp 91.

Deka A., Deka P.C., Mondal T.K. (2006) Tea. In: Parthasarathy, V.A., Chattopadhyay, P.K., Bose, T.K. (eds) Plantation Crops, Kolkata: Naya Udyog. pp. 1–147.

Deka M.K., Saikia G.K. (2011) Antifeedant and Growth Regulatory Effects of *Clerodendron Inerme* (Linn.) Gaertn. and *Lantana Camara* L. Extracts on Tea Mosquito Bug, *Helopeltis theivora* Waterhouse (Hemiptera: Miridae). *Pesticide Research Journal* 23(1), 32-35.

DeLong D. M. (1971) The bionomics of leafhoppers. *Annu. Rev. Entomol.*16: 179–210.

Gogoi B., Choudhury K., Sharma M., Rahman A., Borthakur M. (2012) Studies on the host range of *Helopeltis theivora. Two a Bud* 59: 31-34.

Gurusubramanian G., Rahman A., Sarmah M., Roy S. and Bora S. (2008) Pesticide usage pattern in tea ecosystem: their retrospects and alternative measures. *J. Environ. Biol.* 29 (6): 813-826.

Hazarika L. K., Bhuyan M. and Hazarika B. N. (2009) Insect pests of tea and their management. *Ann. Rev. Entomol.* 54: 267-284.

Hazarika L.K., Borthakur M., Singh K., Sannigrahi S. (1994) Present status and future prospects of biological control of tea pests in North East India. In: Proceedings of the 32nd Tocklai conference on towards sustainable productivity and quality, TRA, Tocklai Tea Research Institute, Jorhat, Assam, India, pp. 169-176.

Hazarika M. and Muraleedharan N. (2011) Tea in India: An overview. Two a Bud 58: 3-9. Jin M. and Baoyu H. (2007) Probing behaviour of tea green leafhopper on different tea plant cultivars. *Acta Ecologica Sinica* 27: 3973–3982.

Kalita H., Deka M.K., Singh K. (2000) Diet breadth of tea mosquito bug, *Helopeltis theivora* Waterhouse (Hemiptera: Miridae). *Crop Res.* 19 (1): 122-124.

Kalita H., Handique R., Singh K. and Hazarika L.K. (1995) Feeding behaviour of tea mosquito bug. Two a Bud 42:34-35.

Light S .S. (1927) Mites as Pests of the Tea Plant. *Trop. Agriculturist.* 68 (4): 229-238.

Loganathan S. (1992) Studies on the pink mite, *Acaphylla theae* (Watt) (Acarina: Eriophyidae) infesting tea in Southern India. Ph. D. Thesis, Bharathiar University, Coimbatore, India. pp. 137.

Mamun M.S.A. and Ahmed M. (2011). Integrated pest management in tea: prospects and future strategies in Bangladesh. *J. Plant Prot. Sci* 3 (2): 1-13.

Mathew G., Shamsudeen R.S.M. and Chandran R. (2005) Insect fauna of Peechi-vazhani wild life sanctuary, Kerela, India. *Zoos'Print Journal*, 20(8): 1955-1960.

Mound L. A. and Palmer G. M. (1981) Identification, distribution and host plants of the pest species of *Scirtothrips* (Thysanoptera: Thripidae). *Bull. Entomol. Res.*71: 467–479.

Mukherjee T. D. (1960) Destructive eelworms of tea seedlings. *Two a Bud* 7 (2): 5-6.

Mukhopadhyay A. and Roy S. (2009) Changing dimensions of IPM in the tea plantations of the north eastern sub-Himalayan region. pp. 290-302. In: V.V. Ramamurthy, G. P.Gupta and S. N. Puri (Eds.) *IPM strategies to combat emerging pests in the current scenario of climate change*. New Delhi, India: Entomological Society of India, IARI.

Muraleedharan N. and Cheng Z.M. (1997) Pests and diseases of tea and their management. *J. Plant. Crops* 25 (1): 15-43.

Muraleedharan N. (1991b) Entomology. pp. 35-48. In: Annual Report of UPASI Scientific Department. UPASI Tea Research Institute, Valparai.

Muraleedharan N. (2010) An IPM package for tea in India In: Glimpses of tea research, published by United Planters' Association of Southern India, Coonoor, pp. 147-169.

Muraleedharan N. and Radhakrishanan, B. (1989b). Recent studies on pest management in South India. *Bull UPASI Tea Sci. Dep.* 43: 16-29.

Muraleedharan N. and Selvasundaram R. (2002) An IPM package for tea in India. *Planters' Chron.* 4: 107-124.

Murthy R. L. N. and Chandrasekaran R. (1979) Effect of pest damage on quality of made tea. *Proc. PLACROSYM II* : 275-284.

Rabindra R. J. 2012 Sustainable pest management in tea: Prospects and challenges. *Two a Bud* 59: 1-10.

Ragesh G. (2013) New record of soapbush (*Clidemia hirta* (L.) Don, 1823) as a host and the biology of Tea Mosquito Bug, *Helopeltis theivora* Waterhouse (Miridae: Hemiptera) In: 2nd Global Conference on Entomology, November 8-12, 2013, Kuching, Sarawak, Malaysia.

Rahman A., Sarmah M., Phukan A.K., Roy S., Sannigrahi S., Borthakur M. and Gurusubramanian G. (2005) Approaches for the management of Tea Mosquito Bug, *Helopeltis theivora* Waterhouse (Miridae: Heteroptera). Proceedings, 34th Tocklai Conference.

Rao G.N. and Murthy R. L. N. (1976) Economics of tea pest control. *Bull. UPASI Tea Sci. Dep.* 33: 88-100.

Rao G.N. and Subramanian H. (1968).Control of mites in tea results in increased yields. *Planters' Chron.* 63: 412-413.

Rao G.N. (1969) Alternate hosts of tea mites. *Planters' Chron.* 64: 139.

Rau A.S. (1940) Helopeltis. *UPASI Tea Sci Dept. Misc.* Paper No. 2, pp. 1-8.

Roy S. and Rahman A. (2013) Distribution pattern of eelworm incidence in nursery tea soils of South Bank region of Assam. *Two a Bud*, 60 (1): 41-45.

Roy S. and Mukhupadhyay A. (2012) Bioefficacy assessment of *Melia azedarach* (L.) seed extract on tea red spider mite, *Oligonychus coffeae* (Nietner) (Acari: Tetranychidae). *Int. J. Acarology* 38 (1): 79-86.

Roy S., Gurusubramaniam G. and Mukhopadhyay A. (2008) A preliminary toxicological study of commonly used acaricides of tea mosquito bug (*Oligonychus coffeae* Neetner) of North Bengal, India. *Resistance Pest Management Newsletter* 18 (1): 7-8.

Roy S., Gurusubramaniam G. and Mukhopadhyay A. (2010) Nem based integrated approaches for the management of tea mosquito bug, *Helopeltis theivora* Waterhouse (Miridae: Heteroptera) in tea. *J. Pest Sci.* 83:143-148.

Roy S., Mukhopadhyay A. and Gurusubramaniam G. (2010a) Anti-mite activities of *Clerodendrum viscosum* Vemtemat (Verbenaceae) extracts on tea red spider mite, *Oligonychus coffeae* Nietner (Acarina:Tetranychidae), *Arch. Phytopathology Plant Prot.* 44 (16): 1150-1159.

Roy S., Mukhopadhyay A. and Gurusubramanian G. (2009b) Biology of *Helopeltis theivora* (Hemiptera: Miridae) on tea (*Camellia sinensis*) in sub Himalayan region. *J. Plant. Crops* 37 (3): 226-228.

Roy S., Mukhopadhyay A. and Gurusubramanian G. (2009a) Varietal preferences and feeding behavior of tea mosquito bug (*Helopeltis theivora* Waterhouse) on tea plants (*Camellia sinensis*). *Academic J. Entomol.* 2 (1): 01-09.

Saha D., Roy S. and Mukhopadhyay A. (2012) Seasonaln incidence and enzyme-based usceptibility to synthetic insecticides in two upcoming sucking insect pests of tea. *Phytoparasitica* 40: 105–115.

Saha D., Mukhopadhyaya A. (2013) Insecticide resistance mechanisms in three sucking insect pests of tea with reference to North-East India: an appraisal. *Int. J. Trop. Insect Sci.* 33: 46-70.

Sannigrahi S. and Talukdar T. (2003) Pesticie use patterns in Dooars tea industry. *Two a Bud* 50: 35-38.

Sarmah M., Basit A., Hazarika L.K. (1999) Effect of *Polygonum hydropiper* L. and *Lantana camara* L. on tea red spider mite, *Oligonychus coffeae*. *Two a Bud*. 46: 20-22.

Sarmah M., Rahman A., Phukan A.K. and Gurusubramanian G. (2009) Effect of aqueous plant extract on tea red spider mite, *Oligonychus coffeae* Nietner (Tetranychidae: Acarina) and *Stethorus gilvifrons* Mulsant. *Afr. J. Biotechnol.* 8 (3): 417-423.

Sasidhar R Sanjay R. (2000) Cultural control of tea mosquito bug. *Newsletter of UPASI Tea Research Institute* 10 (1): 4.

Seal D. R., Klassen W. and Kumar V. (2010) Biological parameters of Scirtothrips dorsalis Hood, on selected hosts. Environ. *Entomol*. 39: 1389–1398.

Selvasundaram R. and Muraleedharan N. (2000) Occurrence of the entomogenous fungus *Beauveria bassiana* on the shot hole borer of tea. *J. Plant. Crops* 28 (3): 229-230.

Selvasundaram R. and Muraleedharan N. (2003) Red spider mite-biology and Control. Hand Book of Tea Culture; UPASI Tea Research Institute, Valparai; pp. 18.

Sen A. R. and Chakrabarty R.P. (1964) Estimation of loss of crop from pest pests and diseases of tea from sample surveys. *Biometrics* 20: 492-504.

Somchowdhury A.K., Samanta A.K., Dhar, P.P. (1993) Approach to integrated control of tea mosquito bug. In: Proc. of International Symposium of Tea Science and Human Health, pp. 330-338.

Srikumar K.K., Bhat P.S. (2013) Biology of the tea mosquito bug (*Helopeltis theivora* Waterhouse) on *Chromolaena odorata* (L.) R.M. King & H. Rob. *Chilean J. Agric. Res.* 73 (3): 309-314.

Sudhakaran R. and Selvasundaram R. (2000) Incidence of tea mosquito bug on *Maesa indica*. UPASI Tea Res. Inst. *Newsl*. 8: 3.

Sudhakaran R., Muraleedharan N. (2006) Biology of *Helopeltis theivora* (Hemiptera: Miridae) infesting tea. *Entomon* 31:165-80.

Sundararaju D. (1993) Studies on the parasitoids of Tea Mosquito Bug, *Helopeltis antonii* Sign. (Heteroptera: Miridae) on cashew with special reference to *Telenomous* sp. (Hymenoptera: Scelionidae). *J. Biol. Control* 7:6-8.

Sundararaju D., Sundarababu P.C. (1999) *Helopeltis* sp. (Heteroptera: Miridae) and their management in plantation and horticultural crops of India. *J. Plant. Crops* 27:155-74.

Thu P.Q., Griffiths M.W., Pegg G.S., McDonald J.M., Wylie F.R., King J., Lawson S.A. (2010) Healthy plantations: a field guide to pests and pathogens of *Acacia* and *Eucalyptus* in Vietnam. Brisbane, Australia, Queensland Department of Employment, Economic Development and Innovation.

Watt G. and Mann H.N. (1903) *The Pests and Blights of the Tea Plant*. Calcutta: Gov. Printing Press. 429 p.

Winotai A., Wright T. and Goolsly J.A. (2009) Herbivores in Thiland on *Rhodomyrtus tomentosa* (Myrtacae), an invasive weed in Florida. *Florida Entomologist* 88(1):104-105. www.mothsofborneo.com

Zeiss M. R. and Braber K. D. (2001) Tea: Integrated Pest Management Ecological Guide. E-book. CIDSE, Vietnam.

Zhu J. G., Kuang R. P., Hu H. M., Tao T. and Zhu Q. Z. (1993) The development, reproduction and spatial distribution of lesser green leafhopper (*Empoasca flavescens*) on different tea cultivars. *Zoological Research* 14, 241–245.

Appendix

Stage wise IPM practices to be adopted

Crop stage and pest	Stage-wise IPM practice	
1. Nursery pests		
Aphids	Cultural practice	Manual removal of infested shoots
	Biological control	Allow predator populations to build up If pest persists spray neem formulation
	Chemical control	If pest persists spray recommended insecticides
Flushworm, Leaf roller and tea tortrix	Cultural practice	Manual removal of infested shoots
	Biological control	Allow populations of natural enemies to build up
	Chemical control	If pest persists spray recommended insecticides
Eelworm	Mechanical control	Heat treatment of soil at 60-65 ^{0}C
	Chemical control	Apply recommended chemicals to the soil
2. Young tea		
Aphids	Cultural practice	Removal of infested shoots by hand
	Biological control	Allow populations of natural enemies to build up If pest persists spray neem formulation
Flushworm, Leaf roller and tea tortrix	Cultural practice	Removal of infested shoots by hand
	Biological control	Allow populations of natural enemies to build up
	Chemical control	If pest persists spray recommended insecticides
Shot hole borer	Cultural practice	Avoid planting susceptible clones in the shot hole borer prone area Use cut stems of *Montanova* plants to trap adult beetles
	Biological control	Spray the formulations of the entomogenous fungus *Beauveria bassiana*
	Chemical control	Spray recommended insecticides
3. Mature tea		
Tea mosquito bug	Monitoring	Regular field assessment of percentage infestation
	Cultural practice	Black plucking, weed control Removal of stalks containing eggs while plucking
	Chemical control	Spraying in the early morning or evening hours with recommended insecticide

Tea: Technological Initiatives, pp. 195-239
New India Publishing Agency, New Delhi, India
Edited by Niladri Bag, Arundhati Bag and L.M.S. Palni

7

Tea Research for Darjeeling Tea Industry - Various Aspects

Anil Kumar Singh, J.S. Bisen, R.K. Chauhan, Mrityunjay Choubey, R. Kumar and N. Kumar

Abstract

*Tea (*Camellia sinensis *L.) is a plantation crop having a wide distribution extending from low elevation to high exceeding 2000 m. The demands and consumption of organically grown residue free Darjeeling tea is increasing day by day due to health consciousness and environmental awareness. The average productivity of Darjeeling tea is far below than the average national tea productivity. Reasons for low productivity of Darjeeling tea may be due to high altitude/elevation; temperature/climate and geographical location in addition to the old age of tea bushes and low plant density. In the past indiscriminate use of synthetic fertilizers has led to the pollution and contamination of the soil, water basins, destroyed micro-organisms and friendly insects, making the crop more prone to diseases and reduced soil fertility. Following Good Agriculture Practices (GAP), sustainable and quality production of Darjeeling tea may be increased through eco-forming without any deterioration of surroundings of plants. Pruning and plucking are important agro-cultural operations for maintenance of the tea crop production and productivity. The physiological parameters such as photosynthesis, stomatal conductance, vapor pressure deficit etc. also play a significant role in maintaining the tea crop production. The photosynthesis process continues during winter also though tea bushes become dormant when minimum temperature falls below 12 ^{0}C. Higher rate of photosynthesis is observed in cool conditions. Although shoot growth diminishes as the temperature falls and ceases below 12 ^{0}C but photosynthesis accompanied by slow rate of respiration continues even at lower temperature. Vapour Pressure Deficit is observed high in summer months while leaf water potential was lowest in summer and highest in rainy season. The decrease in stomatal conductance and transpiration were more pronounced in general during moisture stress* *period. Stomatal conductance is observed normally inversely proportionate to VPD. In Darjeeling, tea plantation started through seed propagation. In general, there are three main mode of propagation in tea i.e. through seeds, vegetative propagation and grafting. Flavour retention has been a serious concern among the Darjeeling tea planters for which different breeding approaches are adopted. Conventional breeding is observed the best approach for improving many characteristics for conservation of genetic diversity which is highlighted in this chapter. Quality of Darjeeling tea is unique in the world and known for its muscatel flavour. The concept of quality dominates over the yield in Darjeeling hills. However, agricultural practices like plucking and pruning including application of copper affect the Darjeeling tea flavor and quality. A biochemical perspective of these practices including organic farming has also been described in the chapter. Tea being a perennial crop and the harvesting of new shoots at regular intervals for commercial production of tea, needs continuous supply of nutrients for faster recovery. Hence organic manures, bio-fertilizers and bio-organic formulations are considered the better options to meet the continuous requirements of nutrient as well as to keep the pest population below thresh hold level. This chapter is based on the experimental findings and the technical information gathered in the past on several aspects on tea cultivation such as tea bush physiology, climatic condition, plant improvement, soil management especially through organic resources, plant protection, cultural practices and quality aspects.*

Introduction

Tea plant belongs to the genus *Camellia* of the family *Theaceae*. There are large number (more than 100) recognized *Camellia* species. Tea which is cultivated and produced from the Darjeeling region in West Bengal, India is called Darjeeling tea. The tea is available in the form of black, green, white and oolong. When properly brewed, it yields a thin-bodied, light-colored infusion with a floral aroma. The flavor includes a tinge of astringent tannic characteristics, and a musky spiciness described as "muscatel". Although Darjeeling black teas are marketed commercially as "black tea", almost all of them have incomplete oxidation (<90%), so they are technically more Oolong than black. After the enactment of Geographical Indications (GI) of Goods (Registration & Protection Act, 1999) in 2003, Darjeeling tea became the first Indian product to receive a GI tag, in 2004-05 through the Indian Patent Office. Tea planted in the district of Darjeeling had begun during 1841 by Dr. Campbell, a civil surgeon of the Indian Medical Service. Campbell was transferred to Darjeeling in 1839 and used seeds from China to begin experimental tea planting, a practice that he and others continued during the 1840s. The government also established tea nurseries during that period. Commercial initiatives began during the 1850s. Flavour and quality of Darjeeling tea makes it the 'unique in the world'. Darjeeling teas produced in the 87 tea gardens covering 18000 ha of land. Out of 87 tea gardens more than 60% have been converted to organic. Darjeeling tea contributes only 1% of the total tea produced in India with the annual production of 10.09 Million kg of orthodox tea both black and green tea including a small amount of white tea.

Tea is one of the most popular drinks and also considered as a heritage non-alcoholic beverage of the world. It is one of the economically important small, evergreen woody plantation crop predominantly grown in the humid tropical and subtropical regions. Nestling in the foothills of the snow covered Himalayan range, for proper maintenance of the health of tea bushes and to obtain high yield, a balanced fertilization and manuring is necessary at certain intervals throughout the year. It is also vital to integrate the various factors relating to climate, soil, plant and cultivation for achieving the maximum return from the investment on fertilizers and productivity of the soil. Photosynthetic efficiency of tea leaf in fixing the atmospheric carbon dioxide and partitioning of assimilates are governed principally by the environmental conditions, the management practices and the genetic potential of the plant (Marimuthu *et al.*, 1994). Chlorophyll is essential components for photosynthesis. Glaucousness in leaves is caused by the deposition of epicuticular wax and may be dependent on the changing environment (Ankunda, 1990). The epicuticular wax effectively reduced water loss due to transpiration (Ghosh Hajra, 2001).Tea is a plantation crop having a wide distribution extending from low elevations to levels exceeding 2000 m (Wijeratne *et al.*, 2007). All physiological processes including photosynthesis need suitable temperature. Under non-limiting conditions of light and carbon dioxide, photosynthesis increases with the rise of temperature till it reaches maximum and thereafter declines photosynthesis of tea leaves is influenced by the temperature, day length, carbon dioxide concentration, genetic potential of given cultivar, and physiological maturity (Raj Kumar *et al.* 2002).

1. Physiological Characteristics of Tea plant: Effect of External and Internal factors

Darjeeling tea plants experience various types of climatic conditions in the Darjeeling hills such as low temperature, low soil moisture in winter foggy climate, high humidity and low levels of solar radiation. All physiological processes including photosynthesis need suitable temperature. Under non-limiting conditions of light and carbon dioxide, photosynthesis increases with the rise of temperature till it reaches maximum and thereafter declines.

1.1 Climate

In Darjeeling, mean maximum air temperature ranges from around 16 ^{0}C in February to 24 ^{0}C in July; a mean minimum temperature of 4.5 ^{o}C was observed in January. A rapid increase of temperature takes place during March and April owing to the warmer air from the plains. In May, the southerly winds reach the hills which are at times are very high. November to February are almost rainless and the light showers which fall in December and March occur when shallow depressions are passing eastward over the plains. In October, northerly winds begin, cloud is much less than the previous months and rainfall occurs mainly owing to cyclonic storms that generally re-curve towards North Bengal at the end of the season. Based on the agro-climatic conditions, the month of April is considered as pre-monsoon, June to August as monsoon, October to December as post monsoon and January to February as winter. Further, December to April could be considered as a moisture-stress period. The daily maximum net photosynthetic rate (Pn) was noticed in clones and minimum in old china bush in canopy depth (from the top of bush) 0-10 cm during autumn (September) when humidity was (85%), air temperature (24°C) and photosynthetic photon flux density (PPFD) (1270 μ mol $m^{-2}s^{-1}$) were moderate (Kumar *et al*, 2015). In Darjeeling, the shoot extension growth stops at monthly mean maximum and minimum temperature of 18°C and 10°C respectively in November and it starts flushing in the middle of March when maximum and minimum temperature exceeds 20°C and 12°C respectively. In Kenya, the monthly mean maximum and minimum temperatures rarely exceed 24°C and 11°C respectively at any time of the year, but the tea plants flush throughout the year and produce annual yields of the same order as plants in many warmer regions. The concept of thermal duration (degrees-days) helps tea growers to determine important plucking policies such as plucking rounds for different periods of the year based on their temperature variation. Accuracy of such predictions depends to a large extent on the precision of the estimation of base (threshold) temperature (Tb) and the absence of other limiting factors for growth such as soil water and VPD. Carr (2000) also reported that small differences in Tb can have relatively large effects on rate of shoot development and extension at high altitudes where ambient temperature (Ta) is low. The other practical implication limiting the use of the thermal duration concept for deciding plucking rounds is the presence of a mixture of genotypes in a given tea plantation as Tb may vary between genotypes.

1.2 Light Intensity

Photosynthetic activity mostly depends on photosynthetic photon flux density (PPFD). Pn significantly increased with the increase of PPFD till it reached the light saturation at 1200 μmol $m^{-2}s^{-1}$ (Barman *et al* 2005). At increasing irradiance 1400-1800 μ mol $m^{-2}s^{-1}$, Pn decreased gradually from lower to higher PPFD and effect between them was insignificant. The results thus indicate that beyond 1200 μ mol $m^{-2}s^{-1}$ PPFD, tea leaf undergoes an irradiance stress. The light saturation points for assimilation of maximum CO_2 in tea leaf were 1100-1500 μ mol $m^{-2}s^{-1}$ in Kenya (Smith *et al.*, 1993), 900-1200 μ mol $m^{-2}s^{-1}$ in South India (Raj Kumar *et al.*, 1998), 1340 μ mol $m^{-2}s^{-1}$ in Darjeeling (Ghosh Hajra and Kumar, 1999), 500-800 μ mol $m^{-2}s^{-1}$ in Sri Lanka (Mohotti and Lawlor, 2002).

1.3 Photosynthesis and Related Parameters

Tea bush obtains its food from the atmosphere through a process called photosynthesis occurs using gaseous carbon dioxide in the presence of chlorophyll and light. Carbon dioxide and water react to synthesize sugar in this process. If there is insufficient light, photosynthesis will be reduced or even stopped. An efficient plant manufactures more sugar than is needed for current growth. The excess sugar is converted into starch and stored mainly in the roots. The stored starch is available for use in an emergency such as when new shoots are produced after pruning, or when the rate of photosynthesis drops at night. The maximum rate of photosynthesis (Pn) was showed during autumn followed by winter and summer seasons. Although maximum humidity and maximum soil moisture had a positive influence on Pn, low PPFD and sunshine hours (less than 2.0 h day^{-1}) affect Pn activity. The limiting factor for Pn was probably the low sunshine hours. The value of Pn was higher when PPFD increased from 950 μ mol $m^{-2}s^{-1}$ (winter) to 1350 μ mol $m^{-2}s^{-1}$ (autumn). Lowest PPFD was recorded during rainy season which in turn affected the Pn. A significant positive correlation exists between Pn and PPFD. In winter season, PPFD was moderate, sunshine hours and relative humidity was high. Ghosh Hajra (2001), Ghosh Hajra and Kumar (2002b) indicated that the tea plant effectively conserves water by stomatal closure and was supported by lower levels of stomatal conductance (gs) during the dry season. An instantaneously water use efficiency (WUE) was positively correlated with Pn which is in conformity with the findings of Read *et al.* (1991). The decrease in stomatal conductance and transpiration (E) was pronounced in summer when both atmospheric and soil moisture were low and demand for water was more. Lower values of gs observed in summer and winter. The highest value of gs was recorded in rainy season. The gs at different seasons more or less followed the patterns of changes in transpiration rate. Transpiration in general increased with increase in leaf temperature and PPFD in summer, autumn and winter as it was observed in the diurnal study (Ghosh Hajra and Kumar, 2002a). Stomata normally close in response to increasing vapour pressure deficit (VPD). Maximum VPD was recorded in summer (3.34 kPa) and minimum in winter season (1.32 kPa). In general, a reduction in g_s with higher VPD values was observed which is in conformity with the findings of Squire and Callender (1981).

Maximum leaf water potential ($ø_L$) was noticed in rainy season followed by summer season. Highest $ø_L$ was recorded in the early morning hours in all the seasons. A gradual reduction of $ø_L$ till 12.00 hrs was observed in winter by Kumar *et al.* (2014). However, in general a lower value of $ø_L$ in all seasons was observed during midday. There are some differences in the diurnal leaf water potential patterns of pruned and un-pruned tea clones. The leaf water potential of the pruned plants is constantly higher than the un-pruned tea clones.

During the dry periods of many tea growing regions of the world, vapour pressure deficit could rise to levels which would influence gs, shoot $ø_L$ and the rate of shoot initiation and extension (Squire and Callander. 1981). In addition, VPD influences these key processes of yield attributes/formation of tea even during periods when the soil is wet. The linear relationship between shoot extension rate and temperature breaks down at higher VPD (Squire and Callander. 1981). During wet periods with frequent rain, shoot $ø_L$ of tea has an inverse, linear relationship with VPD (Williams, 1971; Squire, 1976). This probably operates through the influence of VPD on transpiration, which increases with increasing VPD causing a decrease in shoot $ø_L$. During these wet periods, VPD did not exceed 2 kPa and shoot $ø_L$ did not fell below –1 Mpa.

The chlorophyll is the most important plant pigment playing a vital role in determining the photosynthetic efficiency and productivity of the plant. The Chl content in leaves vary with the day length, irradiance and radiation quality, temperature and nutrient status of the soil. Total Chl content was highest during rainy season and lowest in summer season (moisture stress period) as it was observed by Wickremasinghe and Perera (1966). Certain clones had higher value of chlorophyll content and the same was lowest in old china plant. Decline in Chl due to water stress has also been reported in tea (Rajasekar *et al.* 1991). Chl content was positively correlated with Pn in the case of clones and found negative in seasons as reported in South Indian tea by Rajkumar *et al.* (2002). The physiochemical properties of Epicuticular wax (EW) have been well studied due to its unique properties and fundamental role as the barrier between the leaf and its environment (Baker, 1982). In the present study, EW was found highest during summer and lowest in rainy season, which is in conformity with the findings of Rajasekar *et al.* (1991). An increase in the deposition of EW of leaves during summer exhibits higher ability in the retention of tissue water and reduced water loss due to transpiration that is beneficial during drought.

During the growing season from spring to summer (March to August), photosynthates are supplied to the growing shoots by all the leaves of maintenance foliage. By end October, the lower layer of leaves send photosynthates downwards to the roots and the upper layer continues to support the top growth. But with the fall of ambient temperature in November, all the layers send photosynthates downwards to the roots which accumulate as starch. Barman *et al.* (1998), after estimating the root starch of all TV clones, advocated pruning between end December and early January which supports the present study that the maximum yield was from plots pruned during 1st week of December followed by November pruned plots. The starch accumulation increased gradually from October and reached the peak in December. The quantity of starch reserve was highest around 16.0% in December and lowest in September which was observed by Kumar *et al.* 2012). Considering the highest accumulation of starch during December, the pruning operation can be safely done in December. The production of new shoot after pruning, skiffing or plucking is dependent upon the starch reserves of the plants. If the existing foliage on a skiffed or plucked plant is highly efficient, the starch, which is used, will replenish quickly. But if it is inefficient the amount of stored starch decrease every time the bush is plucked to the point when shoot production could cease whenever the plant is subjected to sub optimal growing conditions. It is therefore essential that the tea bush be allowed to retain at all times an adequate amount of efficient foliage. On a plucked tea bush this foliage is called the maintenance foliage or layer. It is the layer of mature leaves below the plucking surface. If the maintenance layer is too shallow or too sparse then the rate of sugar production may be too slow to permit the accumulation of starch reserves and hence there will be a lot of die backs, which decrease the number of plucking shoots and formation of more banjhi and crow's feet.

2. Tea Improvement through Conventional Approaches

In India, most of the tea plantations were established by using seeds. Continuous seed propagation has produced tea populations with different yield and quality properties, reflecting wide genetic variation. However, like in many cross breeding plants, tea is highly heterozygous, with most of its morphological and biochemical traits showing continuous variation, which are influenced by environmental factors, these adversely affecting productivity and quality of tea.

Generally conventional breeding aims to produce a breeding population that is highly variable for traits. This is accomplished by identifying parents having characters that harmonize each other, the strengths of one parent having the capacity to augment the shortcomings of the other, and then cross-pollinating the parents to initiate sexual recombination. The genetic mechanisms that drive sexual recombination operate during gamete formation by means of meiosis. The main characteristic of sexual reproduction is that it allows and assures all of the characters that vary between the parents are free to segregate in new and improved combinations in the offspring. The other step of breeding includes selection among the segregating progeny for individuals that combine the most useful traits of the parents. Hence, conventional breeding is essentially the normal mating process, but it is manipulated through researcher choice of the parents and selection of their offspring so that evolution is directed toward production of plants with characteristics closely suited to human needs (Manshardt, R. 2004). It is relatively inexpensive and technically simple. Conventional breeding is best approaches for improving many traits controlled by many factors. The morphological traits such as shoot length, root length, shoot- root volume, number of leaves and branches can be used to distinguish between the camellia species. The information on morphological diversity can also be used for future breeding programmes. Clone T-78 performed best in respect of shoot length, number of branches, number of leaves and shoot volume respectively as compare to other clones under Darjeeling conditions (Choubey *et al.*, 2013). .

There are three main means of tea propagation: (i) seeds (ii) vegetative propagation and (iii) nursery grafting.

2.1. Seeds

Traditionally tea plantation is established through propagation of seeds. A matured and healthy seed while attached to the plants are collected from the tea orchards. After eliminating the very small sized seeds, the remaining seeds are transferred to a tank filled with water for sinker-floater test. The sinker seeds are taken out of water and examined for mechanical and pest damage. The common practice is to cut open a sample of 100 seeds from the batch to examine starred, shrunken seeds or damaged seeds by pests and diseases. Floater seeds are discarded as these seeds which have dried cotyledons, normally fail to germinate. The seeds are packed properly after grading and sorting. Moisture content of the packing material may varies from 10 to 30 %. Seeds are spread in layers along with some packing materials, and each layer is separated from the one on the top by a thin sheet of paper (Mondal, 2014). One kilogram of graded and sorted seed may contain about 500 seeds depending on the size of the grader used. After 30-45 days, the germinated seeds are transferred to the polythene sleeves and kept under a shaded nursery for rearing.

2.2. Vegetative propagation

In the case of tea, until about 60 years ago seeds were the only method of propagation. Though, due to the cross breeding, seedlings show a wide variability for yield and quality attributes, and ultimately it forced researcher to find some alternatives. However, due to slow growth, vegetative propagation method could not serve the purpose of rapid multiplication. Meanwhile, vegetative propagation was attempted in several parts of the tea-growing areas around the world

(Tunstall 1931a & b; Tubbs 1932; Wellensiek 1933); although, standardization of tea propagation technique of single-leaf cuttings, practised presently, took a long time to be successful. The faster propagation by single leaf cutting was developed simultaneously in India, Sri Lanka and Indonesia (Mondal 2011). This propagation technique was further optimized to fit the commercial venture. Cuttings from about 100 days old shoots with semi-hard wood are usually collected and immediately subjected to fungicide and rooting-hormone treatment and brought to the nursery for root initiation for 40–60 days depending on the location, planting material, etc. Then, the successful rooted cuttings are propagated to polythene sleeves that filled with virgin soils consisting adequate soil pH, water-holding capacity and are kept in the nursery. The plant becomes ready for plating in field in about 15 months after proper hardening.

2.3. Grafting

Grafting is as alternative technique of propagation, gained significant popularity in the recent past. The main objective of grafting is to develop composite plants comprising the shoot system of a good quality and high yielding clone and the root system of drought tolerant clones to enhance yield per unit area. In this technique, nodal cuttings of 3-4 mm diameter with a healthy leaf and axillary bud are generally taken for both rootstock and scion. Generally, Scion of a good quality clone is grafted on rootstock which is drought resistant. On grafting, the scion and stock affect each other and as a result composite plants combine both the characters, resulting in enhance of yield with better quality than either of the un-grafted clone (Mondal, 2014). Recently, a modified improved 'second-generation' grafting had been developed, where a tender *in vitro*-derived shoot was grafted on the young seedlings of tea, which had an additional advantage over conventional grafting due to the presence of tap-root system (Prakash *et al.* 1999). With the escalating demand for high and quality tea planting materials, vegetative propagation with nodal single-leaf cuttings relics the best choice in the tea industry.

In brief, conventional approach for tea improvement can become appropriate depending on the breeding objectives.

3. Organic Tea Farming in Darjeeling Hill

Organic tea cultivation in Darjeeling hills is getting popularity day by day due to growing health awareness of consumers, fetching good prices and high demand in International market. Maintenance of the present tea crop productivity in the changing scenario of climate and mode of cultivation is a big challenge for every stakeholder of Darjeeling tea industry, while it has been established that the 12-30% reduction in yield is quite natural after conversion to organic

(Bisen & Singh, 2012). At present the planters of this region are worried about the changes in climatic pattern and bio-diversity. It is necessary to overcome such problems by maintaining eco-system. The average productivity of Darjeeling tea is far below than the average national tea productivity. Reasons for this low productivity of Darjeeling tea are high altitude/elevation; temperature/climate and geographical location in addition to this age of tea bushes and low plant density are most important factors for low productivity.

Organic agriculture is not just to cut off the application of agrochemical, it is an ecological production management system that promotes and enhances biodiversity, biological cycle and soil biological activity. It is based on minimal use of off farm inputs which restore, maintain and enhance ecological harmony (Gold Mary, 2007) through different management practices.

3.1 Basic tools in Organic farming (Bhattacharya, P. 2004).

3.1.1 Land and crop management

The purpose of soil management in tea area is to develop suitable soil husbandry systems, which will ensure satisfactory root growth, crop yield and arrest the loss of soil tilth.

3.1.2 Farm waste recycling

Farm waste such as cow dung, sheep, goat, poultry manure, leaf fall and livestock litter have considerable amount of organic matter is also added through recycling. The hygienic disposal of organic wastes by recycling is an environmentally sound and economically viable technology resulting valuable input in organic farming.

3.1.3 Weed management

Weeds should be removed manually by hand weeding. If it is not possible bio-herbicides may be used.

3.1.4 Biological control of pests

In organic tea cultivation preventive measure is most important than curative measure. If pest incidence occurs, non-toxic biological agents such as bio-pesticide can be applied.

3.1.5 Integrated plant nutrient management

Animal dung, crop residues, green manure, bio-fertilizer and bio-solids from agro-industries and food processing wastes are some of the potential sources of nutrient. Accumulations of different organic sources available in the farm are useful resources, for maintenance of soil fertility.

3.1.6 Integrated farming system approach

The integral feature of organic farming is livestock management, crop management and other resource management for optimizing farm profit and ecological balance.

Measures of basic tool to achieve the goal of organic cultivation should be strictly followed. Since organic production warrants the cultivation in the absence of agrochemicals, it involves a careful selection of components of farming system keeping the local resources, agro climatic features and socio-economic structure for the development of package of practices. Organic sources availability is limited, but under different constraints of soil and climatic factors, use of organic inputs proved to be more profitable compared to use of agrochemical.

3.2 The Most Often Promoted Issues on Organic Farming are (IFOAM, 1996):

- To produce food of high nutritional quality.
- To work with natural system.
- To encourage and enhance biological cycle within the farming system, involving microorganism, soil flora and fauna, plants and animals.
- To maintain and increase the long term fertility of soils.
- To use as far as possible renewable resources in locally organized agricultural system.
- To work as much as possible within closed system with regard to organic matter and nutrient elements.
- To maintain the genetic diversity of the agricultural system and its surroundings, including the protection of plant and wildlife habitats.

3.3 Advantages of Organic Farming

- Organic manures produce optimal conditions in the soil for high yield and good quality crops.
- It supplies all the nutrients required by the plant.
- It improves plant growth and physiological activities of plants.

- Improves the soil physical properties such as granulation and good filth, giving good aeration, easy root penetration and improved water holding capacity, stabilizes soil structure, lowers bulk density etc.
- It improves the soil chemical properties such as supply and retention of soil nutrients and promotes favorable chemical reactions.
- Supplies energy for soil organisms, increases microbial populations and their activities, source and sink for nutrients, ecosystem resilience etc.
- Organically grown crops are believed to produce healthier and nutritionally superior food.
- There is an increasing consumer demand for agricultural products which are free of toxic chemical residues, hence organically produced food are of great demand.
- Organic farming helps to prevent environmental degradation.

3.4 Constraints in Organic Farming

- Limited availability of sufficient quantity of locally available quality organic inputs like farmyard manure, compost, vermicompost etc.
- Heavy metal content of urban compost.
- Lack of awareness on the impact of organic farming.
- Unavailability of organic package of practice based on locally available inputs.
- Risk of low production in initial years of organic farming.
- High cost of present inspection and certification system is not affordable by most of the farmers.
- Lack of information regarding global demand and supply.
- Slow rate of organic matter mineralization in places like Darjeeling hill.

4. Bio-Fertilizer (BF) in Tea Cultivation

The deleterious impact of continuous use of chemical fertilizers alone in agriculture was realized in past. Application of chemical fertilizer as such without conservation of soil fertility not only results in depletion of soil nutrient reserves but it also disrupts the biological eco balance of soil - plant system, Singh *et al.*, 2011). Due to intensive cultivation of tea and indiscriminate use of nutrient inputs, there is widening gap between removal and supply of nutrients. Bio-fertilizer generates plant nutrients like nitrogen and phosphorus through their activities in soil or rhizosphere and make available to plant in a gradual manner. BFs are gaining momentum recently due to the increasing emphasis on maintenance of soil health, minimize environmental pollution and cut down on the use of chemicals in tea cultivation. In Darjeeling, where tea cultivation is converted to organic amendment has been found more effective to augment the benefits of microbes (Singh *et al*., 2007). Chemical fertilizers are banned so cost of cultivation per ha through organic resources is very high and not affordable for small and marginal farmers. BF is the ideal inputs for reducing the cost of cultivation and for practicing organic farming. BF is low cost, renewable sources of plant nutrients which supplement chemical fertilizer and referred to sustainable or eco-friendly system. Besides above facts, the long term use of bio-fertilizers is economical, eco-friendly, more efficient, productive and accessible to marginal and small farmers over chemical fertilizers.

4.1 Important Bio-fertilizer used in the Tea Cultivation

4.1.1 Nitrogen Bio-fertilizer

4.1.1.1 Azospirillum

It is chemoheterotropic and associative (with some grasses) in nature. It fixes nitrogen in an environment of low oxygen tension. By producing growth regulating substances, fixes 20-40 Kg N ha^{-1}. The commercially available *Azospirillum* can be used at the rate of 5g/seedling. The use of plant growth promoting rhizobacteria such as *Azospirillum* as bio-fertilizers has been inconsistent (Boddey and Dobereiner, 1982; Elmerich, 1984). Use of these bio-fertilizer results in enhanced mineral and water uptake, root development, vegetative growth and 15-30% increase in crop yield. Azospirillum is recommended for rice, millets, maize, wheat, sorghum, sugarcane, tea and co-inoculants for legumes.

4.1.1.2 Azotobacter

This is a non-symbiotic free living nitrogen fixing bacterium which fixes nearly 10-15 Kg N ha^{-1} $year^{-1}$ depending on the availability of carbon sources (Mishustin and Shilnikova, 1969). It is very predominant in most of Indian soils, having pH 5.8 to 7.8. The ability to fix elemental nitrogen is a vital physiological characteristic of *Azotobactor* species. It increases the yields which are partly attributed to the growth promoting substances such as gibberalins, cytokinins, Indole acetic acids etc. (Bolton *et al* 1993), and suppress/controls plant pathogens (Pal and Jalali 1998). It is widely used as a bio-inoculants in India for various crops like rice, wheat, millets, cotton, tea, vegetable, sunflower, mustard and flowers.

4.1.2 Phosphate mobilizing Bio-fertilizer (PMBF)

Symbiotic relationship between phosphate solubilizing bacteria (PSB) and plants is synergistic in nature as bacteria provides soluble phosphate and plants supply root borne carbon compounds (mainly sugars), that can be metabolized for bacterial growth. Use of phosphate solubilizing microorganisms (PSMs) can increase crop yields up to 70 percent. Combined inoculation of arbuscular mycorrhiza and PSB give better uptake of both native P from the soil and P coming from the phosphatic rock. The commercially available PSB can be used at the rate of 5g/seedling of tea. It may be used for all crops by seed treatment, seedling dipping and soil application. Large amount of P applied as fertilizer enters into the immobile pools through precipitation reaction with highly reactive Al^{3+} and Fe^{3+} in acidic, and Ca^{2+} in calcareous or normal soils. Efficiency of P fertilizer throughout the world is around 10-25% and concentration of bio-available P in soil is very low reaching the level of 1.0 mg kg^{-1} soil. Soil microorganisms play a key role in soil P dynamics and subsequent availability of phosphate to plants. Population of PSB depends on different soil properties (physical and chemical properties, organic matter, and P content) and cultural activities. PSB is beneficial with different agrochemicals in tea crop. Bacterial genus *Bacillus* and *Burhkolderia* have been identified as dominant bacterial genera present in tea rhizosphere of Darjeeling hills. The isolated potential phosphate solubilizing bacteria may be used as bacterial fertilizer for phosphate nutrition of organic tea, production of phoph-compost in Darjeeling hills (Panda, *et al.,* 2012).

4.1.2.1 Vesicular Arbuscular Mycorrhiza (VAM)

The fungi which are considered to be obligate symbionts, described as "a universal plant symbiosis". They are found to be present in every taxonomic group of plants and the beneficial effects are highly pronounced in acidic soils. In Darjeeling region most of the soil suffering from low level of available phosphorus. In such soil an efficient mycorrhizal association can increase phosphorus availability and crop yield. VAM enhance uptake of P, Zn, S and water, leading to uniform crop growth and increased yield. It also enhances resistance to root disease and improve hardness of transplant stock. The practical use of VAM fungi seems to be more appropriate as they are effective in overcoming the stress conditions like draught, disease incidences and deficiency of nutrients.

4.1.3 Compost catalyst

Some saprophytic microorganisms which are capable of decomposing organic matter at a faster rate for example *Aspergillus, Penicillium, Trichoderma* are cellulolytic fungi which break down cellulose of plant material. Use of these microorganism accelerate the natural decomposition process and reduce the composting time 4-6 weeks.

4.1.4 Potassium solubilizing bio-fertilizer (KSB)

Some organisms like *Bacillus mucilagenosus* (rod shaped , Gram -ve , spore former), *Frateuria aurantia* (Gram-ve , rod shaped) *Pseuodomonas putida* (Gram - ve, rod shaped) have been found to be useful for weathering of K-bearing minerals (Chandra, 2005; Bhattacharyya and Tandon, 2012). It has been found to be capable of mobilizing potash (e.g. by secretion of slime in case of *B. mucilagenosus*) and may enhance crop productivity.

4.1.5 Zinc Solubilizing bio-fertilizer (ZSB)

Some microorganisms (e.g. species of *Pseudomonas, Bacillus*) are capable of solubilising insoluble Zinc from Zinc sulphide, Zinc Oxide and Zinc Carbonate e.g. *Pseudomonas fluorescence.* Recently these bio-fertilizers are being promoted in zinc deficient soils.

4.1.6 PGPR bio-fertilizer

The group of beneficial root associative bacteria (N-fixers/P- solubilisers/ Siderophore Bacteria/Sulphur Oxidizing bacteria etc.) that stimulates the growth of plant is termed as Plant Growth Promoting Rhizobacteria. PGPRs play significant role in nutrient cycling.

4.2 Methods of Bio-fertilizer application

Bio-fertilizers are carrier based preparations containing beneficial microorganisms in a viable state intended for seed or soil application and designed to improve soil fertility. Methods recommended for bio-fertilizer application are as follows:

4.2.1 Seed inoculation

- About 3-4 kg ha^{-1} BF is applied for seed treatment in case of tea (Bhattacharyya and Tandon, 2012).
- Prepare the slurry (slurry with 5% Jaggery) to stick the bio-fertilizers on seeds
- For homogenous coating mix the seeds with hand and dry them in shade.
- Showing of the inoculated seeds should be done immediately.

4.2.2 Seedling treatment

- Make the suspension of required quantity of bio-fertilizer in water (1: 10 ratio) in a container.
- Soak the roots of seedling in the suspension for 8-10 minutes after that transplant the seedlings immediately.

4.2.3 Soil treatment

- Prepare the mixture of 25 kg of bio-fertilizer in 100-120 kg soil/ compost.
- The bio-fertilizer mixture in tea field may be applied around tea bushes by placement method at a depth of 10 cm (Bagyalakshmi *et al.*, 2012).

4.2.4 Optimum conditions for higher efficacy of bio-fertilizers (Biswas, & Mukherjee, 2003)

- Soil organic matter – Minimum 0.5%
- Soil Temperature – Optimum 28-30 ^{0}C
- Soil moisture – Optimum 30-40%
- Soil Reaction – Usually pH 6.5 to 7.5 is ideal.
- Nutritional status of soil – optimum in P, K, Ca, B, Mo & low in N.

4.2.5 Advantages of bio-fertilizers applications

- Increase availability of nitrogen by biological nitrogen fixation and by mineralization of organic matter.
- Increase plant growth and help to maintain soil fertility by release of enzymes, antibiotics and hormones.
- Increase crop yield by 10-50%, N fixers reduce depletion of soil nutrients and provide sustainability to the farming system.
- Control soil borne crop diseases.
- Improve physical, chemical and biological properties of soil.
- Help to survive and proliferates beneficial micro-organism in soil.
- Provide residual effect for subsequent crops.
- Help in recycling /decomposition of organic wastes.
- Reduce the dependency on chemical fertilizers and save from harmful effect of fossil fuel base chemical fertilizers.

4.2.6 Constraints in the use of bio-fertilizers

- In India, there was lack of sufficient region specific strains and availability of strains for tea crop specific bio-fertilizers.
- Unavailability of suitable carrier due to which shelf life of bio-fertilizers is short which is a major constraint. Indian conditions, where extremes of soil and weather conditions prevail, there is yet no suitable carrier material identified capable of supporting the growth of bio-fertilizers.
- Bio-fertilizers are sensitive to mutation during fermentation due to environmental conditions and ability to survive in broth.
- In recent years, most of the farmers in India are not aware of bio-fertilizers, their usefulness in increasing crop yields sustainably.
- Inadequate knowledge and unavailability of technically qualified person, who can handle the technical problem, technology transfer to the farmers.
- Due to corrupt marketing, there was poor quality and high cost bio-fertilizers are available to the farmers. So there is challenging goal to supply of cheap and best bio-fertilizer to the farmers.
- Most of the region, agriculture input demands are seasonal only in few months of year. So production units were not sure for their demands.
- Due to short shelf life of bio-fertilizer resource generation is limited.
- Soil and climatic factors are more responsible for microbial growth, multiplication and survival such as pH, fertility status, temperature, moisture and presence of toxic elements.

4.2.7 Quality of bio-fertilizer

Ministry of Agriculture, Govt. of India has made quality control specification of different bio-fertilizers under Fertilizer Control Order (FCO). It is always advised to use bio-fertilizers having high population (1 x 10cfu g^{-1}) for getting success. As tea is grown in strongly acidic soil with pH around 5.0, it is essential to use acid resistant strain so that they can survive and maintain biological activity. There is a need to isolate of native strains from local soil and detailed characterization and screening of isolates from tea rhizosphere in terms of N-fixation, phosphate solubilization, secretion of hormones, siderophore etc (Mazumdar *et al.*, 2007).

4.3 Soil and Nutrients Management through Good Agricultural Practicies (GAP)

The Food and Agricultural Organization of the United Nations (FAO) uses good agricultural practices as a collection of principles to apply for on-farm production and post-production processes, resulting in safe and healthy food and non-food agricultural products, while taking into account economical, social and environmental sustainability (FAO GAP Principles, 2012). GAP requires maintaining a common database on integrated production techniques for each of the major agro-ecological area, thus to collect, analyze and disseminate information of good practices in relevant geographical contexts.

4.3.1 Important issues promoted under GAP are

- To work as much as possible with in closed system with regard to organic matter and nutrient elements.
- To give all livestock conditions of life that allows them to perform all aspects of their innate behavior.
- To avoid all forms of pollution that may result from agricultural techniques
- To maintain the genetic diversity of the agricultural system and its surroundings, including the protection of plant and wildlife habitats.
- To use as far as possible renewable resources in locally organized agricultural system.
- To allow agricultural producers an adequate return and satisfaction from their work including a safe working environment.
- To consider the wider social and ecological impact of the farming system.
- To produce food of high nutritional quality in sufficient quantity.
- To work with natural system rather than seeking to dominate them.
- To encourage and enhance biological cycle within the farming system, involving microorganism, soil flora and fauna, plants and animals.
- To maintain and increase the long term fertility of the tea soils.

4.3.2 GAP related to nutrient management planning

Nutrient management is a major component of a soil and crop management system. Nutrient management planning as such is a relatively new term; however, the principles involved are basic, sound fundamentals necessary for good management. Nutrient management plans must be site-specific. They are tailored to the soils, crops and cultivars, landscapes, and management of a particular farm. Some important steps for nutrient management planning are mentioned herewith:

- Soil samples should be collected/obtained and analyzed as per recognized soil fertility analytical procedures in order to have accurate soil fertility information for each field management unit.
- Estimate yield potential for each field based on soil productivity and intended management and then fixes up the yield target.
- Work out the plant nutrient needs to achieve the pre-set yield target. Nutrient uptake and removal data for tea are available from various sources. It is important to distinguish between nutrient removal/uptake by the target crop, or the physical displacement of the nutrients from the field through the crop harvest.
- Determine the amount of the nutrients to be supplied through organics. The best method is to sample the manures to be used in the field. Determine accurately the nutrient contents of the manure and the nutrient release patterns.
- Decide the doses of the nutrients to be supplied through fertilizers considering indigenous nutrient supply. Keep record of the nutrient sources, their rate, method and time of application.
- Use a combination of mulches and fertilizer to maintain the health of the crop.
- Use organic matter and compost to reduce the need for inorganic fertilizer application.
- Fertilizer recommendations should be followed, but always taking into account the actual condition of the crop.
- Where inorganic fertilizer is required, carefully placed compound fertilizer under the tea canopy is likely to give the most efficient utilization by the growing crop.
- Use leguminous species of shade trees which will help to improve biological Nitrogen fixation and availability to the plant.
- Avoid application of fertilizer in heavy rainy season.
- Soil organic matter is an important component of soil to provide optimum condition in the soil for proper plant growth.
- Retain tea pruning litter in the field as complementary source of nutrient.

- Soil should be covered by a crop or mulch, including pruning litter, especially in rainy season.
- Apply organic manure and plant litter after pruning where organic matter is low.
- Grow in situ green manuring by growing Guatemala or legumes like cowpea, sun hemp in vacancies to add organic matter and other nutrients in the soil.

4.3.3 GAP to maintain soil fertility

Soil fertility is the capacity of soil to provide plant nutrients in available form in adequate amounts at a particular time to the plants. Fertility status of cultivated soil declines with time which affects the productivity. Due to prolonged cultivation, the tea soils of Darjeeling region have undergone considerable changes. This decline can be reduced to a considerable extent by recycling of crop residue. Maintaining optimum level of soil organic matter by applying appropriate organic manures in the soil could ensure adequate supply of essential plant nutrients to the plant and also to combat the global warming (Singh *et al.*, 2014). Thus for sustaining a high productivity; soil needs proper management and supplementation of nutrients under following tips.

- Use residues of plants as complimentary source of plant nutrients.
- Grow green manuring crop in vacant places of field and surrounding.
- Use both bulky and concentrated organic manure which help to maintain C: N ratio of soil organic matter.
- Use suitable bio-fertilizers.
- Utilization of industrial wastes with proper treatment to convert them into good organic manures.
- Use leguminous crops for green manuring because it requires less amount of water and also add nitrogen by symbiotic nitrogen fixation and by biomass production as compared to non-leguminous crops.
- Always keep the soil covered with green plants to avoid the direct exposure of the soil to radiation which induces the loss of soil moisture and accelerate the decomposition of added as well as native organic matter.
- Oil cakes should be well- powdered before application, so that they can be spread evenly and are easily decomposed by microorganism.
- Do not broadcast the organic manures, apply through ring placement method.

4.3.4 GAP related to soil water/moisture management

- Since tea cultivation in Darjeeling depends on rainfall and no irrigation facilities are available, hence, water harvesting and conservation of soil moisture practices should be followed.
- Maintain permanent soil covering particularly in monsoon season to avoid soil and nutrient losses.
- Minimize water loss by maintaining proper drainage system.
- Avoid green manure or intercropping crops with high water requirements in a low water availability region like Darjeeling.
- Harvest water in situ by digging catch pits, crescent bunds across the slope.

4.3.5 GAP related to soil reaction

- Tea requires certain soil acidity. The optimum pH for proper growth and development of tea plant ranged between 4.5 to 5.5
- Soil samples should be collected and analyzed to know the reaction of soil and correction measures to be adopted as per recommendation.

4.3.6 GAP related to soil conservation

- Soil erosion is one of the major reasons for decline in agricultural productivity especially in high rainfall and high slope areas.
- High slope tea soil is particularly vulnerable to soil and nutrient erosion at the time of replanting and after pruning.
- Soil for nursery should be taken from areas to be planted so that soil is returned to the field during planting.
- Construct drains to avoid rapid flows which cause erosion, using stones at vulnerable corners and planting grass along the sides to hold the soil.
- Use mulches of tea pruning and other mulches, including litter from Guatemala and *Crotalaria* etc to minimize soil erosion from tea field (Singh *et al.*, 2014).

5. Vermicomposting: An Eco-Friedly Approach for Nutrient Recycling

Vermicomposting is a method of preparing enriched compost with the use of earthworms. In recent years, earthworms have been identified as one of the major tools to process the biodegradable organic materials (Julka and Mukherjee, 1986; Greig Smith *et al.* 1992 and Ansari and Jaikishun, 2010). Slow rate of decomposition and mineralization of organic matters are major limiting factors in adequate nutrient availability to the plant in the Darjeeling tea soils (Bisen *et al.*, 2012). In these circumstances vermicomposting could be one of the easiest methods to recycle agricultural wastes and to produce quality compost in hilly areas like Darjeeling. Earthworms consume biomass and excrete it in digested form called worm casts. Worm casts are popularly called as Black gold. The casts are rich in nutrients, growth promoting substances, beneficial soil micro flora and having properties of inhibiting pathogenic microbes. Vermicompost is stable, fine granular organic manure, which enriches soil quality by improving its physicochemical and biological properties. It is highly useful in raising seedlings and for crop production. Vermicompost is becoming popular as a major component of organic farming system.

5.1. Needs of Vermicomposting

- An important source of organic manure.
- Helps in recycling any organic wastes into a useful bio-fertilizer and leaves no chance of environmental pollution.
- An eco-friendly, non-toxic product, consumes low energy input while processing.
- A preferred balanced nutrient source.
- Improves physical, chemical and biological properties of soil without any residual toxicity.
- Reduces the incidences of pests and diseases in crop production.
- Improves quality of agricultural produce.

5.2 Suitable recyclable material for Vermicomposting

Any biologically degradable and decomposable non toxic organic wastes, both solid and liquid, available from agricultural waste, animal husbandry, forest, industrial wastes including food processing and household activities can be used as source of raw material as earthworm food. In surrounding areas of Darjeeling hills the major green biomass pruning litter from tea bush, tea waste from the factory, Guatemala, weed biomass, forest leaf litter and to certain extent animal wastes are being mostly used as raw materials for vermicomposting (Singh *et al.*, 2013). The food of earthworm also could be small creatures (living or dead) and the microorganisms present in the soil or organic matter. Usually, under ideal conditions, earthworm consumes organic wastes as per their body weight in 24 hours; therefore, daily 1 kg earthworms consume minimum 1 kg of organic waste.

Table 1: The wastes material can be grouped into three categories as per C: N ratio

Organic wastes	C:N ratio	Suitability
Animal manure, oil seed residues, fish manure.	< 19	Most suitable for high nitrogen content
Vegetable waste, food processing waste including pulses, oil seeds, tea etc, kitchen waste, green andsucculent crop waste and weed biomass and green manures	19-27	Moderately suitable
Saw dust, coir wastes, stubbles and crop wastes, twigs and crop foliage with high lingo-cellulose contents.	27-208	Less suitable

5.3 Characteristics of Earthworm and their suitable species for Vermicomposting

Earthworm belongs to the class Oligochaete. They are bilaterally symmetrical, externally segmented, with a corresponding internal segmentation. The earthworm breathes through its skin, has no skeleton and crawls with the help of back and forth movement of setae on its body. Based on dwelling and food habit, three broad categories of earthworms are:

5.3.1. Epigeic (manure worm)

Small size and uniformly pigmented, lives in upper layer of vermibed and consumes. Species: *Eisenia foetida, Perionyx excavatus*, *Eudrillus eugenae* (Nigerian) etc.

5.3.2. Endogeic (Sub soil dweller)

Small to large in size, they inhabit the mineral soil horizons, feed more on soil. They move horizontally in the soil and not helpful in organic matter processing.

5.3.3. Aneceic (Top soil dwellers)

They are large in size, deep burrowing worms and live in vertical burrows in soil.

In a study conducted at Darjeeling Tea Research and Development Centre, Kurseong it has been found that Epigeic category of earthworm is most suitable for vermicomposting in Darjeeling conditions (Singh *et al.*, 2012).

5.4. Method of Vermicomposting

Vermicomposting is done by various methods but most common methods are bed and pit method.

5.4.1. Bed Method

Composting is done on the pucca/ kachcha floor by making bed (6x2x2 or 8x4x1 feet size). This method is easy to maintain and practice.

5.4.1.1 Preparation of vermibed

The beds should be made in open space under the shade. The size of the beds may vary from 6 to 15 ft in length, 2-4 ft in width and 1- 1.5 ft depth as per the feasibility and availability of the space. The beds can be made in the series from 10 to 100 and may be more depending on requirement. The layout has to be made with a provision of space for dumping of waste materials, pre-treatment and bed positions with walking and working space, harvesting and drying floor. The surface of the bed should be well plastered by clay soil or concrete or covered by polythene sheet. All the bed should have at least 2-4 per cent slope which will be helpful in collection of vermibed wash. All the beds should be covered by galvanized iron sheet or asbestos sheet or thatch material to avoid direct interruption of rain and sunlight.

5.4.1.2 Preparation of raw materials

Biodegradable waste materials like farm waste, weed biomass, tea waste and Guatemala etc. have to be collected and chopped in a fine pieces which has to be well mixed with cow dung slurry and has to be kept for 15-20 days for pre-decomposition and can be used as bedding material.

5.4.1.3 Filling/ loading of vermibeds

The waste materials after pre-decomposition are loaded in the bed and make heap with a maximum height of 1-2 ft. A thin layer of cow dung has to be placed on the surface of waste material. After checking of inside bed temperature which should not exceed 30^{0}C of heaped material, the earthworms are released on the surface of the bed.

5.4.2 Pit method

Composting is done in the cemented pits of size 5x5x3 feet. The unit is covered with thatch grass or any other locally available materials. The preparation of raw materials and filling/loading is more or less similar as bed method. In this method time to time gentle turning of bedding material including worms with soft/blunt hoe have to be done at 15-20 days interval to improve aeration. This method is not preferred due to poor aeration, water logging at bottom, and more cost of production.

5.5. Precautions during Vermicomposting

The following precautions are important for vermicompost preparation.

- Vermicomposting unit should be in a cool, moist and shady site.
- Cow dung and chopped dried leafy materials are mixed in the proportion of 3: 1or 1:1and are kept for partial decomposition for 15-20 days.
- A layer of 15-20cm of chopped dried leaves/grasses should be kept as bedding material at the bottom of the bed.
- In vermibed 60-70% of moisture should be maintained by sprinkling the water from time to time.
- Water content of bed should not go beyond saturation which may deplete oxygen level and may cause adverse effect on their survival of earthworm.
- Temperature inside of the bed should not go beyond 35°C and below 2°C which is lethal for earthworm. The most suitable temperature for adequate growth and reproduction is 10°C to 30°C.
- The bed should be covered by moist gunny bag or thatch to maintain the moisture in side and to avoid direct exposure to sunshine/rain etc.
- During winter season particularly at high altitude like Darjeeling, bed should be covered by paddy straw and polythene sheet as insulating material which can give rise the inside temperature of the bed up to 4°C.
- pH of bedding materials is an important factor which has an impact on earthworm activity which may affect their multiplication and distribution in the bed. Most of the species prefer neutral status of waste. Depending on the type of wastes lime can be used to enhance the pH.
- The vermicompost get matured generally in a period of 45-60 days at low altitude and also it depends on waste material used. At high altitude this period can be extended from 70-90 days.
- Maturity of earthworm can be judged by physical appearance as well as chemical parameters. On maturity the compost becomes soft, spongy and dark brown in color with no smell.
- About 10 days before of harvesting, watering should be stopped.
- Compost is harvested at a moisture level of about 60%.
- The separation of worms from vermicompost is generally done manually or by mechanical process.

- In mechanical process, the earthworms are separated by sieving using a wire net of 2-2.5 mm in size.
- After separation of worms bagging of compost can be done.

5.6 Advantages of Vermicompost

- It provides efficient conversion of organic wastes/crop/animal residues. It is a stable and enriched soil conditioner.
- It helps in reducing population of pathogenic microbes.
- It helps in reducing the toxicity of heavy metals.
- It is economically viable and environmentally safe nutrient supplement for organic production.
- It is an easily adoptable low cost technology.

6. Pruning in Darjeeling Teas

As the age of tea bushes from pruning advances, size and weight of harvestable shoot decline due to the reduction in the vascular supply to growing buds (Ghosh Hazra, 2004). Therefore, pruning is considered as the most important operations in tea with a primary objective to replace the old set of maintenance foliage, so that tea bushes remain healthy and continue to provide succulent harvesting shoots to manufacture tea having good quality. Repeated pruning and skiffing under certain intervals keep the tea bushes under vegetative phase and thus encourage shoot growth for optimum harvest. As such, pruning operation becomes imperative to provide a fresh stimulus to the bushes to restart its vegetative growth.

6.1 The Main Objectives of the Pruning

- To divert stored energy for production of fresh shoots.
- To regulate the crop distribution and quality.
- To remove dead/ unproductive wood and renew the actively growing branches which can support sufficient volume of maintenance foliage on it.
- To minimize banjhi formation, remove knots, control crop during rush periods and reduce incidence of pests/diseases.
- To control height for efficient and economic plucking.

6.2 Types of Pruning

6.2.1. Collar prune (CP)

All above ground portion cut leaving a maximum of 10 cm when bush frame becomes unproductive but still having good root system. Appropriate time of pruning should be December – middle January

6.2.2. Heavy prune/ Rejuvenation prune (RP)

RP is a heavy type of pruning and the basic objective is to rejuvenate the frame of old tea by removing almost all the knots in the frame along with dead and diseased woods by pruning right up to the clean wood.

6.2.2.1. Height

For China/ China hybrid bushes pruning at 15 – 30 cm and for Assam /Assam hybrid bushes pruning at 25 – 40 cm above the ground are ideal to be followed.

6.2.2.2.Time of Pruning

December – middle January is considered to be appropriate.

6.2.3. Medium Prune (MP)

This pruning is done at a higher height than the rejuvenation prune to rectify the tall and knotty frames with congested top hamper, but with a reasonably better frame below. The objective is mainly for top frame renewal and height reduction.

6.2.3.1 Height

For China/ China hybrid bushes pruning at 30 – 35 cm and for Assam /Assam hybrid bushes pruning at 50 – 60 cm above the ground.

6.2.3.2 Time of Pruning

December – middle January are considered to be ideal.

6.2.4. Light Prune (LP)

This is a normal prune done at a regular interval at the end of a pruning cycle. LP is for cleaning out the bush and removal of wood for new branches, 4 -5 cm or $1^1/_2$ -2 inches above last prune.

6.2.4.1 Time of Pruning

December – middle January

6.2.4.2 Height

The pruning height may vary depending on the level of knots and availability of healthy and sizeable sticks.

6.2.5 Deep Skiff (DS)

The height of the pruning operation is being determined according to earlier operation and sequence.

6.2.6 Medium Skiff (MS)

This is being done just below majority of crow's feet, in order to remove congestion in the bush frame.

6.2.6.1 Height of skiffing

For China/ China hybrid bushes this is being practiced at 10 - 12 cm above last LP level and low deep scale.

6.2.6.2 Time of skiffing

Last week of November – December is considered to be ideal.

6.2.7 Light Skiff (LS)

Cut through the red wood of the plucking table is the operation followed in this case.

6.2.7.1 Time of skiffing

In the month of January

6.2.8 Level-Off-Skiff (LOS)

Cut through the green wood of the plucking table is the normal practice being followed.

6.2.8.1 Time of skiffing

In the month of January

Table 2: Pruning type and pruning level

Prune Type	Pruning level (cm)
Rejuvenation Prune	< 20
Medium Prune / Cut Back	25-30
Light Prune	35-40
Deep Skiff	45-55

6.3 Pre-requisites for Pruning

The following criteria are considered important while doing the pruning operation in tea bush (Pathak, 2008).

- Low temperature and high humidity.
- Weak bushes should be rested for a period of 3-4 weeks before pruning.
- During resting period SOP should be applied @ 2 % in 2-3 rounds at fortnightly intervals to improve starch reserve in the frames and roots.
- Adequate moisture in the soil should be available.
- Adequate starch reserve in the root for quick recovery of bushes.
- Iodine Test: Dip cut ends of pencil thick roots collected at random for Iodine test. (The Iodine solution contains Iodine crystals 1 gm, potassium iodide 1 gm in 12 ml of distilled water and shaking the container gently then diluted to 100 ml and stored in dark colored bottle and kept away from light in the dark or Tincture iodine diluted 8 times).
- Cut end turning blue black- indicating the presence of adequate starch. If it remains yellow indicating low starch reserve. If starch reserve is low, the bushes may be rested for 1- 2 months.
- Sap rise through cut ends is indicative of bush recovery after drought and minimize splitting of branches during pruning.
- Use 15 cm blade knife for LP of young tea and 8-10 cm blade knife for cleaning operation. The weight of pruning Knife should be at least 450 gm or more.

6.4 Pruning Cycle

A pruning cycle is followed through the interval in years i.e. in between two successive light pruning. Implementation of pruning cycle helps to optimize yield of a section, distribution of crop during a cropping season, formation of wood to prune in next light prune and to balance crop and quality. Hence, duration of a pruning cycle should be adjusted in such a way that at the end of the cycle, the primaries attain the desired thickness to prune again. A pruning cycle is influenced by the following factors.

- Elevation/Altitude
- Planting material
- Age and vigour of bushes
- Frame height
- Bush hygiene
- Climate and Soil texture

- Quality requirement
- Recurrence of drought
- Incidence of pests and diseases.

The length of a pruning cycle may vary from 4 to 7 years depending upon altitude and planting materials. In case of Chinary tea bushes at very high altitude, say above 1650 m, a 6 year cycle of LP-UP-UP-UP-DS-UP-LP would be ideal, where as in case of Assam type bushes at low elevation, say below 1200 m, a 4- year cycle of LP-UP-MS –LS-LP would be ideal (Singh, I. D. 2005). The pruning cycles in Darjeeling tea plantation are as follows.

- 4 years cycle (LP-UP-MS–LS-LP) - Low Altitude (below 5000 feet's from the sea level)
- 5 years cycle (LP-UP-LS -MS-LOS-LP) - Medium Altitude (5000 feet's from the sea level).

Apart from above some Darjeeling tea planters' are following 5, 6, and 7 years pruning cycle at very high altitude as mentioned here under.

- 5 years cycle (LP-UP-UP -UP-UP-LP) - high altitude.
- 6 years cycle (LP-UP-UP-DS-UP-UP-LP) - High Altitude (above 5000 feet's from the sea levels).
- 7 years cycle (LP -UP -UP-UP-DS-UP-UP-LP) - High Altitude (above 5000 feet's from the sea level).

The most common cycle is the 5-year pruning cycle of LP-UP-LS MS/DS-LS-LP covering majority of the hill slopes. The distribution of crop during a cropping season depends on the type of prune/ skiff and the pruning cycle.

6.5 Some Important Considerations in Pruning

- Remove banjhi shoots for more crops from pruned bush soon after pruning but not beyond bud break time.
- Avoid defoliation from pruned bushes for higher crop.
- Any tea left unplucked for 2- 3 weeks, it may be skiffed at last plucking height followed by weekly plucking.
- Skiffing, if restored in the growing season, should be done latest by 15th August in Darjeeling.
- Leave at least two "breathers" (lung shoots) in south-west of the bush following HRP/MP/rejuvenation prune. Cut them after the new flush of shoots which have produced some leaves (2-3).

6.6 Post Pruning Care

- Apply Trichoderma pasting (20%) to large cut ends-rejuvenated and hard pruned teas.
- Provide a thorough lime washing to the cleaned out frames using 12 kg washing- soda, or 4-5 kg caustic soda/soda-ash, 4-5 kg quick lime dissolved in 200 liter of water.
- Clean the dried up mosses and lichens after a week using Hessian cloth.

6.7 Pruning of Young Tea

- Young tea should be pruned during 15th January to 15th February.
- Provide adequate rest, so that starch build up is adequate prior to prune.
- Newly planted tea with 3-5 good laterals should be de-centered in end January at 20 cm above ground.
- The single stemmers should be thumb pruned at 20 cm above ground.
- First frame forming prune (FFP1) should be done within 24 – 36 months after planting followed by FFP II at the end of FFP1-UP-UP.
- Apply additional dose of potash in the form of 10:5:15 mixture of YTD, if soil potash status is low or medium.
- Young tea area should be kept under mulching during winter months.

6.8. Pruning Type and Tipping Height

The height of tipping depends on the type of prune/skiff adopted, and is usually undertaken at a height just below the first "banjhi" horizon usually above 5 leaves in pruned tea. The different tipping measures for different pruning or skiffing are as follows.

Table: 3 Pruning type, pruning height and number of leaves retained in new lateral/primaries

Sl. No.	Type of pruning/ skiffing	Tipping height (in cm)	Number of leaves retained in new lateral/primaries
1	Rejuvenation prune	30-35 cm above pruning level	5-7 leaves
2	Medium pruning	25-30 cm above pruning level	5-7 leaves
3	Light prune	20-25 cm above pruning level	5-6 leaves
4	Deep skiff	10-12 cm above skiffing level	2-3 leaves
5	Medium skiff	4-5 cm above skiffing level	1 leave
6	Light skiff	Just above the level of skiffing	Nil
7	Levelling of skiff	Same level	Nil
8	Young tea (FFP1/ FFP2)	60-65 from ground	8 leaves

6.9 Management of Tipping

- Height of tipping is the average height just above which maximum number of primaries attains banjhi.
- The stick (measured bamboo stick) should be placed at last LP mark or in ground (in case of taking measure from ground) at the central portion of the bush.
- Strict supervision is required to ensure the primary objective of tipping.

7. Plucking

The operation of harvesting of tender shoots of tea bush for the manufacture of tea is known as plucking (Arunachalam, 1995). Besides giving crop, it encourages regeneration of new shoots, checks vertical growth of bushes and keeps them in vegetative phase. Plucking in Darjeeling hill generally starts in early March when shoots emerge and continued throughout the growing season and stopped in late November when shoot production ceases. A few variations may occur in the duration of continuity of plucking depending on the climate and time of pruning and skiffing. The best shoots for plucking of tea are those that have not more than two leaves and a growing bud and are undamaged. Plucking is considered as the most important cultural operation in the tea cultivation because it is the collection of the economic crop of the plant. A round of bad plucking can affect the current round's crop as well as few successive rounds in addition to its far reaching adverse effect on quality. In Darjeeling, plucking is generally is being done by hand. However, shear and machine plucking is also being done but it is not common because it needs certain conditions.

7.1 Maintenance foliage

The shoot left below the plucking table provides nourishment and sustenance to the pluckable shoots, known as maintenance foliage. Depth of maintenance foliage is controlled by tipping.

7.2 Standard of plucking

In a handful of plucked shoot, different proportions of shoots are present. To have an idea about the composition of shoots in the plucked amount of leaf and to indicate whether it is primarily composed of fine or coarse leaf some standards of plucking have been fixed. Thus standard of plucking denotes the types of shoots which may be fine, standard and coarse.

7.3 Plucking round

It refers to the time interval in days between two successive plucking. It is determined on the basis of pruning/ skiffing operation, rate of shoot growth, and system of plucking followed. Normally plucking interval in Darjeeling hill ranged from 5-10 days followed during the whole plucking season.

7.4 Plucking round v/s Yield and Quality

Plucking round has a close impact on yield as well as on quality. Few experimental findings are as follows:-

- A significant increase in crop could be obtained by plucking on appropriate round.
- Shorter plucking round yielded higher crop and produce better quality teas as compared to extended rounds.
- Negative correlation of extended plucking round with yield and quality was observed in some cases.
- Quality of Darjeeling tea may be obtained by varying appropriate plucking cycle from flush to flush throughout the plucking season.
- To achieve the optimum yield and good quality tea in Darjeeling hill, the plucking round in first flush 4-5 days, second flush 5-6 days, third flush/ rain flush 6-7 days and 8-10 days plucking round should be followed in autumn flush.

7.5. Important points regarding plucking

- During plucking season, foliar application of any chemical, if necessary, should be done immediately after plucking to get better efficacy of applied chemicals besides reducing the residue hazard.
- The pluckers should not be allowed to keep their belongings over the plucking surface to avoid heat damage to shoots.
- Maintain a uniform plucking table by effective supervision.
- Do not allow pluckers to dip hands into the plucking table.
- *Banjhi* shoots at plucking table should be removed after each flush, as these take longer period to regenerate "or" removing of banjhi at the time of plucking itself is a good alternative to shorten the banjhi period.
- Weeds within bush retard plucking efficiency. Hence, weeding should be done in proper time.
- Plucked leaves should not be pressed tightly in the plucking basket to accommodate more leaves which may trigger the leaf fermentation earlier than necessary.

8. Pest Management in Organic Tea

The productivity of agro ecosystems depends on the interaction of plant populations with several physical, biological and managerial factors. Pests are one of the important biological components in this ecosystem. Cramer (1967) estimated that tea in Asia sustained a crop loss of 8% through pests, while Banerjee (1976) reported that the estimate shows the loss to be anywhere between 6 and 14%.

Tea pests can be categorized in five groups depending on their nature of feeding:

- Sucking pests: Aphids, Jassids, Thrips, Tea mosquito bug, Red spider mites and mealy bug / scale insects.
- Leaf eating: Bunch caterpillar, Red slug caterpillar, Looper caterpillar, Nettle grub, Jelly grub, Grass hoppers etc.
- Leaf tiers: Flush worm, Tea leaf roller, Tea tortrix, Tea leaf webber, Nest forming caterpillar.
- Borers: Common red borer, Large stem borer etc.
- Root feeder: Cockchafer grub, Nematodes, Weevils etc.

Table 4: Common symptoms and their causes

Common symptoms	Probable cause
• Curled, deformed leaves	• Aphid, mites, mosquito bug
• Leaves with dark brown strips or spots	• Red spider mite, mosquito bug
• Leaves changing colour	• Aphids, mites, mosquito bug, green leaf hopper
• Sooty moulds growing on leaves	• Aphids, scales/ mealy bugs
• Leaves or buds folded or tied with silk	• Bud rollers, leaf tiers and leaf folders
• Holes in leaves	• Caterpillars, grasshopper etc.
• Dried, dead leaves	• Mosquito bugs, mites, red borer, termite
• Pluckers or other workers with burns on their skin	• Nettle caterpillar and saddle back caterpillars
• Tubes made out of sawdust or mud on the surface of trunk or branches	• Termites
• Big pieces of sawdust on the ground under tree	• Red borer

8.1 Control measures

The use of conventional insecticides and acaricides are the most common method to bring down immediate reduction in pest population but in organic cultivation there is a restriction in use of synthetic insecticides. The alternatives of inorganic insecticides are as follows-

8.1.1 Biological control

Numerous bio-control agents are active in tea fields and identified as natural enemies for certain pests and are able to keep their population under control such as Coccinelid beetle larva/ grub, Chrysoperla carnea, Syrphid fly, Praying mentid, Reduvid bug, Spiders, Tachanid fly etc.

8.1.2 Microbial agents

It includes bacteria, fungi, viruses, protozoa and nematodes that are pathogenic or antagonistic to certain pest species. Encouraging results were obtained with introduction of *Bacillus thuringinsis* (Bt) against *Homona magnanima, Caloptilia theivora* and *Adoxophysis* sp. (Kariya, 1977). Its efficacy has also proved against *Andraca bipunctata* and *Buzura suppresaria* (Reddy *et. al.*,1990; Borthakur, 1986) and *Cydia leucostoma* (Bisen and Ghosh Hajra, 1997). Some other microbial agents such as *Verticillium catenulatum, Beauveria bssiana, Paecilomyces fumosoroseus, Verticillium leucanii* etc. are also effective against different pests.

8.1.3 Botanicals

These are derived from extractions, pressing and infusions of any plant part which can be used as antifeedant, repellent, toxicant, growth inhibitor, antiovipository etc. against common pests. Among botanicals, Neem has been found to have multifarious properties mentioned above. The antifeedant properties of the extract derived from the aerial parts of *Artimissia vulgaris Linn., Urtica dioica Linn., Polygonum runcinetum Hm* and *Eupatorium glandulosum* were evaluated in the laboratory against bunch caterpillar (*Andraca bipunctata*) and found effective. (Bisen and Kumar, 1997)

9. Tea Quality: Influence of Agricultural Practices

A brief description of tea quality as influenced by the main agro-cultural practices is being presented here. Darjeeling tea also known as 'The Champagne of Teas' originated from Darjeeling region of West Bengal, India is world famous for its muscatel flavour. The biogenesis of Darjeeling tea flavour is a consequence of type of genotype (tea cultivar-Chinary seedling jat) prevailing climatic conditions- short day low temperature coupled with cold night and foggy days, interaction between genotype and environment, agro-cultural and manufacturing practices. In Darjeeling hills orthodox black tea is produced.

9.1 Plucking cycle in relation to quality of Darjeeling tea

The objective of harvesting tea flush is the maximum production of high quality tea maintaining the tea bush in good health. Harvesting of tea crop (two leaf and a bud) is done on definite intervals. This periodic phenomenon of plucking is referred to as Plucking Round or Plucking Cycle. In Darjeeling hills tea shoot comprising two leaf and a Bud is plucked which produce the flavor and quality of Darjeeling tea. In Darjeeling hills the general practice of harvesting tea flush at seven days interval prevails since time immorial. This is perhaps after Wight

(1932) who suggested a plucking round of 7 days for chinary jat based on phyllocron using formula (2* leaf period -1). He further argued that plucking rounds should be adjusted in accordance to the leaf period so that maximum number of good quality shoots is available. Besides, a number of workers have studied the impact of harvesting intervals on the yield and quality (Grice, 1979, Owuor, 1987, Kabir 1996). In Darjeeling Hills also an experiment was conducted on the old chinary teas in to evaluate the effect of plucking cycle on chemical composition and the quality of tea. The tea bushes were maintained at 4, 5,6,7,8 and 10 days plucking rounds. A distinct pattern was found in the chemical composition of teas. Chlorophyll a, Chlorophyll b, total carotenes, lipids, and crude fibre content were found to increase along with the length of plucking rounds. However, total polyphenols declined in the same order (Kumar *et. al.*, 2000). No plucking cycle produced consistent quality score throughout the season. During the first and second flush a five and six days plucking cycle produced better quality teas as compared to the conventional seven days cycle, which, however, produced better quality teas in the second post second flush period. Flavour index of such teas was also observed high. Consistent increase in chlorophylls and carotenoids up to seven days plucking rounds and thereafter depletion in their content for longer plucking cycle perhaps suggests that the plucking cycle in Darjeeling hills should not be extended beyond seven days. However, in colder climate when growth rate of shoot is slow and the harvesting of crop on 7 days cycle becomes impractical, it may be extended to 8-10 days rounds keeping in view the convenience of pluckers. Further, it is concluded that quality of Darjeeling teas may be maintained by varying appropriate plucking cycle from the flush to flush throughout the season.

9.2 Plucking rounds in relation to Yield and quality

5 DAYS	:	1st FLUSH
6 DAYS	:	2nd FLUSH
7 DAYS	:	RAINY FLUSH; POST 2nd FLUSH
8-10 AYS	:	AUTUMN FLUSH

9.3 Pruning operations and the Tea quality

The prime objective of pruning is to keep the tea bush in vegetative phase to facilitate the harvesting operations for sustenance of productivity. This agronomic practice influences the crop yield and quality of made tea. In Darjeeling hills pruning operations are carried out with varying length of pruning cycle from 4 to 6 years depending upon the health of the tea bush and elevation. In a study conducted on old chinary seedling tea plantation in order to investigate the effects of pruning operations on the quality attributes of Darjeeling teas a distinct variation was recorded in tea quality constituents with the type of pruning. Total polyphenols and total chlorophylls declined in the pruned teas (LP) and, thereafter, progressively increased up to the 5th year of prune in UP / LOS operation. Similar pattern was found for chlorophyll a and chlorophyll b. The level of carotenes was low in pruned teas showing little increase in skiffed teas and the unpruned teas. Total soluble solids and dry matter content were low in pruned teas. After tipping of pruned teas the level of caffeine decreased and thereafter progressively increased with the age from pruning. The Group I compound mainly non-terpenoids enhanced in pruned teas whereas Group II compounds mainly terpenoids declined. The progressive improvement in Flavour Index (FI) with the passage of time from pruning was noticed as evident also with the organoleptic evaluation of the professional tea tasters. Liquor characteristics of orthodox black teas varied with the type of prune during a pruning cycle affecting tea quality. The level of theaflavins (TF) declined in pruned teas. But at the same time thearubigins (TR) that impart color to tea liquor increased. Also total liquor color (TLC) considered detrimental to Darjeeling tea quality was more in pruned teas whereas the brightness of liquor camouflaged. Seasonal variations in biochemical constituents in a year of harvest were pronounced irrespective of the type of prune. Total polyphenols, dry matter content, total soluble solids and group II volatile compounds declined with increase in group I compounds in monsoon season as compared to pre and post monsoon season affecting the quality of Darjeeling orthodox black teas (Kumar *et. al.*, 2014b). The age from pruning was also observed significant in the recovery of these chemical constituents in Darjeeling teas.

10. Copper in Tea

Copper as copper oxychloride is abundantly used in tea crop to control the Blister Blight disease. Copper is present in tea leaf as prosthetic group of Polyphenol Oxidase. Copper (Cu) is needed for chlorophyll production, respiration, and protein synthesis (Solberg *et al.*, 1999). However, copper being heavy metal draws special attention of tea planters. Next to pesticide residue, heavy metals have been identified as a non- tariff trade barrier under W.T.O. The new tea distribution Export Order 2005 under section 'Additional requirements- metallic contaminants' set permissible limits of 150ppm for copper in tea. European Union has also recently set temporary MRL limits for copper at 40 ppm effective from September 2008. In an analysis of 244 India tea samples for copper content 94% teas were found below 30 ppm including 6% containing in between 31-40 ppm. In general Cu content in plant dry matter is between 5 and 20 ppm. Below 4 ppm Cu in plant dry matter is considered as critical level below which the plant may suffer from Cu deficiency. The critical level of Cu varies from crop to crop but not as widely as in other micronutrients (Tandon, 2009).

In a geochemical survey done recently in Darjeeling hills copper content in tea soils reported to vary 14.1 to 60.4 ppm, in fresh leaf 7.7 to 17.6 ppm and made tea 11-20.4 ppm. (Solveig, 2012). In Turkey, Horuz and Korkmaz (2006) determined Cu content of tea leaves harvested in different offshoot periods varying between 13.00 to 23.60 mg kg^{-1}. In addition, Kacar *et al.* (1979) reported Cu amounts of tea leaves collected during the second offshoot period as 14 to 21 mg kg^{-1}. In the Sri Lanka (Ceylon) teas, Cu contents normally varied between 25 to 30 mg kg^{-1}. On the other hand, Ramaswamy (1960) reported that the Cu amounts of tea leaves vary within quiet wide ranges such as 19 to 62 mg kg^{-1}. Availability of copper is reported directly related with pH values. Also negative relationship is observed between Cu and organic matter (Mehmet *et al.*, 2011). It is, therefore, expected that with the increasing use of organic inputs copper requirements will be increased and more copper will be required to maintain the copper level by organic tea estates in future. In the study conducted at the DTRDC experimental Farm both crop suppressive and crop promoter role of copper were observed. The foliar application of copper as copper sulphate at higher concentration @ 3 percent suppressed the crop growth by 22 percent whereas application of copper sulphate at lower concentration @ 0.5 percent increased the yield up to 9 percent in the agro-climatic conditions of Darjeeling hills (Annonymous, 1993). Suppression of tea crop with the application of copper in excess was also noticed by Shaikh (2007). He further opined that quality of Darjeeling tea is better for the high content of copper in those soils. However, no data of copper content or quantum of crop loss was reported by him. The reason of crop suppression in tea due to copper application in excess was explained by Guha *et al.* (2001) which he termed as 'action spectrum of photosynthesis' According to him red and blue color of sunlight are utilized for the process of photosynthesis. Due to application of copper in excess red and blue color rays are transmitted back hampering the process of photosynthesis and, hence, the crop yield. On the contrary negative relationship between net photosynthetic rate and yield of tea crop in Darjeeling was observed by Ghosh Hajra (1999).

A distinct behaviour of copper treated tea flush was observed during fermentation. The color of such teas turned to dark coppery (blackish) at much faster pace as compared to untreated teas which changed to coppery during fermentation. The foliar application of copper as copper sulphate on crop enhanced the rate of fermentation and it increased with the increase in level of concentration of copper application (Annonymous,1994). Faster rate of fermentation may be attributed to the increased activity of polyphenol oxidase.

During the process of fermentation degradation of chlorophylls leads to the formation of pheophytin and pheophorbide (Wickremasinghe and Perera, 1966). A reverse pattern of pigment composition as influenced by the copper application was observed. Copper application at higher concentration on tea crop also catalyzed the degradation of chlorophylls. On the other hand application of copper at lower concentrations helped in the synthesis of chlorophylls. The biochemical composition of Darjeeling teas was appreciably affected with the foliar application of copper at higher concentrations. Residual polyphenols of made teas declined significantly. Liquor characters of copper treated teas were also affected and it was reflected with change in the level of theaflavins, thearubigins, brightness and color. But no residual metallic character was noticed by the professional Darjeeling tea tasters during organoleptic evaluation of copper treated teas. (Kumar *et al.*, 2012a). These observations are corroborated with the findings of Keegel (1952) who also observed deterioration in liquor characters and quality of Sri Lanka teas due to high concentration of copper application. Further, Balasuriya (2001) observed that in tea brew not more than 25% of total copper of tea shoot, whether inherent or otherwise is passed on to the liquor and no taints were found in tea samples containing up to 15ppm of copper.

It may be concluded that the application of copper sulphate at lower concentrations increases the tea crop yield and also improves quality. Crop suppression occurs only at higher concentration above 0.5percent affecting the liquor characters and vis-à-vis quality. Further, no odor or taint of residual metallic character is noticed by the professional tea tasters during organoleptic evaluation of copper treated teas. Application of copper as copper sulphate at much lower concentrations may be considered for increasing fermentability of slow fermenting cultivars. Judicious Application of copper sulphate in tea is safe and may be used as preventive measure towards incidence of Blister Blight Disease in Darjeeling.

11. Impact of Organic Farming on the Tea Quality

Organic farming is a 'back to nature' approach based on specified traditional farming philosophy. It is independent of conventional agriculture and independent of chemicals and fertilizers. The demand of organically grown food is increasing worldwide and the concept of organic farming is gaining ground. In Darjeeling hills also, organic tea cultivation is gaining momentum where 60 % of the tea gardens have converted to organic practices. The organic tea is a value-added product and such teas command a premium of 30-40% over conventionally grown teas. Today Darjeeling tea is a major foreign exchange earning commodity. To understand the effect of the organic tea cultivation practices on the flavour precursors leading to the variation in quality of Darjeeling teas, a study was conducted at the Darjeeling Tea Research and Development Centre (DTR&DC) ofTea Board, Kurseong. The study suggests that the biosynthesis of polyphenols, proline and lipids are greatly influenced under the organic cultivation of tea. Total polyphenols and proline contents were relatively more in organic teas. However, total lipids were found less in such teas. Pigment profile such as chlorophyll a, chlorophyll b and total carotenes were also observed low in organically grown teas as compared to conventional farming. In non-quality season the level of total chlorophylls, carotenes and lipids were observed high in both type of cultivation (Kumar *et. al.*, 2012b). Also the dry matter content that affects the recovery of made tea from tea flush and hence, the net profit was high in organic teas. The high antioxidant activity of organic tea as compared to inorganically produced teas makes the former a healthy drink. Thus organically produced teas have greater health benefits and longer shelf-life. Further, (Kumar *et al.*, 2014a) observed that application of organic inputs did not adversely affect the quality. It rather improved the quality of organic teas. Organic inputs like de-oiled cakes and compost as well increased the polyphenols, proline content and chlorophylls. Mishra *et al.*, (2009) observed higher antioxidant activity that improves the defense against pest and better bush health of organic teas. Recovery percentage of made tea increased enhancing profitability with the application of organic inputs. Mellowness in organic tea increased affecting the palatability during Organoleptic testing of teas. Application of organic inputs containing high nitrogen contents responded positively in increasing the polyphenolic contents of organic teas. It may be, therefore, concluded that organic farming does not have adverse effect on the quality of Darjeeling orthodox black teas. It rather improves the quality parameters as well as shelf-life. Organic teas, devoid of chemicals with high antioxidant capacity, make them a perfect health drink.

12. Future line of Research for Darjeeling Tea Industry

Development of superior cultivars using classical and molecular approaches for sustaining Darjeeling tea quality is the demand of time. Data base of volatile flavoury compounds in various tea cultivars released for Darjeeling tea industry needs to be prepared. Quality mapping of Darjeeling teas covering different valleys is an important aspect of study. Development of bio-pesticides using abundantly available local herbs, screening and identification of bio-control agents for efficient pest management in organic tea cultivation is required. Preparation of data-base for Darjeeling tea germplasm. Generation of estate wise accurate data-base for soil to develop soil health card, soil quality index, soil fertility map and establishment of critical limit of micronutrient as a diagnostic tool is essential to adopt the site specific nutrient management practices. Improving nutrient use efficiency (both applied and native) to minimize the cost of cultivation, efficient recycling of organic wastes/crop residues through introduction of natural decomposers may improve the soil health and productivity. The hygienic disposal of organic wastes by recycling is an environmentally sound and economically viable technology resulting valuable input in organic farming. Development of enriched compost and vermicompost in terms of beneficial microbes and essential nutrients like K, Ca and Mg etc. may ensure balanced nutrient supply. Identification of potential microbes from Darjeeling tea soils to develop bio-fertilizers (macro and micronutrient) using local strain, free living (non symbiotic) N fixing microbes, K and Zn solubilizer and mobilizer etc. need to be identified for better and cost effective nutrient management. Development of effective soil conservation measures to control/minimize soil erosion from tea growing areas of Darjeeling and creating awareness among planters about the value of soil and negative effect of its improper use. Heavy metal study in detail of Darjeeling tea soils to avoid any contamination of such elements in the final produce also needs to be initiated.

13. Conclusion

Adoption of good agriculture practice (GAP) through the judicious application of fertilizers, pesticides and micronutrients may be useful to maintain the sustainability of tea crop production and quality in Darjeeling hills. Appropriate agricultural practices in tea cultivation mainly pruning and plucking have great impact on the quality parameters and also help in maintaining the crop productivity. The genetic diversity and flavour retention may be achieved through intensive programmes of conventional breeding.

14. References

Ankunda E.M.D. (1990) Physiological aspects of productivity in coffee: Some aspects of water relations and dry matter production of coffee L. in Kenya. pp. 404-425. In: Proceedings of the international congress of Plant physiology, 15-20 Feb., New Delhi, India.

Annonymous (1993) Annual Scientific Report of DTRDC, 1992-93. Biochemistry.Effect of Copper and Zinc on the yield and quality of Darjeeling teas, pp 16-17.

Annonymous (1994) Annual Scientific Report of DTRDC, 1993-94. Biochemistry. Effect of Copper and Zinc on the yield and quality of Darjeeling teas, pp 19-20.

Ansari A. A. and Jaikishun S. (2010) An investigation into the vermicomposting of sugarcane bagasse and rice straw and its subsequent utilization in cultivation of Phaseolus vulgaris L. in Guyana. American-Eurasian *J. Agric. & Environ. Sci.* 8 (6): 666-671.

Arunachalam K. (1995) A handbook on Indian Tea.Arunachalam Associates (Plantation consultants). Wellington Bazar, The Nilgiris, Tamil Nadu, India, pp-64.

Bagyalakshmi B., Balamurugan A., Ponmurugan P. and Premkumar R. (2012) Compatability study of indigenous plant growth promoting rhizobacteria with inorganic and organic fertilizers used in Tea (*Camellia sinensis*). *International Journal of Agricultural Research* 7: 144–151.

Baker E.A. (1982) Chemistry and morphology of plant epicuticularwaxes.PlantCuticle 62: 139-166.

Balasuriya A. (2001) Control of blister blight disease of tea in Sri Lanka. A historical overview; milestones and future directions. *Bull. of UPASI TRF*, 54: 146-158.

Banerjee B. (1976) Pesticides and pesticide residues in tea. *Two and a Bud* 23(2): 35-42.

Barman T. S., Baruah U. and Saikia J. K. (2005) Effect of light intensity on tea leaf physiology under field conditions. *Two and a Bud* 52: 22-27.

Barman T. S., Saikia J. K. and Pathak S. K. (1998) Starch reserve and growth of tea. *Two and a Bud* 45: 14-8.

Bhattacharya P. (2004) Organic food production in India status, strategy and scope. Agrobios (India).

Bhattacharyya P. and Tandon H. L. S. (2012) Biofertiliser Handbook – research – production – application. Fertiliser Development and Consultation Organisation, New Delhi, pp. 190.

Bisen J. S. and Singh A.K. (2012) Impact of inorganic to organic cultivation practices on yield of tea in Darjeeling hills-A case study. *Indian Journal of Horticulture* 69 (2): 288-291.

Bisen J.S. and Ghosh Hajra N. (1997) Testing of delfin with some commonly used insecticides to control flushworm in young tea of Darjeeling. *Petology*, 30(3): 12-14.

Bisen J.S. and Kumar R. (1997) Studies on the antifeedant properties of some plnt extracts against bunch caterpillar (*Andraca bipunctata*) on tea (*Camellia sinensis*). *Petology* 21(10): 13-15.

Bisen J.S., Singh A. K., Kumar R., Bora D.K. and Bera B. (2011) Vermicompost quality as influenced by different species of earthworm and bedding material. *Two and a Bud* 58: 173-140.

Biswas T. D. and Mukherjee S. K. (2003) Text Book of Soil Science. Tata McGraw-Hill Publishing Company Ltd., New Delhi.

Boddey R.M., Dobereiner J., (1982) Non-symbiotic nitrogen fixation and organic matter in the tropic. In: Abstracts of 12th International Congress of Soil Science, New Delhi, India. pp. 28-47.

Bolton H. J., Fredrikson J.K. and Elliot L.F. (1993) Microbiology of the rhizosphere. *Soil Microbial Ecology* pp. 27-63.

Borthakur M.C. (1986) Role of entomopathogenic bacteria for the control of tea looper caterpillar. *Two and a Bud* 33: 1-3.

Carr M.K.V. (2000) Shoot growth plus plucking equals profit. TRIT Occasional Publication No. 1, April 2000, Tea Research Institute of Tanzania pp. 1-7.

Chandra K. (2005) Liquid biofertilizers. Regional Centre of Organic Farming, Bangalore Publication 85p.

Choubey M., Kumar R., Chakraborty A., Bisen, J. S., Singh A.K. and Singh M. (2013) Performance of tea clones in the nursery through vegetative propagation in Darjeeling. International *Journal of Scientific& Research Publications* 3 (11): 1-4.

Cramer H.H. (1967) Plant protection and world crop production. Bayer, Levekusen pp. 524.

Elmerich C. (1984) Molecular biology and ecology of diazotrophs with non leguminous plants. *Bio-Technology* 2: 967-978.

FAO GAP Principles (2012) Food and Agricultural Organization of the United Nations.

Ghosh Hajra N. (1999) Seasonal variation in photosynthesis and productivity of young tea. In: Proceedings of the scientific seminar on Darjeeling tea organised by Darjeeling tea Research and Development Centre, Tea Board, Kuseong, Darjeeling pp. 5-15.

Ghosh Hajra N. (2001) Tea cultivation: Comprehensive Treatise. International Book Distributing Co. Lucknow, India, 520p.

Ghosh Hajra N. and Kumar R. (1999) Seasonal variation in photosynthesis and productivity of young tea. *Expl. Agric*. 35: 71–85.

Ghosh Hajra N. and Kumar R. (2002a) Diurnal seasonal variations in gas exchangeproperty of tea leaves. *J Plant Biol.* 29(2): 169-173

Ghosh Hajra N. and Kumar R. (2002b) Responses of young tea clones to Subtropical climate: Effects on photosynthetic and biochemical Characteristics. *J Plant Biol.* 29(3): 257-264.

Ghosh Hazra N. (2006) Organic Tea Cultivation (cultivation and marketing). International Book Distribution Company Lucknow, UP. pp. 80.

Gold Mary V. (2007) What is organic production? National Agricultural Library. USDA.

Grice W.J. (1979) Quarterly Newsletter, Tea Research Foundation, Central Africa 54: 4-7.

Guha J., Das Gupta B., Dulal C. (2001) In Biology. Publisher, Shree Mayapur Chandra press, Kolkata, India.

Horuz A., Korkmaz A. (2006) Yield, nitrogen and mineral matter composition of tea plant harvested in different periods. J. Fac. Agric., OMU 21(1): 49-54.

Julka J.M. and Mukharjee R.N. (1986) Preliminary observations on the effect of Amynthusdiffringens (Baired) on the C/N ratio of the soil.Pro. Nat. Sem. Org. Waste Utility, Vermicompost. Part B: Verms and Vermicomposting pp. 66-68.

Kabir S.E., Ghosh Hajra N. and Kumar R. (1996) *J. Plantation Crops* 24(S): 758-762.

Kacar B., Przemeck E., Ozgumus A., Turan C., Katkat A.V., Kayýkçýoglu I. (1979) A research about micronutrient status of tea plant and its soils in Turkey. *TUBITAK, TOAG Report* 321: 1-67 (in Turkish).

Kariya A. (1977) Control of tea pests with *Bacillus thuringiensis*. *JARQ*. 11(3): 173-178.

Keegel E.L. (1952) Studies in blister blight control IX. The effect of spray residue on the quality of manufactured tea. *Tea Quarterly* 23(1): PP. 2-6.

Kumar N., Biswas P., Singh M. and Bera B. (2014a) Effect of organic inputs on the quality of Darjeeling tea. *Journal of Plantation Crop*, pp. 95-96.

Kumar N., Rai R. and Ghosh Hajra N. (2000) Plucking cycle in relation to quality of Darjeeling teas.Plantation Crop Research and Development in New Millennium, pp. 236-441.

Kumar N., Rai R. and Bera B. (2012a) Impact of organic farming on the biochemical constituents and quality parameters of Darjeeling teas [*Camellia sinensis* (L) O. Kuntze.]. *Two and a Bud* 59(2): 50-55.

Kumar N., Rai R., Singh M. and Bera B. (2012b) Effect of Copper on quality and productivity of Darjeeling teas [*Camellia sinensis* (L) O. Kuntze.]. PLACROSYM XX pp. 132.

Kumar N., Rai R., Singh M. and Bera B. (2014b) Pruning operations in Darjeeling Tea [*Camellia sinensis* (L) O. Kuntze.]- A biochemical Perspective. *Agrotechnology* 2(4): 96.

Kumar R., Bisen J. S. and Bera B. (2014) The effect of pruned and un- pruned tea clones on the diurnal leaf water potential patterns in Darjeeling hill. *J. of plantation Crop* 42 (1): 75-79.

Kumar R., Bisen J.S., Choubey M., Singh M. and Bera B. (2015) Studies on effect of altitude and environment on physiological activities and yield of Darjeeling tea(*Camellia sinensis* L.) plantation. *J. of Crop and Weed* 11: 71-79.

Kumar R., Bisen J.S., Singh M. and Bera B. (2012) Effect of pruning and skiffing on growth and productivity of Darjeeling tea. *PLACROSYM – XIX*-2010, pp. 78.

Manshardt R., (2004) Crop improvement by conventional breeding or genetic engineering : how different are they ?; cooperative extension service, University of Hawai.

Marimuthu S., Kumar R. and Manivel L. (1994) Factors affecting partitioning of assimilates in tea. *J. Nuclear Agric. Biol.* 23: 219-223.

Mazumdar T., Goswami C. and Talukdar N. C. (2007) Charecterization and screening of beneficial bacteria obtained on King's B agar from tea rhizosphere. *Indian Journal of Biotechnology* 16: 490-497.

Mehmet A., Gulen O. and Orhan D. (2011) Determination of micronutrients in tea plantations in the eastern Black Sea Region, Turkey. *African J. of Agri. Res.* 6(22): 5174-5180.

Mishra T.K., Saha A., Nanda A.K. and Mandal P. (2009) Study of Free-Radical Scavenging Activity in different grades of organically and non-organically produced tea. In: Improving Productivity and Quality of Tea through Traditional Agricultural Practices. Proceedings of National Seminar, pp. 87-97.

Mishustin E. N. and Shilnikova V.K. (1969) The biological fixation of atmospheric nitrogen by free-leaving bacteria.Soil Biology. Review of Research pp. 72-109.

Mohotti A.J., Lawlor D.W. (2002) Diurnal variation of photosynthesis and photoinhibition in tea: effects of irradiance and nitrogen supply during growth in the field. *J. Exp. Bot.* 53: 313-322.

Mondal T.K. (2011) *Camellia.* In: Kole C. (ed.) Wild crop relatives: genomics and breeding resources plantation and ornamental crops. Springer, USA, pp. 15–40.

Mondal T.K. (2014) Breeding and biotechnology of tea and its wild species. Springer New Delhi, Dordrecht, Heidelberg, London, New York. Vol. XVI.

Owuor P.O. Obanda A.M., Othino C.O., Horita H., Tsushida T. and Murali T. (1987) *Agric. Biol. Chem.*,51 3383-3384.

Pal V. and Jalali I. (1998) Rhizosphere bacteria for biocontrol of plant disease. *Ind. J. Microbiol.* 38: 187-204.

Panda P., Choudhury A., Singh A. K., Chauhan R. K., Singh M. and Bera B. (2012) Isolation and screening of phosphorus transforming bacteria from acidic tea soils of Darjeeling hills. Souvenir Extended Summary & Abstracts Silver Jubilee Year of IISS, Bhopal, pp.126-129.

Pathak S.K. (2008) Pruning and skiffing of tea in the hills. In: Field Management in Tea, Tea Research Association, Tocklai Experimental Station, Jorhat (Assam), India, pp.113-121.

Prakash O., Sood A., Sharma M., Ahuja P.S. (1999) Grafting micropropagated tea (*C. sinensis* (L.) O. Kuntze) shoots on tea seedling- a new approach to tea propagation. *Plant Cell Rep.* 18: 137-142

Rajasekar R., Cox S. and Satyanarayana N. (1991) Evaluation of certain morphological and physiological factors in tea (*Camellia sinensis*) cultivars under water stress. *J Plant Crops* 18: 83-89.

Rajkumar R., Manivel L., Marimuthu S. (1998) Longevity and factors influencing photosynthesis in tea leaves. *Photosynthetica* 35: 41-46.

Rajkumar R., Marimuthu S., Jayakumar D. and Jeyaramraja P. R. (2002) *In situ* estimation of leaf chlorophyll and their relation with photosynthesis in tea. *Indian J. Plant Physiol.* 7(4): 367-371.

Ramaswamy V. (1960) Copper in Ceylon teas. *Tea Q.*, 31: 76-80.

Read J.J., Johnson D.A., Asay K.H., Tieszen L.L. (1991) Carbon isotope discrimination gas exchange and water use efficiency in crested wheat grass clones. *Crop Science* 31: 1203-1208.

Reddy K.S.P., Chaudhury T.C. and Ghosh Hajra N. (1990) Infectivity of *Bacillus thuringiensis var. kurustaki* against bunch caterpillar. *Two and a Bud* 37: 15-16.

Shaikh H. (2007) Effect of copper (Cu+) on the quality and quantity of made tea and the exact reason for suppression of tea crop for the application of bivalent copper products to the tea bushes by spraying. The Assam Review & Tea News, Oct. pp. 12-14.

Singh A. K., Bisen J. S., Bora D. K., Kumar R. and Bera B. (2011) Comparative study of organic, inorganic and integrated plant nutrient supply on the yield of Darjeeling tea and soil health. *Two and a Bud* 58: 58-61.

Singh A. K., Bisen J. S., Chauhan R.K., Singh M. and Bera B. (2013) Development of Vermicompost Technology for Organic Cultivation in Darjeeling hills. 'National Seminar on Organic Tea Cultivation' Organized by TRA, Jorhat, Assam, pp. 11-12.

Singh A. K., Chauhan R. K. and Bisen, J. S. (2014) Role of soil organic matter in soil health sustainability. *International Journal on Agricultural Sciences*, 5 (II): 219-227.

Singh A. K., Chauhan R. K., Bisen J. S., Singh M., Bhagat R.M. and Bera B. (2014) Development of soil conservation measures for mitigation of soil erosion problems from tea growing areas of Darjeeling hill. COBACAS, National Conference on "Adaptation and mitigation strategies of climate change for sustainable livelihood", UBKV, Pundibari, Cooch Behar, West Bengal. pp. 18.

Singh A. K., Kumar R.P. and Singh A.K. (2007). Significance of microbes in agriculture and human life. *Agriculture Update*, 2 (1): 44-45.

Singh I.D. (2005) The Planter's guide to tea culture and manufacture. N.B. Modern Agencies, Siliguri, West Bengal.

Smith B.G, Stephens W., Burgess P.J., Carr M.K.V. (1993) Effects of light, temperature, irrigation and fertilizer on photosynthetic rate in tea (*Camellia sinensis*). *Exp. Agric*. 29: 291-306.

Smith Greig P.W., Becker H., Edwards P.J. and Heimabach F. (1992). Eco toxicology of earthworms.Athenacum Press Ltd., Newcastle. U.K., 269P.

Solberg E., Evans I. and Penny D. (1999) Copper Deficiency: Diagnosis and Correction. Agriculture, Food, and Rural Development. http://www.agric.govagdex/500/532-3.

Solveig P. (2012) Geochemical investigation on the tea plant *Camellia sinensis*: Implications from element soil-plant transfer factors for provenance studies – the case of Darjeeling Tea from NE-India, Diploma thesis 111P.

Squire G.R. (1976) Xylem water potential and yield of tea (*Camellia sinensis*) clones in Malawi. *Exp. Agric*. 12: 289-297.

Squire G.R. and Calendar B.A. (1998) Tea plantations In; Water deficits and plant growth,(Ed). Kozlowski T.T., Academic Press, Newyork, Vol. VI, pp. 471-510.

Tandon H.L.S. (2009) Copper requirements and its critical levels. In: Tandon H.L.S. (Ed.) Micronutrient Handbook, Publishers Fertilization Development and Consultation Organization, New Delhi, pp. 70-71.

Tubbs F.R. (1932) A note on vegetative propagation in tea by green shoot cuttings. Tea Q-5: 154-156.

Tunstall A.C. (1931a) Experiment on vegetative propagation of tea by green shoot cuttings. Bull, Tocklai Experimental Station, pp. 113-114.

Tunstall A.C. (1931b) A note on the propagation of tea by green shoot cuttings. *Quart J Indian Tea Assoc* 4: 49–51.

Wellensiek S.J. (1933) Floral biology and technique of crossing with tea. *Arch Thea Cult* 12: 27-40.

Wickremasinghe R. and Perera V.H. (1966) The blackness of tea and the colour of tip. *Tea Quart* 37: 75-79.

Wight W. (1932) Quart. J. Indian Tea Assoc., pp. 200-217.

Wijeratne M.A., Anandacoomaraswamy A., Amarathunga M.K.S.L.D., Ratnasiri, J., Masnayake B.R.S.B. and Kalra N. (2007) Assessment of impact of climate change on productivity of tea (*Camellia sinensis* L.) plantation in Sri Lanka. *J. Natl. Sci. Found. S.* L. 35(2): 119-126.

Williams E.N.D. (1971) Investigations into certain aspects of water stress in tea. In: Carr M.K.V. and Carr S. (Eds.) Water and the Tea Plant, Tea Research Institute of East Africa, Kericho, Kenya, pp.79-87.

Tea: Technological Initiatives, pp. 241-270
New India Publishing Agency, New Delhi, India
Edited by Niladri Bag, Arundhati Bag and L.M.S. Palni

8

The Flavour of Tea: Chemistry and Advanced Analysis

Prosenjit Biswas and Biswajit Bera

Abstract

In this chapter the origin of flavour in tea have been explained under the light of chemistry of green leaves and made tea. The dependency of flavor in the final product on several factors involved in every step of its manufacturing process has been explained briefly to draw a back ground picture of the subject. On the other hand in-depth explanations on the chemical basis of diversity and complex nature of tea flavour have been given following a simple manner.

Moreover, traditional and recently developed analytical methods have also been reviewed critically. Analytical strategies and details of instrumentation involved to study or quantify the flavour in tea have been depicted in details with their prose and cones under the light of complicacy of the subject. Potential and loop holes of the new technology viz. E-Nose have been explained critically. Finally a hypothesis with little experimental proofs has been explained to meet a probable solution to the present problems in flavour analysis of tea.

Keywords: *Tea, flavour, chemistry, reactions, analysis, instruments, methods, e-nose, integrated solution.*

Introduction

Tea has been accepted globally as the most preferred noncarbonated, nonalcoholic beverage. By convention it is mostly consumed as a warm or hot beverage from the very beginning of civilization. Presently some other forms, in the category of value added products, like ice tea, bottled ready to drink (RTD) tea, instant tea, tea tablet, tea cola etc are also gaining popularity to certain extent. Though tea is commercially recognized as a beverage commodity but due to its health beneficial properties it has been considered as a health drink intended to maintain good health and wellness. With time the trend of its consumption for this propose is increasing enormously due to several new health beneficial properties of tea through extensive research globally. The countries primarily known to have major tea consumption are China, India, Rusia, U.K., Germany, Japan etc. Besides, several other countries also considered tea as a very popular drink. Earlier USA was known to have the lowest consumption of tea but at present tea has become very popular in this country also. As per the recent reports, more than 3 million hectares of the world is now under use for planting tea which is increasing continuously (Pripdeevech *et al.* 2011).The countries known to have the largest tea plantation are China, India, Sri Lanka and Kenya. Most of the tea plants belong to *Camellia sinensis*. It is again sub-divided into Sinensissinensis from China and *Sinensis assamica* from India. Broadly there are three classic types of tea namely Black tea, Oolong tea and Green tea (Wang *et al.* 2008). Among these Black tea is most popular and usually produced in India, China, Sri Lanka, Kenya etc. Oolong tea is produced predominantly in China and Taiwan and Green tea comes mostly from China and Japan (Pripdeevech *et al.* 2011).

Tea as a beverage have gained popularity mainly due to its flavor (Kawakami *et al.* 1995). In general, quality of a tea is evaluated based on the colour, aroma and taste. But a tea with a good aroma always attracts the consumers over others with fewer aromas (Togari *et al.* 1995). Thus in a simple manner it is said that the attractive flavour or aroma boosts up the quality of a tea (Ullah 1967, 1985).

Undamaged green tea leaf on the bush has no flavour but it develops during the processing, especially in the fermentation stage of tea manufacture. According to the origin aroma of a made tea may be classified into two groups. One being the "basic type", developed from the inherent essential oils of green tea leaves, and the other is "acquired" developed from the transformation of certain other substances present in the leaf. The flavour of "basic type" is found to be dependent on genetic character of tea plant and agroclimatic condition. For example, flavour is more pronounced in China type *(Camellia sinensis* var. *sinensis*) and China hybrids specially when cultivated at high altitude to grow slowly in cool condition. The "acquired aroma" develops in post harvest process, largely during withering and fermentation of the leaves where the major chemical transformations takes place into the leaves (Ullah 1967, 1985).

In general, during Black tea manufacturing after plucking the two leaves and a bud *i.e.* young tender shoots from the tea bushes it is kept under shade to remove moisture by high air flow, which is called "withering". When the removal of moisture reaches its desired level the and leaves become spongy. These spongy leaves are then rolled to curl into small sizes. In case of orthodox tea manufacturing system the finely plucked shoots get their final shape during rolling. In case of CTC manufacturing system there is another step called cutting; in which the curled leaves are cut into very small sized fine particles. The next step is same for both CTC and Orthodox manufacturing process which is called fermentation. Practically in this step the curled or cut leaves are kept spread over floor or table or a specially designed machine for a certain period of time. Majority of chemical transformation in tea leaves occur during this step. The chemical components of green tea leaves get oxidized by aerial oxygen catalyzed by some enzyme present in leaves. The original purpose of fermentation was to enhance or retain the flavour of tea leaves. In practice flavour differs in accordance with the degree of fermentation at standard set of other conditions. The details of dependence of flavour on different conditions are discussed in subsequent sections (Ullah 1985, Mukherjee 1966).

In earlier days of tea research attention was paid mainly to the components contributing to tea taste such as polyphenols, caffeine, amino acids etc compared to the components responsible for flavour. In those days the first consideration for setting discrimination of standards of tea quality was level of theaflavins, thearubigins and un-reacted polyphenols *i.e.* catechins present in tea. The ratio of theaflavins to thearubigins near to 0.1 was considered to represent a tea of good quality. The deviation of this ratio from 0.1 was considered deviation of standard good quality.

The HPLC analysis of catechins is practiced in most of the tea laboratories whereas the quantification of theaflavins and thearubigins is being done by the spectrophotometric method of analysis. However, due to the lower accuracy and lengthy multistage sample preparation involved in spectrophotometric method of analysis it is not always appropriate for accurate analysis of large number of samples. Although the HPLC method has been developed to quantify few of the theaflavins but its actual application may still depend on the universality of the commercial authentic compounds. Therefore, presently it has been accepted by the scientists that discrimination of tea of different quality only by the analysis of theaflavins, thearubigins as well as other polyphenols would be inappropriate. The analysis may be called to be complete if and only if the chemical components responsible for flavour i.e.; VFCs (volatile flavour component) are also analyzed simultaneously (Wang *et al.* 2008).

But in reality there are numerous factors which complicate the task of analysis of VFCs. The compounds that contribute to the aroma of any food are usually present in trace to ultra-trace amounts comprising a diverse range of chemical compounds. This chemical diversity may be further enhanced by subsequent processing. Physical properties of these compounds are also very much diverse in nature viz. boiling points varies from very low to very high. This diversity facilitates separations but complicates simultaneous recovery of the full range of aroma compounds. Despite of their differences tea has a common feature that they are characterized by very low levels of aroma compounds (Togari 1995).

The expensive traditional assays for VFC in terms of time and labor might be one of the major reasons for which they have not been assessed properly till date. In addition, the multiplicity of VFC detected in tea, amounting to hundreds, complicates the results. Presently due to the development of more efficient extraction procedures and detection systems research hasnow become possible to be focused on assessment of flavor through analysis of odor active compounds *i.e.* VFCs (Wang *et al.* 2008). Finally, the expression and proper interpretation of chemical analysis data in terms of quality again depends on perfectly established correlations among the two parameters mentioned. This step, again, complicates the whole procedure few folds more due to presence of hundreds and thousands of types of flavours and their constituent chemical composition in tea all over the world (Ullah 1967).

The Chemistry of Flavor

In principle the term 'flavour' describes an integrated perception resulted due to combined effect of the senses of taste and odor and not the perception experienced by accumulation of odor or taste alone (Wang *et al.* 2008). So, more specifically, it results from a multifaceted series of sensory responses of tastes like sweet, sour, salty, bitter, umami etc accompanied by the olfactory responses. Olfactory responses involve large number of descriptors. In addition oral sensations like coolness, astringency and pressure also contribute to the final sense of flavour. Instead of these entire factors odor plays the leading role to sense the flavour. For example, if a food has no odor then its "flavour" is never felt and it is experienced primarily in terms of bitter, sweet, sour, salty etc. For the view point of chemical composition, non-volatile components may impart only taste. While the volatile components give the odour (Kawakami *et al.* 1995).

In general, constituents of the naturally occurring essential oils are volatile in nature and are known to be primarily responsible for the characteristic odour of any food. Eessential oils, in general terms, are lipophilic liquids of plants containing volatile chemical constituents. The nomenclature of the substance is based on the fact that the oil contains "essence" of fragrance of its origin plant. Tea is a naturally flavoured beverage due to the content of its unique essential oil. The exact details about chemical nature of the constituents present in the essential oil of tea are still open to question although the percentage yield is relatively fixed. "Takei and Sakats (1932) obtained a yield of 0.0140% from fresh tea leaves harvested in the spring, and 0.007% from leaves harvested in the summer.

Sen (1960) found 0.016% of steam distillable oil in Darjeeling tea (black) and 0.018% in Assam tea (black). The yield appears to be comparable whether it is isolated from fresh leaves or from manufactured tea. Further, isolation of these chemical constituents has been described in many scientific literature, particularly in short abstracts, but in many cases the sources of these constituents, whether from green leaf or made tea, are not adequately described (Mukherjee 1966).

In a crude manner the main group of compounds prevailing in tea essential oil may be classified into two major groups as Group I & Group II. Group I consists of mainly non-terpenoids and Group II consists of terpenoids. The presence of Group II compounds is highly desirable for a good tea aroma. The classical approach to identify flavour potential of a tea was to use the ratio of concentration of Group II to Group I constituents. The ratio was termed as "Flavour Index" (Yamanishi *et al.* 1986).

Table 1: Terpenoids and Non-Terpenoids responsible for tea flavour

VFC Group I: Non-Terpenoids	VFC Group II: Terpenoids
1-Penten-3-ol	Linalool and Oxides
n-Hexanal	α and β - Ionone
n-Hexanol	Methyl salicylate
cis-3-Hexenal	Phenylacetaldehyde
trans-2-Hexenal	Geraniol
cis-3-Hexenol	Benzyl alcohol
trans-2-Hexenol	2-Phenylethanol
n-Pentanol	Benzaldehyde

Table 2: Flavour of individual compounds

Compounds	Flavour
Hexenals	Green and contributes to mouthfeel
Acetaldehyde	Cooked flavour
β-Ionone	Fresh flavour
Phenyl ethanol	Fruity Note
Phenyl acetaldehyde	Fruity Note
Linalool and its oxides and Geraniol	Floral Sweet note; Rosy
Hexanoic acid	Fatty

According to Takei *et al* the VFC of both green and black tea consist of the

Table 3: Structures of few important VFCs of tea

Geraniol	Linalool
α – Ionone	β – Ionone
Methyl salicylate	Benzyl alcohol
2-Phenylethanol	Phenylacetaldehyde
Hexanoic acid	cis-3-Hexenal

condensation products of large number of aldehydes and primary alcohols like hexenol and hexenal, phenyl ethyl alcohol and citronellol etc. But later on this view was contradicted by many other researchers. Till date 3-Hexen-l ol, 2-Hexenal, Butyl alcohol, Iso amyl alcohol, Iso butyl alcohol, 2-hexen-l ol, 1-Hexanol, Octyl alcohol, Benzyl alcohol, Geraniol, 3-Methyl alcohol, Phenyl ethyl alcohol, Methyl salicylate have been found to be most abundant in tea essential oil. In addition to those many other flavouring compounds have been isolated in trace and ultra-trace amounts. The concentration of 3-Hexan-l-ol in tea is remarkably high and often characterised by a 'fresh green odour'. But when 3-Hexan-1-ol was synthetically prepared and assessed for flavour it did show the same fresh green odour as the natural analogue. There after it was found that the synthetic hexenol had the same chemical formula as that of the natural analogue but it did in fact have a different structure (spatial configuration) and this was the reason for the difference in smell. This is an example which shows that the odour of a compound depends upon its stereo chemical structure also (Takei *et al.* 1937).

On the other hand, in an experiment carried out at Tocklai Tea Research Institute, it was shown that the addition of three volatile alcohols separately to a series of tea infusions in small amounts produced "rawness", moderate amount produced "briskness" and the larger quantity produced a flavour similar to that of "green apples". All of these flavours were different from the actual aroma of tea. So, the conclusion of this experiment was that a single flavour compound though it may be of very good flavour can only form a very small part of the chemical mixture responsible for tea flavour and not at all can sufficiently explain the tea aroma (Mukherjee 1966).

It is already explained before that in terms of origin the aroma complex of tea appears to be classifiable into two groups: (1) Basic aroma which is developed from the inherent essential oils of the green leaf. (2) Acquired aroma which is developed from the transformation of substances in the leaf other than essen-tial oils.

So, hypothetically flavour of made tea might be artificially produced by mixing these two categories along with other, as yet unknown, chemical components for any desired effect. But the question now arises that why Darjeeling and Assam leaf with approximately the same percentage of essential oil content should have such different flavours in the made tea. Had the basic and acquired aroma played the decisive role in causing the flavour of tea? If so, then all made tea irrespective of their origin and place of growth would have been equally flavoury tea. The implication is that flavour is ultimately determined by other chemical components present in the mixture. This anomalous situation leads one to assume that the basic aroma as well as the acquired aroma become latent or submerged in Assam tea because of their higher polyphenolic content and oxidation products which comprise the single majority group amongst all the chemi-cal constituents present in the tea liquor. Besides, the volatility alone is not enough for a compound to produce a sensation in our odour sensor cells. The compound must have to be lipid soluble to fulfil the criterion of being sensed by animal nose. Moreover, acidic/basic or oxidizing/reducing substances generate a sensation of pain because of its chemical reactivity and this will usually hide any true odour if it possess. These are called odour inhibitory chemical constituents. Tea liquor is a mixture of many chemical ingredients having such properties. For example polyphenols which are well known for their reduction properties and the thearubigins which have characteristic acidic properties. Thus the presence of these components in the tea liquor in large amounts may impose an inhibitory effect on its flavour. This may be one of the factors for which Assam teas are less flavoury than Darjeeling tea, because former one usually contains larger amount of polyphenols and their oxidized products than the latter (Mukherjee 1966).

Diversity in Tea Aroma

Quantitative studies on VFCs and their correlation with the aroma patterns of different tea revealed that the chemistry is full of diversity both in number and relative abundance of the VFCs. In many cases it has also been found, interestingly, that a same group of VFC can manifest different aromas in their different set of relative abundances of the components. Few representative cases for some world famous tea have been discussed below.

The most important role in flavor of Darjeeling black tea is played by linalool and geraniol, besides, vanillin (vanilla-like), 4-hydroxy-2,5-dimethyl-3(2H)-furanone (caramel), 2-phenylethanol (flowery), and (E,E,Z)-2,4,6-nonatrienal have also been identified to have the positive impact. Black tea of Assam mostly contains linalool, geraniol and α-terpineol as the key flavour constituents. From China, Chin Shin Oolong tea is dominated by linalool, indole and cis-jasmone whilst the major flavour volatiles of Chin Hsuan Oolong tea are trans-nerolidol, cis-jasmone and geraniol. Green/Oolong tea of Thailand are found to have hotrienol, geraniol and linalool to be the major flavoury components (Mahanta *et al.* 1995).

Different types of made tea from same plants shows remarkable difference in aroma pattern and VFC composition. For example, three different types of Fennel tea, namely, conventional black tea, pre-packaged teabags and instant tea of same origin shows a remarkable difference in VFC composition. Conventional teas from crushed fruits and teas prepared from the other instant tea showed the highest levels of volatile constituents.Anethole and/or anisaldehyde are found in this case to be the main constituents. Methychavicol, eugenol and fenchone are found in most samples. Carvone and camphor presents only in teabag samples and volatile constituents of instant tea includes limonene and α-terpineol (Bilia *et al.* 2002).

In case of value added products the chemistry becomes more complicated as in this case VFC from other source also gets imparted into the system. For example, a study on odorants in Chinese jasmine green tea scented with jasmine flowers (*Jasminum sambac*) found to contain linalool, methyl anthranilate , 4-hexanolide, 4-nonanolide, (E)-2-hexenyl hexanoate, and 4-hydroxy-2,5-dimethyl-3(2H)-furanone as the principle odorants. The enantiomeric ratios of linalool in jasmine tea and *Jasminum sambac* determined by a chiral analysis showed to be 81.6% ee and 100% ee for the (R)-(-)-configuration, respectively. It was reported that the jasmine tea flavor could be closely duplicated by a model mixture containing these six compounds on the basis of a sensory analysis. As per their claim, omission of methyl anthranilate and the replacement of (R)-(-)-linalool by (S)-(+)-linalool may lead to great changes in the odor of the model. Based upon results these two compounds were concluded to be the key odorants of the jasmine tea flavor. It seems to be a story completely different from any generalization in chemistry of tea in its original form (Ito *et al.* 2002).

Formation and Flavour Stability

As mentioned earlier, the fresh green tea leaves usually do not show any significant flavor but after manufacturing it shows different flavours depending upon the different set of conditions (Pripdeevech *et al.,* 2011). Apart from preserving the inherent flavour potentials of the tea leaf, the methods of black tea manufacture is always targeted to influence the development of more pleasant aroma in tea. Orthodox tea manufacture appears to facilitate the production of geraniol, linalool, trans linalool oxide, trans-2 hexenol and cis 3 hexenol type of compounds which contribute more towards the flavour of black tea. Whereas, trans 2 hexenal, n hexanal and phenylacetalde-hyde are found to predominant in C.T.C. tea (Ullah 1985).

In CTC method of tea manufacturing plyphenolic compounds of tea leaves are oxidised fully letting the enzymes to act as per their full potential whereas in Orthodox process the oxidation of polyphenols is restricted and slowed down by giving the leaf a hard wither, which reduces the polyphenol oxidase activity (Ullah and Roy, 1982), followed by controlled cell rupture by only rolling the withered leaf. In case of C.T.C. process of manufacturing light withered leaf with extensive cell rupture, fast and drastic oxidation of the polyphenols dominates all through the fermentation process. Un-like flavoury, light liquored Orthodox tea, C.T.C. produces strong heavy li-quors with less flavour. It appears that, slow Orthodox process with its contro-lled oxidation might facilitate the associated chemical changes accompanying the fermentation process for which the tea flavoury components might be be-tter developed in Orthodox than in C.T.C. process (Ullah 1985).

Degree of fermentation, irrespective of manufacturing procedure, causes a large change in volatile flavour components resulting different characteristic flavours. Both the number and quantity of the flavoury compounds suffer a large change during this step. But in practice, increased flavour is felt only in optimally fermented tea as in this case concentration of flavour inhibitor remains under the optimum condition. The change in flavour continues during the process of drying and takes the final form (Wang *et al.,* 1994). In particular, VFCs with positive impact significantly increase in the final product as compared to that in green leaves (Pripdeevech *et al.*, 2011). The concentration of the compounds like trans-2-hexenal, benzaldehyde, methyl-5-hepten-2-one, methyl salicylate, and indole, alters remarkably during fermentation. The change is so prominent that the profile of these compounds is able to discriminate non-fermented, semi-fermented and fermented tea (Wang *et al.*, 2008). Higher amount of trans-2-hexenol is usually found in non-fermented tea whereas semi-fermentation tea contains significantly lower amounts. Though this result was contradicted with the study of Borse *et al.* (2002) who reported that trans-2-hexenol only can appear significantly in heavily fermented tea. Green tea are not processed through fermentation, in that case it is assumed that the high temperature during the process of steaming or pan-firing bring about higher appearance of VFCs (Borse *et al.* 2002). The same assumption fits for the observation that Green Oolong tea contains a higher number of volatile flavour components than Green tea. Seven constituents, heptanoic acid, nerol oxide, isoborneol, c-terpineol, 2, 6, 6-trimethyl-1-cyclohexene-1-carboxaldehyde, nonanoic acid and trans-β-damascone are the foremost flavour constituents of Green Oolong tea only. Ten components are mainly formed in the semi-fermented tea process. These are palmitic acid, methyl jasmonate, methyl 9, 12, 15-octadecatrienoate, methyl linoleate, bicyclo (3.3.0) oct-1(2)-en-3-one, methyl palmitate, 6, 10, 14-trimethyl-2-pentadecanone, cis-2-(1- pentenyl) furan, geranyl linalyl ester, 2(3H)-furanone and geranyl phenylacetate. In few cases the increase in concentration of phenylacetaldehyde with fermentation time has also been noticed. Presumably, as fermentation proceeds, conditions become favorable for the Strecker degradation of phenylalanine, which is a known precursor of phenylacetaldehyde in tea (Siajo *et al.*, 1972).

Heat processing during manufacturing and packaging of tea is also one of the responsible factors for the change in flavor of black tea. Evidence is also growing that substantial amounts of some compounds are formed during the drying (Giaturco *et al.*, 1974). VFCs like 3-methylbutanal, methional, α-damascenone, dimethyl trisulfide, and 2-methoxy-4-vinylphenolare usually found to show higher concentrations after heat processing of the black tea, therefore, these odorants are involved in changing the black tea odor during heat processing (Kumazawa *et al.*, 2001).

It is noteworthy from a phytochemical viewpoint that flavor of tea as well as aroma constituents are dependent on genetic character genotype of tea plant and the agroclimatic condition of the region of cultivation. Always flavour is found to be more pronounced in China type (*Camellia sinensis* var. sinensis) and China hybrids of tea plants specially when cultivated at high altitude to grow slowly in cool condition (Wickremasinghe, 1974). Tea bushes planted at both of high and medium elevations produce cis-3-hexen-1-01, linalool oxides and linalool more than the bushes planted at lower elevations (Gianturco *et al.* 1974). The China hybrid tea verities produce, in general, a higher concentration of phenylacetaldehyde, linalool, trans lina-lool oxide than the Assam varieties. Trans 2 hexenol, trans 2 hexenal, n hexanal and methyl butanal are produced almost in same extent in both of China and Assam varieties. On the other hand Cis 3 hexenol which imparts a fresh green leafy odour is less in China hybrid (Hazarika *et al.,* 1984).

Precursors of Tea Flavor

In the preceding sections it has been stated that the green leaves of tea usually do not manifest any aroma but after manufacturing the made tea gets its characteristic flavour. In practice, flavour remains within the green leaves even before manufacturing and the process of manufacturing is nothing but the act of unleashing the incumbent potential inherent to it. It is practically impossible to produce a tea with an excellent flavour from the leaves which is poor in its flavour potential. Now the question arises that scientifically what should be called the descriptors of the term "flavour potential". There are few group of compounds viz. amino acids, carotenes, lipids, glycosides etc are called in general "flavour precursor". Here also only few things have been generalized and rest are still in ambiguity (Britton *et al.,* 1995).

Some school of thoughts believe that amino acids play the most important role in production of aroma in tea. There are about twenty five amino acids (including the amides) present in tea and they contribute between 4 and 5 per cent of the total dry weight of fresh tea leaf. Among them Aspartic, glutamic, serine, glutamine, tyrosine, valine, phenylalanine, leucine, isoleucine and theanine (5-N-ethylglutamine) are found to be most abundant in general. Out of all these, thea-nine (ethyl amine of glutamic acid), which is specific for tea only is present in the highest concentration. This alone accounts for 60 to 70 per cent of the total amino acids present in tea. High concentration of theanine is usually present in high quality green tea. These are mainly responsible for the 'acquired' aroma of tea. This assumption is based on the fact that different types of aroma are produced by different am-ino acids when added to hot aqueous solutions of tea catechols (polyphenols), different aromas are produced, which are different for different amino acids. Addition of phenylalanine is found to produce a rose like odour, threonine, a wine like odour and nor leucine, a spicy odour. Unpleasant odour is induced by the addition of tryptophan and tyrosine. There are certain amino acids which produce no odour on chemical transformation. Glutamic acid or alanine produces a flower like odour. All of these artificially produced aromas in the experiment have been found to mars upon addition of aldehyde binding agents in those solutions. It may be due to the fact that amino acids upon reaction with other substrates present in the system giving result to produce some secondary compounds with aldehyde groups having property to impart flavour in the system (Giantcurco *et al.* 1974).

During withering of green leaves, proteins breaks down into free amino acids which gets oxidized to aldehydes by quinones produced from polyphenols in other sequence of reactions. Finally, these aldehydes contribute to the flavour. Few potent flavoring aldehydes viz. coniferyl aldehyde, cinnamic aldehyde, p hydroxybenzaldehyde and vanillin present in manufactured tea are assumed to be produced from reactions of amino acids with quinines during manufacturing (Mukherjee 1966).

Polyphenol No Flavour	→	Quinones No Flavour		
Protein No Flavour	→	Amino acids No Flavour		
Quinones No Flavour	+	Amino acids No Flavour	→	Aldehydes Flavoury substances

The following Volatile Carbonyl Compounds formed from the amino acids during processing:

Glycine	→	Formaldehyde
Alanine	→	Acetaldehyde
Valine	→	Isobutyraldehyde
Leucine	→	Isovaleraldehyde
Isoleucine	→	2-methylbutanol
Methionine	→	Methional
Phenyl alanine	→	Phenylacetaldehyde

Sometimes during manufacturing process, it has been found that the amino acids interact with the VFCs resulting in modification of flavor of the final product due to formation of addition compounds. For example, Glycine has been found to contribute only to significantly decrease some of the carbonyl aroma compounds, that is, 1-octen-3-one and geranial. In a model reaction containing a mixture of 1-octen-3-one and glycine indicated the formation of the tentatively identified *N*-1-(3-oxooctyl) glycine resulting from a 1,4-addition reaction (Britton *et al.* 1995).

The perceived aroma of green tea infusion is likely to be affected by the formation of new aroma compounds and the changes in release of aroma may be affected by interactions with tea nonvolatile components. Though documents on this type of observations are very less in number and deserve further investigations on the sensory level too (Cheng *et al.,* 2008).

Carotenoids are another important group of compounds which contribute largely to the accrued aroma of tea. Carotenoids represent one of the most widely distributed and structurally diverse classes of natural pigments (Britton *et al.*, 1995). Apart from crucial functions in photosynthesis, photoprotection and nutrition, carotenoids are considered to be important aroma precursors. During fermentation of tea leaves these pigments readily degrades into volatile flavour unsaturated aldehydes (Mukherjee, 1966).

Aroma generation from carotenoids is thought to proceed via both of enzymatic and non-enzymatic pathways. Non-enzymatic cleavage includes photo-oxygenation, auto-oxidationas well as thermal degradation processes. The biodegradation of carotenoids is assumed to be catalyzed bydioxygenase systems. Whereas detailed knowledge is available concerning the biosynthesis of carotenoids in plants little is known about the metabolic pathways that lead to the formation of aroma compounds (Isoe 1969; Hoa 2015). In general three steps are required to generate an aroma compound from the parent carotenoid viz (i) the initial dioxygenase cleavage, (ii) subsequent enzymatic transformations of the initial cleavage product giving rise to polar intermediates and (iii) acid-catalyzed conversions of the non-volatile precursors into the aroma-active form (Winterhalter *et al.* 1990).

Step-I (Oxidative Cleavage):

Carotenoid → Primary Cleavage Product.

Step II (Enzymatic Transformation):

Primary Cleavage Product → Non-volatile metabolite.

Step-III (Acid catalysed conversion) :

Non-volatile metabolite → Aroma Product.

One example illustrating these reactions is the formation of â-damascenone from neoxanthin. The primary oxidative cleavage product of neoxanthin, grass-hopper ketone, has to be enzymatically reduced before finally being acid-cata-lyzed converted into the odoriferous ketone. In very limited cases, i.e. formation of α-and β-ionone, the target compound is already obtained after the initial cleavage step (Winterhalter *et al.*, 1990, Winterhalter *et al.*, 2002).

For step I and II of carotenoid cleavage extremely little knowledge is available. Some insight into the enantioselectivity of the enzymatic conversion of the primary cleavage products (step II) has been obtained through determination of the chiral composition of norisoprenoids in natural substrates. But almost no information is available concerning the postulated dioxygenase systems (step I) (Winterhalter *et al.*, 2002).

Presently it is universally established that Orthodox tea manifest superior flavour quality over the CTC tea produced from same green leaves at standard set of conditions. Results of investigation revealed that lipids / fatty acids play the main role to cause such discrimination. Production of VFCs is related to the changes in lipid/fatty acids during manufacturing process. Generally Glycolipids consists of nearly 50% of the total lipids and are rich in linolenic acid. Phospholipids are present in the least amount (15%) and have a high proportion of oleic, linoleic and palmitic acids. Neutral lipids are found in moderate amounts (35%) and have a high content of lauric, myristic, palmitic, stearic, oleic and linoleic acids. Major losses of lipids/fatty acids are observed in the withering process. The other stages of processing *viz.* rolling, fermentation and drying register only a minor change in lipid/fatty acid content (Ramaswamy *et al.,* 2000).

Fatty acid composition in relation to lipolytic and lipoxygenase activities in tea leaf shoots at different stages of manufacturing process have been investigated in a series of black tea manufacturing experiments. More degradation of fatty acids and higher lipoxygenase activity have been found to be characteristics of tea leaves allowed to undergo the withering for longer periods. The oxidative breakdown of fatty acids are usually higher in the later part of the rolling process, while lipolytic activity are more prominent in the withering process as well as the earlier stage of rolling. A gradual decline of lipoxygenase activity as well as fatty acid content has been recorded during the fermentation and drying processes. Although the lipids and fatty acid degradation are directly related to the degree of withering, the volatile content of tea manufactured from hard withered leaves shows unexpectedly lower than that in black tea from that of optimally withered tea leaves (Mahanta *et al.,* 1993). Sequential hydrolysis and lipoxygenase mediated reactions of lipids during withering and rolling shows at least 3-fold increase of hexenol, hexenal, and other related volatiles (Saikia *et al.*, 2002).

Variations of fatty acid compositions, glycosides precursors and activity of enzymes viz. lipoxygenase, glycosidase are used to differentiate seasonal changes of tea flavour. The muscatel flavor of second-flush tea are found to be associated with increased activities of glycosidase in association with several terpenes, phenolics, and aliphatic compounds bound to glycosides, whereas high levels of fatty acids and lipoxygenase activity synthesize more green volatiles in monsoon (Saikia *et al.,* 2002).

Nearly 4-fold increase in floral bend of volatiles is found to be developed due to the hydrolysis of the glycoside precursors throughout the processing stages. Glycosides are known to be precursors of the alcoholic aroma compounds of black tea. They are hydrolyzed by endogenous glycosidase enzymes during the manufacturing process. It is found that primeverosides are 3-fold more abundant than glucosides in fresh leaves but they decrease greatly during the manufacturing process especially during the stage of rolling. After the final stage of fermentation, primeverosides almost disappeared whereas glucosides substantially remain unchanged. These observations reaches to a conclusion that hydrolysis of the glycosides mainly occur during the stage of rolling and confirm that primeverosides are one of the main black tea aroma precursors. It is also supported by the changes in glycosidase activities in tea leaves. The glycosidase activities usually remain at a high level during withering but decreases drastically after rolling (Wang *et al.,* 2001).

During manufacturing of oolong tea, it is found that the amounts of most of the glycosides increase from the solar-withering stage reaching to the highest level at the final stage of manufacturing. It has been noted that no glycoside decreases in its content during the manufacturing process, which quite different from the manufacture of black tea. In addition, the content of alcoholic aroma compounds in the free aroma concentrate remains almost unchanged or slightly decrease and constitute only about 12 and 17% in amount of the whole oolong tea aroma compounds. However, jasmine, lactone and indole are found markedly higher in the final oolong tea products (Wang *et al.,* 2001).

Most of the above mentioned chemical transformations leading to production of VFCs from their precursors involve several enzymes in their reaction pathways of which the role of carotenase and glycosidaseenzymes have already been explained.

In case of the reactions converting glycosides to VFCs glycosidase enzymes plays the leading role. At an acidic pH, β-galactosidase has been found to be the most active among these enzymes at fermentation temperature needed for black tea processing. The enzyme is optimally active at 50°C and at pH 4.0. Monovalent and divalent cations had no effect on the activity of the enzyme. Galactose, a hydrolytic product of this enzyme was found to be a competitive inhibitor. Galactose (5mmol/L) could provide partial, but significant protection against thermal inactivation. Several thiol modibing agents at low concentrations could rapidly inactivate the enzyme; inactivation could be reversed with dithiothreitol. Significant protection against inactivation by methyl methane thiosulphonate (MMTS) could be observed in presence of galactose indicating the presence of an essential thiol at the active site of the enzyme (Mahanta *et al.,* 1995, Halder *et al.,* 1997).

Studies on the activity of tea shoot lipoxygenase showed that with the maturation of tea shoot components lipoxygenase activity increases parallel with the increase in lipid content. The relative proportions of neutral and structural lipids such as phospho- and glycolipids also increase throughout maturation. At pH ranging from 6.5 to 9.5 the enzyme have been found to be present in three forms of isoenzymes. Details studies are still required to gather more information regarding functioning of this enzyme in tea processing system (Saikia *et al.*, 2002).

Finally, it will never be a wise decision to give all credit of flavour to the volatile components only because the non-volatiles also contribute to flavour. Conventionally non-volatiles are not well known to contribute to the tea aroma but few reports are available which reveal their role in producing flavor also. The major non volatiles in tea are polyphenols like theaflavins, thearubigins, epicatechin (EC), epicatechin gallate (ECG), epigallocatechin (EGC) and epigallocatechin-3-gallate (EGCG). These non-volatile compounds contribute indirectly towards the flavour. Sometime they combine with other flavour precursor and give volatile products and sometimes they are degraded into smaller molecules which are volatile in nature. The roles of quinones have already been discussed in earlier sections to produce VFC by reacting with amino acids. One example of the second category of non-volatile producing flavour is Epigallocatechin, which is known to break down into volatiles like pyrogallol (1, 2, 3-trihydroxybenzene). The non-volatiles contribute in several other ways also like maintaining pH balance during processing, inhibition of undesired enzymes etc (Pongsuwan *et al.*, 2008).

Analysis of Flavour

From the very beginning major emphasis in flavor studies has always been given on the identification and quantification of volatiles and their correlation with sensory results. Quantitative measurements have received a little less attention over that of qualitative, although it is well known that the same chemicals can contribute to flavors or off-flavors depending on their concentration (Togari *et al.*, 1995). Identification and quantification of volatile compounds are essential but the correlation of these data with sensory properties is equally important and this requires a large number of quantitative data. But the tusk is complicated by the inherent differences in the response of the olfactory system and instrumental detectors. The latter are designed for maximum linear responses, whereas the human olfactory system is nonlinear, which means sensitization in human olfactory system doesn't depend upon the concentration or amount of the odorant linearly. Some of the odorants sensitize it extremely even in very concentration and few others do the opposite (Schuh *et al.*, 2006). Thus analysis, correlation of analytical and sensory data and identification of compounds contributing to flavor are complex problems (Pripdeevech *et al.*, 2011).

The first challenge, during analysis of flavour, faced by a chemist is the efficient extraction of the essential oils from the sample. As discussed earlier the compounds present in aroma extract of tea are of diverse nature both physically and chemically. Many of them are very close in nature and some are completely different. Both these similarity and dissimilarity simultaneously facilitates and complicates the perfect extraction of them (Zeller *et al.,* 2006).

The oldest method for volatile extraction from tea is hydro distillation followed by solvent extraction using organic non-polar solvent. Dichloromethane is conventionally used in this purpose. Though this is the oldest method of VFC extraction from tea but still it is in use in several laboratories working in this area of research. Till date, few more developed methods have been established and used worldwide. Among them, simultaneous Steam Distillation and Extraction method (SDE), Solid-Phase Micro-Extraction (SPME) and Pervaporation may also be mentioned to be remarkable. Amount and composition of the volatile components extracted by the different methods usually differ considerably. In this section we will discuss briefly the comparative efficiency of different extraction methods (Kawakami *et al.,* 1993, Vitzthum *et al.,* 1975).

Hydro-distillation: In this method water or steam is introduced into the flask containing the samples. During boiling the generated water vapor takes out the vaporized volatile flavour compounds to the condensation flask forming an aziotropic mixture. At the condensation flask the aziotropic mixture of water vapor and VFCs condenses into liquid phase and forms two different layers, one of water which remains bellow and other of VFCs at the top. The VFCs from the two-phase system of water and the organic distillate is extracted by partitioning with suitable solvent and then dried for analysis (Rawat *et al.,* 2007).

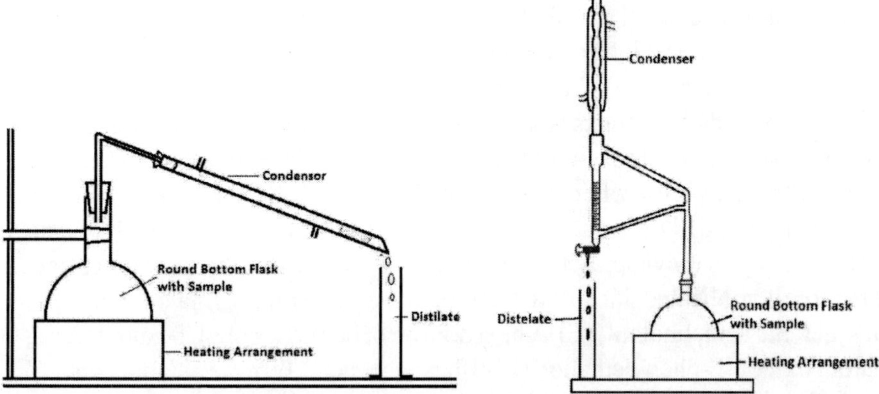

Fig. 1: Hydrodistillation

Fig. 2: Hydrodistillation

Simultaneous Steam Distillation and Extraction method (SDE); the basic principle of this method of extraction is same as that of the hydro-distillation. The exception in this case is that the apparatus is designed/fitted in such an arrangement that the distillation and extraction of VFCs by water vapor and organic solvent, respectively, occurs simultaneously. SDE has been found to be more efficient technique than the hydro-distillation processes for extraction of volatile components in terms of sample size and number of components extracted (Rawat *et al.*, 2007).

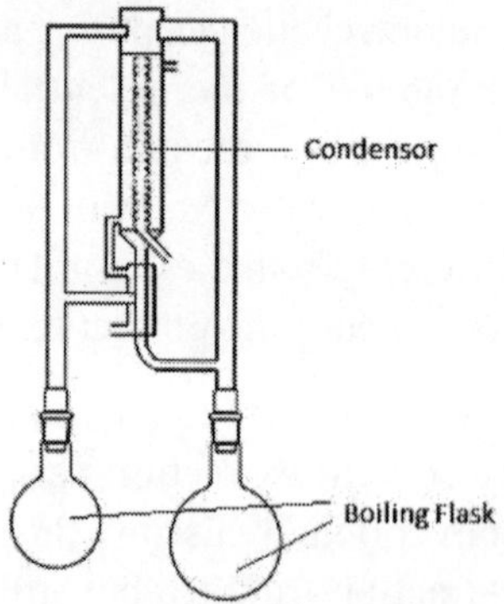

Fig. 3: Simultaneous Distillation and Extraction Apparatus

The main drawback of Hydro-distillation as well as Simultaneous Distillation and Extraction is that it requires a large quantity of sample for flavour extraction. It has also been reported that during the process of extraction in both the cases some reactions among the chemical components present in the sample may initiate due to the harshness of the process conditions. As a result few new volatile compounds may generate which never exists naturally in tea. When such aroma extract is analyzed, the results very unfortunately, only mislead the analyst to a wrong conclusion or great confusion (Masuda *et al.* 2000).

Experimental results has proven the above hypothesis to be true when the chemical profile of a simple tea brew extracted with dichloromethane was compared with that of SDE extract of same samples of black and oolong tea. The volatile components of the two kinds of extracts were analyzed by GC-MS to study their chemical compositions. Results of dichloromethane extract of brewed tea were found to differ greatly from that of the extract prepared by simultaneous steam distillation and extraction method (SDE). The brewed extract was found to contain higher amounts of acids, aromatic alcohols, and mono-terpene alcohols than the SDE extract. The total number of compounds identified in these two kinds of extracts was also found to differ widely. For example, in case of Darjeeling tea thirty-eight compounds were identified from the extract collected from tea brew whereas fifty-two from that of SDE extract. The main components in the SDE extract were geraniol, linalool, linalool oxides 11, I, and IV, and methyl salicylate, while the main components in the brewed extract were more complicated, including linalool oxide 11, geraniol, linalool oxide IV, trans-geranic acid, linalool, (E)-2-hexenoic acid, benzyl alcohol, hexanoic acid, linalool oxide I, 2-phenylethanol, (Z)-3-hexenoic acid. Terpene alcohols such as geraniol and linalool, which are the main Darjeeling tea flavour, were found to be decreased in amount in the brewed extract (Kawami *et al.,* 1995).

Experimental results have also proven that not only the step of isolation of aroma extract is crucial but the efficiency of extraction of VFCs from the aziotropic mixture also is crucial for the analysis. The liquid-liquid partition by solvent extraction followed by drying is very crude in nature with respect to the concentration of the VFCs present in tea. Most recently, due to the development of Solid-Phase Micro extraction (SPME) sampling technique, isolation of VFC from samples has become much more simple process in comparison to the other conventional methods. Solid-Phase Micro extraction (SPME) is a very simple and efficient, solvent free sample preparation method, invented by Pawliszyn in 1989. SPME has been widely used in different fields of analytical chemistry since its first applications to environmental and food analysis and is ideally suited for coupling with mass spectrometry (MS). All steps of the conventional liquid– liquid extraction (LLE) such as extraction, concentration, derivatization and transfer to the chromatograph are integrated into one step and one device considerably simplifying the sample preparation procedure. It uses a fused-silica fiber that is coated on the outside with an appropriate stationary phase. The analyte in the sample are directly extracted to the fiber coating. The SPME technique can be routinely used in combination with gas chromatography, high-performance liquid chromatography and capillary electrophoresis and have no restriction to Mass Spectrometry. SPME reduces the time necessary for sample preparation, decreases purchase and disposal costs of solvents and can improve detection limits. The SPME technique is ideally suited for MS applications, combining a simple and efficient sample preparation with versatile and sensitive detection. SPME relies upon the extraction of solutes from a sample into the SPME absorptive layer. After a sampling period, during which extraction has ideally reached equilibrium, the absorbed solutes are transferred with the SPME layer into an inlet system that desorbs the solutes into a gas (for GC) or liquid (for LC) mobile phase. Success relies upon choosing conditions such that solutes favor the SPME absorptive layer as much as possible in the presence of bulk sample and then are subsequently released as quickly and completely as possible for chromatographic analysis by changing the conditions to favor solute release from the absorptive layer. Secondary trapping and release of desorbed solutes after SPME is sometimes necessary when desorption is too slow to permit full use of column resolving power. This trapping and release can be accomplished using a discrete thermal trap or with column stationary-phase trapping by injection onto a cold column and subsequently temperature programming for solute elution (Wang *et al.* 2008, Aprea *et al.* 2009).

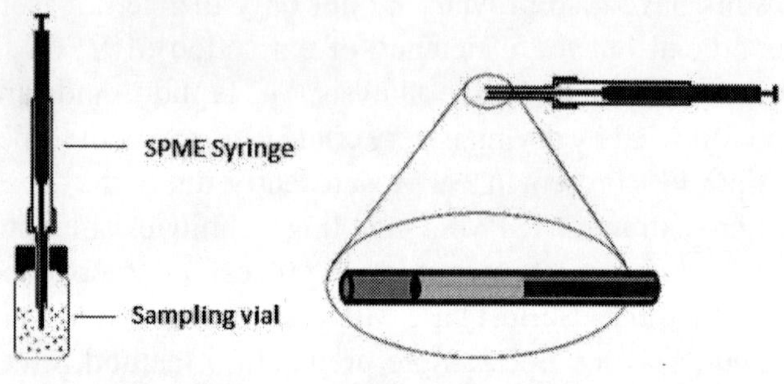

Fig. 4a: SPME Syringe **Fig. 4b:** SPME Fiber

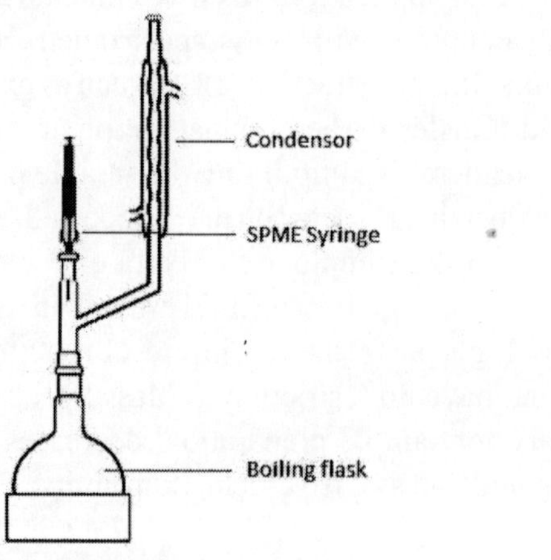

Fig. 4c: Working of a SPME Sampler

Very recently, scientists are trying to explore the possibility of using pervaporation to recover the tea VFCs from tea aroma condensate generated during manufacturing of tea as well as directly from the aziotropic mixture formed during different methods of aroma distillation for VFC analysis (Kanani *et al.* 2003).

The pervaporation method consists of two different steps: (a) permeation of components through the membrane, then (b) their evaporation into the vapour phase. Due to its simplicity and in-line nature this method of purification and analysis of extracts is becoming very much popular in number of industries and research fields. In this method a membrane is used to act as a selective barrier between the two phases viz. the liquid phase and the vapor phase. It allows the desired component(s) of the mixture in liquid phase to transfer through it by vaporization to get purified fractions of single components. Separation of components is based on a difference in transport rate of individual components through the membrane. Typically, the collection side of the membrane is kept at ambient pressure and the feeding side is kept under vacuum. The pressure gradient between the two sides makes the selective components to evaporate after permeation through the membrane. Driving force for the separation is the difference in transport rate and partial pressures of individual components in the feeding mixture. Here it may be noted that the volatility difference of the components never plays any role to separate the components in this process (Kanani *et al.,* 2003, Li *et al.,* 2008).

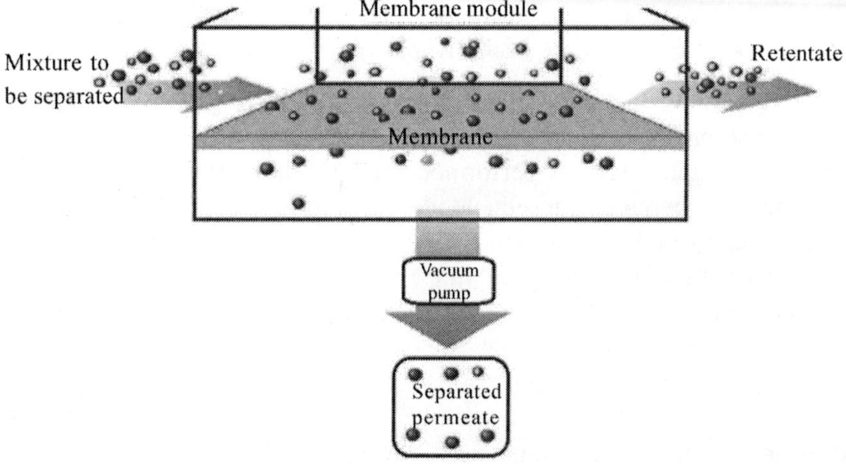

Fig. 5: Pervaporation of FVCs through membrane

The mechanism described above is a typical model mechanism involving polydimethylsiloxane based hydrophobic membranes as the pervaporation membrane. Hydrophilic membranes, instead, are more widely used. Commercially the most successful pervaporation membrane system to date is based on polyvinyl alcohol. More recently polyimide, ceramic, amorphous silica layer *etc.* based membranes found to be used or in the process of development, for example, recently Titania and Zirconia have been introduced into the research to develop novel hydrophilic ceramic membranes. More recently a break-through in hydrothermal stability of the membranes has been achieved through the development of an organic-inorganic hybrid material (Li *et al.,* 2008).

Assuming the potential of this method to extract the aroma condensate prepared during isolation of VFCs from tea for their analysis few scientists have focused their work in its application. Among them the work of Pangarkar *et al.* (2003) may be described as a typical example. Eight important FVCs contributing significantly to tea aroma, namely, *trans*-2-hexenal, linalool, *cis*-3-hexenol, 3-methylbutanal, 2-methylpropanal, benzyl alcohol, phenylacetaldehyde, and α–ionone, were studied in this work. Poly (octyl methyl siloxane) (POMS) and poly (dimethylsiloxane) (PDMS) membranes were used for separation of tea

VFCs by pervaporation system. The study revealed that α-ionone, *trans*-2-hexenal, linalool, *cis*-3-hexenol, and 3-methylbutanal offered very good selectivity, whereas phenylacetaldehyde, 2-methylpropanal, and benzyl alcohol gave moderate selectivity. The results indicate that pervaporation is an attractive technology for the recovery of VFCs from tea flavour condensate as it (i) affords excellent separation and (ii) works under comparatively mild conditions. However, it should be noted that, selectivity of this method varies considerably from compound to compound. Presently, this is only the beginning stage of the method in this area of application and the final form of application method still needs studies on case-by-case basis (Kanani *et al.*, 2003).

Finally, the extracted mixture of VFCs reaches to the stage of identification and quantification of individual components. As the extracts are volatile in nature gas chromatography (GC) is performed for their separation into individual components and thereafter subsequent identification and quantification. A Gas Chromatograph fitted with suitable detector is designed to separate, identify and quantify the components of the extracted VFC mixture. The main components of the instrument are sampler, separation module (column and column compartment) and detector. The column is the narrow tube filled with different packing materials having suitable physical property to separate gaseous mixtures when they will be made to pass through it. During analysis the extracted mixture is fed into the column head, called sampler, through which the components of the mixture pass into a gas stream, called carrier gas, coming into the column from separate arrangement like gas generator or cylinder. In terms of chromatography the stream of carrier gas is called the "mobile phase" and the packing material of the column is called the "stationary phase". The components of the mixture having diverse chemical and physical properties interact differently with the stationary phase of the column during the journey through it. During Gas Chromatography columns are usually kept into a thermostatic chamber capable of heating the column to acquire any desired temperate necessary to perform the separation efficiently. This arrangement is called column compartment. Due to the dissimilar interactions with the stationary phase of the column the individual components acquire different speed during their journey. As a result the components elute from the column end at different times depending on their properties as well as interaction with stationary phase. At the column end there exists the detector(s) which works to identify and quantify the components with the help of the data analysis software (Wright *et al.*, 2007).

The most commonly used detector for VFC analysis is the flame ionization detector (FID). The most efficient detector system coupled with GC is Mass Spectrometer. A Gas Chromatograph fitted with Mass Spectrometer is abbreviated as GC-MS. In this instrument a completely configured mass spectrometer acts as the detector. Some GC-MS are using together with an NMR spectrometer acting as a backup detector. This assembly is called GC-MS-NMR. Sometimes GC-MS-NMR instruments are connected to an Infrared Spectrophotometer which acts as a backup detector. This combination is known as GC-MS-NMR-IR. However, these arrangements are very rare as most analyses can be done via purely GC-MS, more specifically by GC-MS-MS where two mass spectrometers work in a combined mode for higher resolution in chromatograph resulting in more accuracy (Pripdeevech *et al.* 2011).

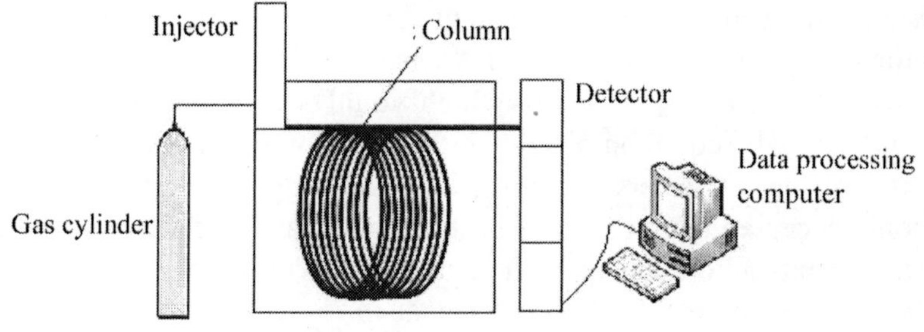

Fig. 6: Typical assembly of a GC system

Few typical examples of columns used are phenyl-polymethylsiloxane capillary column, RTX- 5 ms column, fused silica column, DB-Wax fused silica capillary column etc. Column temperature is generally varied as per the method used by the analyst. Commonly the column temperature ranges from minimum of 40°C to maximum of 250°C, with an increase rate 2-5°C/min are used but it also varies according to the analyst's method. Helium is generally used as the carrier gas in a flow rate usually of 1 to 5ml/min which may be selected any other as per the requirement of the analysis's method (Saikia *et al.* 2002, Kim *et al.* 2011).

Currently Aroma Extract Dilution Analysis is most frequently being used for evaluating ordor-active compounds in tea research. At first the volatile concentrate from a made tea sample or a tea infusion is stepwise diluted with an organic solvent. Diluted fractions are then subjected to GC - Olfactometry with the help a glass sniffing port. Each of the volatile components is then simultaneously evaluated by a panel of human aroma evaluation experts and the glass sniffing port. Finally the dilution factor of each odor-active component is used as an index of its contribution to tea aroma. This technique has been used to identify the most potent odorant compounds in tea and to evaluate the flavour changes that occur during the manufacturing of tea beverages (Kumazawa *et al.* 1999, Kumazawa *et al.* 2001, Kumazawa *et al.* 2002).

Electronic Nose: A New Technology

Though in the above section use of high end equipments e.g. GC, GCMS to determine tea flavour has been described till the most accepted method of flavour determination of tea is the primitive one i.e. judgment by tea tasters or organoleptic analysis is in practice commercially. Tea Tasters, in general, simply measure the aroma of a tea by assessment through their nose and give score in 1 to 10 numeric scales or in some terms. This is perhaps due to the fact that the instruments used for chemical analysis can measure the composition of the flavour extract of the sample but when the term 'aroma' in simple and straight meaning is concerned it fails to give a simple and straight forward result understandable to any common person having no knowledge about chemistry of tea flavour. In this case it is felt that the results of these analysis needs to be down scaled to more simple formats which would be analogues with the 'terms of flavours' used by common people. Further, it should be reported in a simple scale as like the tea tasters do. Now, if these analytical data are not being commonly accepted in commercial sector then what was the driving force to make the scientists to develop such technology? The answer is the accuracy and standard measurement which lacks in the method of flavour analysis by tea tasters. It has been found in most of the cases that the results of tea tasters' vary enormously from one taster to other. Some time it has also been found that the results become biased to any brand, region or season of production over other. Moreover chemical analysis has always been a very costly and time consuming procedure which additionally needs specially trained manpower also (Chatterjee *et al.* 2012).

Now, to find solution further investigations are still going on in defferent field of research oriented to detection and quantification of flavour. Recently to find a single solution of all of the problems stated above application of electronic gas sensors fabricated in a suitable form are attracting the attention of tea researchers. The technology is typically named as Electronic Nose (E-Nose). The technique used is an array of electronic gas sensors selected or newly developed having sensitivity to VFCs of tea is fitted with an user interface viz. a PC or specially designed microprocessor based embedded systems configured to give result in simple format like Aroma type, Aroma Index, Flavour Index, Flavour Score etc (Biswas *et al.* 2014).

The E-Nose systems are usually designed to mimic the functioning of mammalian olfactory systems. These systems gives standard results in simple format and in comparison to the chemical analysis are fast and cost effective solution to the problem concerned. During analysis the electronic sensors suffer conductivity change upon expose to the VFCs. In simple language due to the conductivity change in them a corresponding voltage change is recorded in the data recorder, which in-tern calculated into result of aroma content in desired format through application of different data analysis techniques. A standard data base is usually needs to be fed into the data processing system of the instrument assembly as like the calibration table. This data base is used during analysis to generate the result out-put from the voltage change data. The process of feeding the standard data into the data processing system is called 'training' of the instrument (Zeller *et al.* 2006).

Gas sensors of the kind conductivity sensors, conducting polymer composite sensors, intrinsically conducting polymers, metal oxide sensors, piezoelectric sensors, surface acoustic wave sensor, quartz crystal microbalance sensor, optical sensors, metal-oxide-semiconductor field-effect transistor sensor etc. are usually used in a modern E-Nose system based on the type on chemicals present in the sample (Masuda *et al.* 2000).

The common data analysis techniques generally used in the E-Nose system are Graphical analysis, Multivariate data analysis (MDA) and most recent Artificial Neural Network (ANN) and Radial Basis Function (RBF).The choice of method always depends on the type of responses obtained and purpose of its use (Biswas *et al.* 2014).

Today, reports claiming efficiency of Electronic Nose to determine tea quality are increasing. Various new properties of this instrument like capability of discriminating the flavours of major categories of tea, fermented and non-fermented tea, seasonal tea, regional tea, value added tea etc are increasing. At least, to a conclusion that - an Electronic Nose assembly having correct sensors and suitably fitted with a perfect data processing algorithm can act as a potential instrument in the field of tea quality analysis (Dutta *et al.,* 2003, Dutta *et al.* 2003).

But the most serious problem with this method is that in case of improper selection of sensors and data processing algorithms the results will be anomalous. In general Electronic Nose systems whenever have been tried to determine tea flavour was found to lack in precision in results. Sometimes they have shown the results far apart from the reality (Chatterjee *et al.* 2012). So, till date, though it has been established that this method have benefits over the traditional methods, but the use of the traditional methods is still predominant in the industry. The technology still needs careful research and development works in proper selection of sensors, application of proper data processing algorithms and proper instrument training strategies to give a precise analytical technology (Biswas *et al.* 2014).

Integrated Solution

As stated earlier, tea aroma is basically nothing but the stimulation of the VFCs to the human nose sensors and its reading by brain. So a clear knowledge is basically the primary need for this time on exact count, composition and contribution of VFCs towards the stimulation of human senses generated during sniffing of tea aroma. To reach to any solution for the above said problem, in our laboratory we have formulated a strategy for the study towards a hypothetical model of on integration of the three methods of flavour analyses. During this study tea samples of known quality have been analyzed simultaneously by tea tasters', chemical analysis and by equipments based on electronic sensors. Three set of results produced in this way have been correlated with each other following feature extraction technique to generate a data base correlations among - features of tea in terms of chemical composition and its response to human and electronic sensors. Finally, the correlations found in this approach have been utilized to mimic the human readings on different aromas in electronic sensing instruments. The work is a very much laborious and truly complicated in practice but results are surprisingly hopeful (Chatterjee *et al.* 2012, Biswas *et al.* 2013).

For example, the correlation equation drawn by applying multiple linear regression from the data set generated from CTC tea samples when used for validation by Darjeeling Orthodox Black tea samples it was found to give incorrect results, and a new equation prepared from data set of Orthodox samples by feature

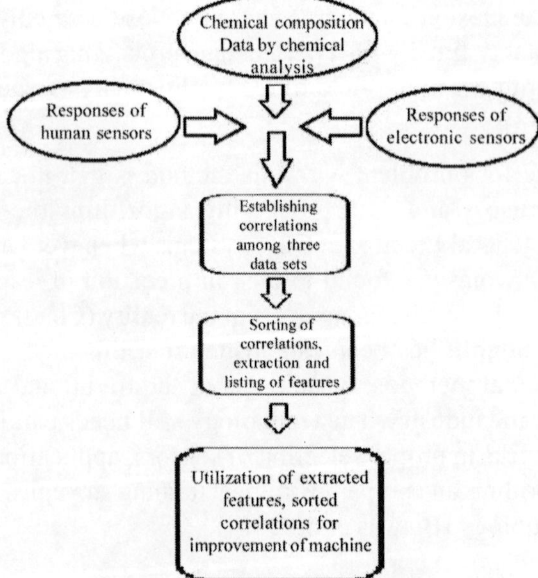

Fig. 7: Flow diagram of integrated approach

extraction technique showed best fit to the results of human sense. In case of sensor selection also this technique worked well. This way, improvement in preinstalled algorithms, creation of new algorithms based on need and for the improvement of hardware part also this strategy is showing its success. This procedure is found to be lengthy to complete collection of a good number of features which would finally contribute to the process of development of this technology in a collective manner. So, it is now only the matter of time and success of the hypothesis (Biswas *et al.,* 2014).

Conclusion

Though the Electronic Nose technique seems to be the simplest solution of the problems, but it still have to go many more miles to reach the ultimate solution. Now electronic sensors specific to individual VFCs are to be developed, algorithms are also to be developed to map every possible combination of VFC composition towards human nose response. Moreover elaborative study on identification and quantification of VFCs still remains a challenge, which needs many more developments in equipments/methods used to get the best resolution. A clear understanding between stimulations of chemical components to human sensors and electronic sensors are still left to be resolved to the final stage.

The hypothesis is based on integration of the three methods viz. tea tasters' evaluation, chemical analysis and use of electronic sensors seems to be promising at this stage towards reaching the ultimate goal for study, analysis and evaluation / identification of tea flavor in more effective and efficient way.

References

Aprea E., Biasioli F., Carlin S., Endrizzi I., Gasperi F. (2009) Investigation of Volatile Compounds in Two Raspberry Cultivars by Two Headspace Techniques: Solid-Phase Microextraction/ Gas Chromatography"Mass Spectrometry (SPME/GC"MS) and Proton-Transfer Reaction"Mass Spectrometry (PTR"MS). *J. Agric. Food Chem 57*(10): 4011-4018.

Bilia A.R., Flamini G., Taglioli V., Morelli I., Vincieri F.F. (2002) GC–MS analysis of essential oil of some commercial Fennel teas. *Food Chemistry* 76: 307-310.

Biswas P., Chatterjee S., Kumar N., Singh M., Majumder A.B., Bera B. (2012) The taste-Appearance relationship of Darjeeling Tea: Cognition Through Human and Electronic Perception Analysis by Chemistry. Proceedings of National Seminar cum Workshop on sensor and Sensing System for Taste Characterization of Food and Agro Produces, IIT, Kharagpur, West Bengal pp. 49.

Biswas P., Kumar N., Singh M., Chatterjee S., Majumder A.B., Bera B. (2013) Chemistry behind uniqueness of Darjeeling tea liquor colour. Proceedings of National symposium on recent Trends in Plant and Microbial Research, University of North Bengal, Siliguri, West Bengal, pp. 21.

Biswas P., Kumar N., Singh M., Majumder A.B., Bera B., Karki I.B., Chatterjee S. (2012) Correlation of tasters scores with biochemical and electronic sensor data for Darjeeling orthodox black tea, IEEE Conference Publication, Sixth International Conference on Sensing Technology (ICST): 769-774.

Biswas P., Chatterjee S., Kumar N., Singh M., Basu Majumder A. and Bera B. (2014) Integrated Determination of Tea Quality Based on Taster's Evaluation, Biochemical Characterization and Use of Electronics, *Chapter 5, Sensing Technology: Current Status and Future Trends II, Smart Sensors,* Measurement and Instrumentation 8:95-117.

Borse B.B., Rao L.J.M., Nagalakshmi S., Krishnamurthy N. (2002) Fingerprint of black teas from India: Identification the regio-specific characteristics. *Food Chemistry* 79: 419-424.

Chatterjee S., Karki I. B., Biswas P., Majumder A. B., Kumar N., Singh M., Bera B. (2012) Zinc oxide nanorod sensing element for detection of tea aroma. IEEE Conference Publication, Sixth International Conference on Sensing Technology (ICST), pp. 567- 570.

Cheng Y., Huynh-Ba T., Blank I., Robert F. (2008) Temporal Changes in Aroma Release of Longjing Tea Infusion: Interaction of Volatile and Nonvolatile Tea Components and Formation of 2-Butyl-2-octenal upon Aging. J. Agric. *Food Chem.* 56: 2160-2169.

Dutta R., Hines E. L., Gardner J.W., Kashwan K.R., Bhuyan M. (2003) Tea quality prediction using a tin oxide-based electronic nose: an artificial intelligence approach Sensors and Actuators 94: 228–237.

Dutta R., Kashwan K. R., Bhuyan M., Hines E. L., Gardner J.W. (2003) Electronic nose based tea quality standardization. *Neural Networks* 16: 847-853.

Gianturco M.A., Biggers R. E., Ridling B.H. (1974) Seasonal variations in the composition of the volatile constituents of black tea. Numerical approach to the correlation between composition and quality of tea aroma. *J. Agric. Food Chem.* 22 (5): 758–764.

Halder J., Bbaduri A. (1997) Nutritional Glycosidases from tea leaf (Camellia sinensis) and characterization of β-galactosidase Biochemistry 8: 378-384.

Hazarika M., Mahanta P. K., Takeo T. (1984) Studies on some volatile flavour constituents in Orthodox Black Tea of various clones and flushes in North East India. *Journal of the Science of Food and Agriculture.* 35: 298-303.

Hoa C.T., Zhenga X. and Lib S. (2015) Tea aroma formation. Food Science and Human Wellness 4(1): 9-27.

Isoe S., Hyeon S.B. and Skan T. (1969) Photo-oxygenation of carotenoids I The formation of dihydroactinidiolide and â-ionone from â-carotene. Tetrahedron Lett. pp. 279-281.

Ito Y., Sugimoto A., Kakuda T. and Kubota K. (2002) Identification of Potent Odorants in Chinese Jasmine Green Tea Scented with Flowers of *Jasminum sambac. J. Agric. Food Chem* 50: 4878-4884.

Kanani D.M., Nikhade B,P., Balakrishnan P., Singh G. and Pangarkar V.G. (2003) Recovery of Valuable Tea Aroma Components by Pervaporation. *Ind. Eng. Chem. Res. 42*(26): 6924-6932.

Kanani D.M., Nikhade B.P., Balakrishnan P., Singh G. and Pangarkar V.G. (2003) Recovery of Valuable Tea Aroma Components by Pervaporation Ind. *Eng. Chem. Res*. 42: 6924-6932.

Kawakami M., Ganguly S.N., Banerjee P.J. and Kobayashi (1995) A Aroma Composition of Oolong Tea and Black Tea by Brewed Extraction Method and Characterizing Compounds of Darjeeling Tea Aroma *J. Agric. Food Chem.* 43: 200-207.

Kawakami M., Kobayashi A. and Kunihiko K. (1993) Volatile constituents of Rooibos tea (Aspalathus linearis) as affected by extraction process. *J. Agric. Food Chem. 41*(4): 633–636.

Kim Y., Goodner L.K., Park J.D., Choi J. and Talcott S.T. (2011) Changes in antioxidant phytochemicals and volatile composition of Camellia sinensis by oxidation during tea fermentation. Food Chemistry 129(4): 1331-1342.

Kumazawa K. and Masuda H. (1999) Identification of Potent Odorants in Japanese Green Tea (Sen-cha). *J. Agric. Food Chem.* 47(12): 5169-5172.

Kumazawa K. and Masuda H. (2001) Change in the Flavor of Black Tea Drink during Heat Processing. *J. Agric. Food Chem*. 49: 3304-3309.

Kumazawa K. and Masuda H. (2001) Change in the Flavor of Black Tea Drink during Heat Processing. *J. Agric. Food Chem.* 49(7): 3304-3309.

Kumazawa K. and Masuda H. (2002) Identification of Potent Odorants in Different Green Tea Varieties Using Flavor Dilution Technique. *J. Agric. Food Chem.* 50(20): 5660-5663.

Li X., Basko M., Prez F.D. and Vankelecom I.F.J. (2008) Multifunctional Membranes for Solvent Resistant Nanofiltration and Pervaporation Applications Based on Segmented Polymer Networks. *J. Phys. Chem.* B 112(51): 16539-16545.

Mahanta P.K., Tamuli P. and Bhuyan L.P. (1993) Changes of Fatty Acid Contents, Lipoxygenase Activities, and Volatiles during Black Tea Manufacture. *J. Agric. Food Chem.* 41: 1677-1683.

Mahanta P.K., Tamuly P. and Bhuyan L. P. (1995) Comparison of Lipoxygenase Activity and Lipid Composition inVarious Harvests of Northeastern Indian Tea. *J. Agric. Food Chem.* 43: 208-214.

Masuda H. and Kumazawa K. (2000) The Change in the Flavor of Green and Black Tea Drinks by the Retorting Process, *ACS Symposium Series* 754(34): 337-346.

Mukherjee S.L. (1966) Aroma complex of tea. *Two and a Bud*, 13(2): 67-69.

Pongsuwan W., Bamba T., Yonetani T., Kobayashi A., Fukusaki E. (2008) Quality Prediction of Japanese Green Tea Using Pyrolyzer Coupled GC/MS Based Metabolic Fingerprinting *J. Agric. Food Chem.* 56(3): 744-750.

Pripdeevech P. and Machan T. (2011) Fingerprint of volatile flavour constituents and antioxidant activities of teas from Thailand. *Food Chemistry* 125: 797-802.

Pripdeevech P. and Machan T. (2011) Fingerprint of volatile flavour constituents and antioxidant activities of teas from Thailand. *Food Chemistry* 125: 797-802.

Ramaswamy R. and Ramaswamy P. (2000) Lipid occurrence, distribution and degradation to flavour volatiles during tea processing. *Food Chemistry* 68: 7-13.

Rawat R., Gulati A., Babu G.D.K., Acharya R., Kaul V.K. and Singh B. (2007) Characterization of volatile components of Kangra orthodox black tea by gas chromatography-mass spectrometry. *Food Chemistry* 105: 229-235.

Saijo R. and Takeo T. (1972) Volatile and non-volatile forms of aroma compounds in tea leaves and their changes due to injury. *Agricultural and Biological Chemistry* 37: 1367-73.

Saikia P. and Mahanta P.K. (2002) Specific Fluctuations in the Composition of Lipoxygenase- and Glycosidase-Generated Flavours in Some Cultivated Teas of Assam *J. Agric. Food Chem.* 50: 7691-7699.

Saikia P., Mahanta P.K. (2002) Specific Fluctuations in the Composition of Lipoxygenase- and Glycosidase-Generated Flavours in Some Cultivated Teas of Assam. *J. Agric. Food Chem.* 50: 7691-7699.

Saikia P., Mahanta P.K. (2002) Specific Fluctuations in the Composition of Lipoxygenase- and Glycosidase-Generated Flavors in Some Cultivated Teas of Assam. *J. Agric. Food Chem.* *50*(26): 7691-7699.

Schuh C. and Schieberle P. (2006) Characterization of the Key Aroma Compounds in the Beverage Prepared from Darjeeling Black Tea: Quantitative Differences between Tea Leaves and Infusion. *J. Agric. Food Chem.* 54: 916-924.

Takei S., Sakato Y. and Ohno M. (1937) Odoriferous substances from green tea. IX. Carbonyl compounds of black tea oil. *Bull. Inst. Phys. Chem. Res.* 16; 773-781.

Togari N., Kobayashi A. and Aishima T. (1995) Pattern recognition applied to gas chromatographic profiles of volatile components in three tea categories. Food Research International 28(5): 495-502.

Togari N., Kobayashi A. and Aishima T. (1995) Relating sensory properties of tea aroma to gaschromatographic data by chemometric calibration methods. *Food Research International* 28(5): 485-493.

Ullah M.R. (1967) Amino acids and tea aroma. *Two and a Bud* 14(4): 163-164.

Ullah M.R. (1985) Aroma constituents of Assam and China hybrid teas and their manifestation during processing. *Two and a Bud* 32(1&2): 60-62.

Vitzthum O.G., Werkhoff P. and Hubert P. (1975) New volatile constituents of black tea aroma. *J. Agric. Food Chem. 23*(5): 999–1003.

Wang D., Ando K., Morita K., Kubota K. and Kobayashi A. (1994) Optical Isomers of Linalool and Linalool oxides in Tea Aroma. *Biosci. Biotechnol. Bwchem*. 58: 2050-2053.

Wang D., Kubota K., Kobayashi A. and Juan I.M. (2001) Analysis of Glycosidically Bound Aroma Precursors in Tea Leaves. 3. Change in the Glycoside Content of Tea Leaves during the Oolong Tea Manufacturing Process. *J. Agric. Food Chem.* 49: 5391-5396.

Wang D., Kurasawa E., Yamaguchi Y., Kubota K. and Kobayashi A. (2001) Analysis of Glycosidically Bound Aroma Precursors in Tea Leaves. 2. Changes in Glycoside Contents and Glycosidase Activities in Tea Leaves during the Black Tea Manufacturing Process. *J. Agric. Food Chem*. 49: 1900-1903.

Wang L. F., Lee J.Y., Chung J. O, Baik J.H, So S. and Park S.K. (2008) Discrimination of teas with different degrees of fermentation by SPME–GC analysis of the characteristic volatile flavour compounds. *Food Chemistry* 109: 196-206.

Wickremasinghe R. (1974) The mechanism of operation of climatic factors in the biogenesis of tea flavour. *Phytochemistry* 13: 2057-2063.

Winterhalter P. and Rouseff R. (2002) Carotenoid-Derived Aroma Compounds: An Introduction, *Carotenoid-Derived Aroma Compounds. ACS Symposium Series* 802(1):1–17.

Winterhalter P., Sefton M.A. and Williams P.J. (1990) Two-dimensional GC-DCCC analysis of the glycoconjugates of monoterpenes, norisoprenoids, and shikimate-derived metabolites from Riesling wine. *J. Agric. Food Chem.* 38: 1041-1048.

Wright J., Wulfert F., Hort J. and Taylor A.J. (2007) Effect of Preparation Conditions on Release of Selected Volatiles in Tea Headspace. *J. Agric. Food Chem*. 55: 1445-1453.

Yamanishi T., Kobayashi A. and Nakamura H. (1986) Flavour of Black tea, V. Comparison of aroma of various types of black tea. *Agriculture and Biological Chemistry* 32: 379-86.

Zeller A. and Rychlik M. (2006) Character Impact Odorants of Fennel Fruits and Fennel Tea. *J. Agric. Food Chem.* 54(10): 3686-3692.

Tea: Technological Initiatives, pp. 271-300
New India Publishing Agency, New Delhi, India
Edited by Niladri Bag, Arundhati Bag and L.M.S. Palni

9

Health Benefits of Tea and Environmental Impacts of Its Cultivation

Jigisha Anand and Nishant Rai

Abstract

Tea is the most consumed, rejuvenating and thirst-quenching beverage. It is bestowed with 'miraculous' health benefits which includes promotion of cardio-vascular health, skin protection, antioxidant activity, defence against microbial infection, prevention from cancer, obesity, and many more. Among the four types of tea, green tea has attained much attention of tea consumers due to its magical health potentials.Owing to increased knowledge and marketing of its health benefits, there has been astonishing changes in its consumption pattern and increase in its production. The tea industry drives economies of the tea growing regions and contributes greatly to the country's GDP growth as well as foreign exchange earnings. However, in recent years, tea farming has become subject to environmental deterioration because of exploitation in cultivation practices. To meet the increase commodities, unsustainable agriculture practices are being followed at the cost of environmental degradation. Desertification of vast habitat converts bare lands into tea farms which consequently lead to unfortunate loss of biodiversity. Monoculture farming is often in tea cultivation and is related with soil erosion, decrease in water retention and increase in water runoff which accompanies indiscriminate removal of wetland habitats. The soil degradation associated with tea plantation is a cycling process which feeds upon itself and increases the environmental degradation associated with it. These degradation processes compel the farmers to rely on chemical inputs to maintain the productivity which further exaggerate soil's degradation. Several research investigation have reported frightening outcomes of pesticides used in tea estates, however it is very unfortunate that still some tea companies permits implementation of unapproved hazardous chemicals beyond their recommended limits.

In the present chapter, disquieting factors such as habitat conversion, land deforestation, soil erosion, and health benefits pertaining to green tea have been highlighted. Also, biodiversity extinctions associated with tea cultivation have been discussed to highlight some alarming environmental and health concerns pertaining to unsustainability of ecosystem. Foresight and stringent consideration of decisive good manufacturing practices should be stringently followed to reduce the global environmental impact of tea cultivation and to increase sustainability.

Keywords: *Camellia sinensis, tea cultivation, environmental impacts, pesticides, sustainability.*

Introduction

Tea is a nature's treasure for the mankind and is one of the most consumed beverages throughout the world. Everyday, a cup of brisk full flavours and beautiful aromatic essence of tea, accompanied with miraculous fragrance and solace in its taste, vanishes the laden moments in our life. The history of tea cultivation and brewing begins in China which is credited to introduce the world with this magical beverage. The principal producing countries of green tea are India, China, Sri Lanka, Kenya, Indonesia and Turkey (Schramm, 2013). An herbal drink with multiple health benefits is derived from the leaves of *Camellia sinensis*, which is an evergreen dicot shrub (Anand *et al.*, 2014).

There are basically four types of tea which are commercially available for drinking. Green tea, white tea, oolong tea and black tea, all of which come from the leaves of *Camellia sinensis* plant. These varieties of tea have varied processing which determines their polyphenolic content and thus their relative health well beings (Table 1). Of the worldwide tea production, 78% is black tea which is usually consumed in western countries, while 20% of green tea (GT) is normally consumed in Asian countries, and 2% is Oolong tea which is produced by partial fermentation in Southern China (Parley *et al.*, 2012).

Table1: Different variety of tea and their processing

S.No.	Types of Tea	Processing
1.	Green tea	Unwilted, unoxidized
2.	White tea	Wilted, unoxidized
3.	Oolong tea	Wilted, partially oxidized
4.	Black tea	Wilted, fully oxidized

Over the last years, numerous epidemiological and clinical studies have revealed enormous health benefits of tea and among the four types of tea; GT known as 'queen of tea', has gained maximum appreciation relevant to promotion of health and prevention or treatment of some chronic diseases. GT not only captures the taste, aroma and colour of spring, but delivers its qualities along with the highest concentration of beneficial phytonutrients and the least caffeine content of all the teas. Its rich source of catechins polyphenols confers the tea with powerful antioxidant, anticarcinogenic and antimutagenic properties (Archana and Jayanthi, 2011) (Figure 1). Due to increasing health concerns and consumers health awareness of magical GT, it is assumed that worldwide consumption of GT is expected to exceed 1.2 million tons by 2015 (San, 2010).

The main tea-producing countries in terms of area planted are China (898,000 ha), India (438,000 ha), and Sri Lanka (180,000 ha) which represents more than 65 % of all land planted to tea globally and nearly 61 % of all production (Figure 2). Other significant producers of tea are Kenya (113,000 ha), Indonesia (110,000 ha), Turkey (76,800 ha), Myanmar (66,908 ha), and Vietnam (7,300 ha). These countries hold 19 % of all land planted to tea and produce 20.4 % of the world's tea (FAO, 2001). With progression of tea cultivation at larger scale in former British colonies of India, Sri Lanka and Kenya, tea can now be brought to market at cheaper rates, and as a result, tea has become widespread in the Middle East, which is a traditional coffee stronghold (Aoyama and Akiyama, 2013).

Tea ranks as one of the most profitable commodities of Indian export market. However, over the last 20 years, India's world ranking as an exporter of tea has fallen down from number one to number three, in the face of stiff competition from Sri Lanka, Kenya, and China (Basu *et al.*, 2010). Currently, India produces 23% of total world tea and consumes about 21% of total world tea (Table 2) with about 2.3 million hectare of area comes under tea cultivation with the global production of 3.0 million metric tons.

Table 2. List of World largest exporter of Tea

Rank	Country	Export value 1000 US$
1.	Sri Lanka	1,476,881
2.	China, Mainland	965,080
3.	India	867,143
4.	Kenya	858,250
5.	United Kingdom	262,959
6.	Germany	229,383
7.	Viet Nam	204,018
8.	United Arab Emirates	191,814
9.	Indonesia	166,717
10.	Poland	139,393

(*Source:* FAOSTAT Database, 2011)

The active polyphenols are the powerful antioxidant which contributes one third of nearly 4000 bioactive compounds present in tea (Tariq *et al.*, 2010). Catechins are the major polyphenols (flavonoids) that are greatly responsible for the antioxidant activities and majorly, epicatechin (EC), epigallocatechin gallate

(EGCG), epigallocatechins (EGC) and epicatechin gallate (ECG) are present in green tea. Among the health claims of green tea includes removal of free radicals, skin protection, prevention of cancer, cardiovascular health, kidney disorders, curing of neurogenerative disorders, diarrhoea, fatigue and inhibition of microbial infection etc. (Anand *et al.*, 2012).

Fig. 1: Multi-potential benefits of Green tea (Anand *et al.*, 2012).

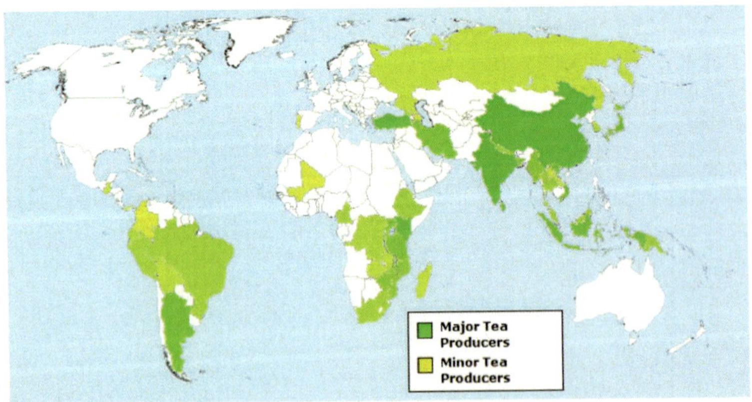

Fig. 2: Main tea producing countries in the World. (Source: FAOSTAT Database, 2008).

Green tea is said to contain over four times the concentration of antioxidant catechins than black tea, about 70 mg catechins per 100 ml compared to 15 mg per 100 ml of black tea. EGCG being the most efficient component of green tea makes it the best candidate for availing maximum health benefits. Antioxidant activity of EGCG is about 25-100 times more potent than vitamin C and E and is the single most studied catechins in relation to health contributing potential (Carmen *et al.*, 2006). Apart from catechins, other green tea polyphenols (GTPs) are flavanols such as kaempferol, quercetin, isoquercetin, myricetin, myricitrin, rutin, kaempferitrin, and their glycosides, as chlorogenic acid, coumaroylquinic acid besides the water soluble anthocyanins (Anand *et al.*, 2014).

Health Benefits of Green Tea

Antioxidant properties

Phenolics are suggested to be the major bioactive compounds responsible for the health benefits of tea (Yang and Liu, 2013). A plethora of evidence suggests strong antioxidant potentials of tea flavonoids. Major catechins present in GT i.e. epicatechin (EC), epigallocatechin gallate (EGCG), epigallocatechins (EGC) and epicatechin gallate (ECG) have strong potentials to protect the body from toxic free radicals. Henceforth, it is appeared that these antioxidants slows or halts the initiation of cancer, heart disease, suppresses immune function and accelerated aging (Hamilton, 2001). It has been reported that EGCG possesses 10 times higher potential than Vitamin C and β- carotene in scavenging the allyl peroxyl radical. However, at the same time evidences in a study suggests a reverse correlation between the amount of phenolic compound in green tea and its antioxidant potentials (Armoskaite *et al.*, 2011). Research investigations have been done which have confirmed rich antioxidant activity of green and black tea leaves (Kopjar *et al.*, 2015).

Application as Nanoparticles

Nanotechnology has emerged as a promising technology that has been advocated for the delivery of antimicrobial phenolic compound extracts. The phenolic compounds can be used as natural and safer substitute to chemical disinfectants in delivery of antimicrobial agents (Ravichandran *et al.*, 2011). There have been some recent efforts to enhance antimicrobials bioavailability by delivering EGCG as lipid nanocapsules and liposome encapsulation, suggesting the possibility of this molecule being developed further by medicinal chemists (Barras *et al.*, 2009).

Synthesis of nanoparticles using biological entities has great interest due to their unusual optical (Lin *et al.*, 2000), electronic properties (Chandrasekharan and Kamat, 2000), chemical (Krolikowska *et al.*, 2003) and photoelectro-chemical (Ahmad *et al.*, 2003). The synthesis and assembly of such nanoparticles is assisted with clean, nontoxic and environmentally acceptable 'green chemistry' procedures, which operates organisms ranging from bacteria, fungi and even plants (Roh *et al.*, 2001; Bhattacharya and Rajinder, 2005).

These days, assimilating green chemistry principles in nanotechnology is a developing area of nanoscience research. Thus, there is an increasing demand for developing environmentally friendly and sustainable methods for the synthesis of nanoparticles that utilize nontoxic chemicals, environmentally benign solvents, and renewable materials to avoid their adverse effects (Sharma *et al.*, 2012).

Silver and gold nanoparticles have been synthesized using various natural products like GT (*Camellia sinensis*) which is non-polluted, environmentally acceptable, and safer for human health (Gardea- Toreesdey *et al.*, 2003; Sharma *et al.*, 2012). The complicated nanoparticle mediated delivery system accompanying phenolic compounds can effectively decontaminate food borne pathogens and improve food safety. Iron (II,III) nanoparticles synthesized by green tea extract (GT-Fe nanoparticles) were found to have negative ecotoxicological impacts on important aquatic organisms (Markova *et al.*, 2014) while causing degradation of aqueous pollutants such as organic dyes, chlorinated organic, or arsenic as well as degrading malachite green (MG) in aqueous solution (Weng *et al.*, 2013).

Effectiveness in Skin Damages

Several *in vitro* and *in vivo* animal and human studies have suggested that GT Polyphenols (GTP) are photo protective in nature, and can be used as pharmacological agents against solar UVB light-induced skin disorders such as photo aging, melanoma and non-melanoma skin cancers (McKay and Blumberg, 2002; Wu and Wei, 2002; Lee *et al.*, 2004; Katiyar, 2011; Godic *et al.*, 2014).

Tea is effective in the area of skin care, particularly in alleviating the symptoms of acne and eczema. When used in a combination with sunscreen, GT enhances sun protection. Due to the presence of antioxidants, GT is also effective in impeding aging process. GT catechin protects from oxidative stress-induced cell death in fibroblasts and thus has potential as a therapeutic agent against skin aging (Tanigawa *et al.*, 2014).

GT has been found to reduce the release of pro- inflammatory cytokines such as IL-1β, IL-6, IL-8, TNF- α and prostaglandin E-2 (PGE-2) in human white blood cells in culture. Anti inflammatory properties of GT have been reported at University of Rochester Medical Centre, USA, where effective treatment of patients with severe skin deterioration following cancer radiotherapies has been reported (Pajonk *et al.*, 2006).

Oral Health

Tea has a rich diversity of minerals and is thus a natural source of fluoride. It acts as a natural vehicle for delivery of fluoride to oral cavities and promotes healthy teeth and gums (Kushiyama *et al.*, 2009). The mean fluoride concentration in GT is ~2.1 ppm, which lies within the acceptable daily intake. It is believed that by applying GT as a mouth cleanser; approximately 34% of the fluoride is retained and interacts with oral tissues, and their surface integuments (Simpson *et al.*, 2001). The fluoride prevents tooth decay, tooth loss and oral cancer (Lee *et al.*, 2004; Okamoto *et al.*, 2004) and helps in whitening of teeth's (Tahir and Moeen, 2011).

The frequent consumption of GT greatly reduces halitosis, i.e., bad breath by inhibiting growth of periodontopathic bacteria which produces volatile sulfur compounds associated with halitosis (Liao *et al.*, 2001). GT has proved to have anti- *Streptococcus mutans* (Neturi *et al.*, 2014) and anti-*Porphyromonas gingivalis* (Chatterjee *et al.*, 2012) activity which helps in reducing dental caries and periodontal implications. In consequence, GT has been considered as functional food for oral health and is widely used in toothpaste formulation. Greater is the concentration of catechins, better will be the health benefits. So, the consumption of GT in comparison to other beverages may be widely recommended (Venkateswara *et al.*, 2011).

Green tea have been shown to inhibit the growth and activity of bacteria associated with mouth infections. Green tea mouthwash containing 1% tannin could reduce the aerobic mouth bacterial load about 26%-32% and eventually it may prevent plaque formation on teeth and therefore, halitosis (Moghbel *et al.*, 2011). GT can be used as adjunct to conventional therapy of oral candidiasis and can prevent oral candida infection (Doddanna *et al.*, 2013).

Anti-Obesity Effect

Many researchers have investigated association of obesity with major risk factor like diabetes, cardiovascular diseases, metabolic diseases and several forms of cancer such as breast, colon and prostate, and many more. Several evidences suggests the efficient role of GT in fat metabolism by reducing food intake, interrupting lipid emulsification and absorption, suppressing adipogenesis, lipid synthesis and increasing fat oxidation, fecal lipid excretion and energy expenditure via thermogenesis, (Huang *et al.*, 2014).

The consumption of GT extract is associated with a significant reduction in total and low density lipoprotein cholesterol levels (Kim *et al.*, 2011). Regular consumption of GT along with routine exercise induces abdominal fat loss in overweight and obese adults (Maki *et al.*, 2009).

Epigallocatechin gallate (EGCG) have been shown to reduce adipocytes differentiation and proliferation, lipogenesis i.e., birth of new fat cells; fat mass, body weight, fat oxidation, plasma levels of triglyceride, free fatty acids, cholesterol, glucose, insulin leptin and increase beta oxidation and thermogenesis (Wolfram *et al.*, 2006). The meta-analysis suggested that green tea consumption had a favorable effect on decrease of BP (Li *et al.*, 2014).

Prevents Vision Loss and Hair Loss

GT "catechins" are among a number of antioxidants likewise vitamin C, vitamin E, lutein, *etc.*, which are capable of protecting the eye (Chu *et al.*, 2010). A study conducted at Chosun University College of Medicine in Korea discovered that the GT antioxidant EGCG have photo protective effect and can protect human retina against UV damage. They concluded that EGCG increases the cell count and the cell activity after UV irradiation in cultured human retinal pigment epithelial cells (Yang *et al.*, 2007).

In an animal trial, it was found that GT may acts as a putative anti-cataract agent (Gupta *et al.*, 2009) and possesses significant antioxidant defence system against age related macular degeneration and glaucoma (Zhang *et al.*, 2006). In a recent *in vivo* study, epigallocatechin gallate catechin (EGCG) eye drops were found effective in preventing UVB radiation–induced corneal oxidative stress. The study showed that the protective effects of EGCG may be due to an increase in the activity of the antioxidant defence system and inhibition of lipid per oxidation and protein oxidative modification (Mu-Hsin *et al.*, 2014).

GTPs were understood as positive factors in hair growth and follicle health, adopting inhibitory mechanism like apoptosis (programmed cell death), radioprotection of follicle cells, profound antioxidant activity, and potential follicular inhibition of TGF-beta (Charles *et al.*, 2008). GTP has been recognized as a reliable herbal inhibitor of dihydrotestosterone, the hormone linked to hair loss in men (Patil *et al.*, 2010). A high intake of GT correlates to higher levels of sex hormone-binding protein globulin (SBGH) which carries hormones like testosterone around the body in a bound, unusable form so that tissues cannot use it directly. Testosterone is usually carried around the body by this binding protein, therefore, the reducing levels of free testosterone, so that it cannot be converted to dihydrotestosterone (DHT) in the hair follicle, which is responsible for hair cycle shortening and loss of hair in men. GT is thought to affect 5α-reductase type I enzyme which converts testosterone to DHT (Patil *et al.*, 2010). Although these findings are at preliminary stage, but these studies suggest that further analysis in this regards can prove to be promising in future. In a recent investigation, a hair tonic formulate of GT ethanolic extract at varied concentrations (2.5%, 5%, and 7.5%) was found stable, safe and effective towards hair growth (Amin *et al.*, 2014).

Antimicrobial Activity

Polyphenols in tea possesses efficient inhibitory potentials against opportunistic as well as pathogenic microorganism. In several investigations, antimicrobial activity of tea extracts have been reported which represent them as a powerful candidate of herbal drug therapy. The first documented report on an antibacterial action of tea was made in 1906, when McNaught, a British Army surgeon, showed that tea killed the causal organisms of typhoid fever (*Salmonella typhi*) and brucellosis (*Brucella melitensis*) (McNaught, 1906). Antimicrobial effect of tea polyphenols have been studied against *Candida albicans* (Doddanna *et al.*, 2013), *Staphylococcus aureus, Vibrio parahemolyticus, Clostridium perfringens, Bacillus cereus, Pleisomonas shigelloides* and *Aeromonas sobria* (Tiwari *et al.*, 2005), *Escherichia coli*, Pseudomonas aeruginosa (Anand *et al.*, 2015).

The GTPs have been found to be inhibitory against, *Enterococcus faecalis* (Garg *et al.*, 2014), *Salmonella typhi, Staphylococcus aureus* and *Pseudomonas* (Archana and Jayanthi, 2011). Inhibitory effect of aqueous and ethanolic extracts of GT was reported against *Streptococcus mutans* and *Lactobacillus acidophilus* (Tahir and Moeen, 2011) and food borne pathogens in microbiological media and food (Kim *et al.*, 2001).

GT is considered as a prebiotic since the active polyphenols provides benefit to the host by inhibiting pathogens growth and regulates commensal bacteria including probiotics (Yang *et al.*, 2003). In a similar way, GT inclusion had shown to enhance growth of lactic acid bacteria and aerobic bacteria counts in ruminants, which again helps in improving and maintaining intestinal microbial balance (Bureenok *et al.*, 2007).

GT is also known to inhibit the reproduction and growth of medically important bacteria, like *Salmonella*, *Clostridium* and *Bacillus* (Takabayashi *et al.*, 2004). Inhibitory effect of GT catechins on *Helicobacter pylori* infection has been reported (Mowafy *et al.*, 2011). Antifungal activity of GT catechins against *Candida albicans* and *Aspergillus fumigates* has been explored (Archana and Jayanthi, 2011). These findings suggest that regular consumption of GT can help us to combat with frequent microbial infections.

Antiviral potentials of GT have also been reported which suggest that GTP prevents viral attachment and entry into cells. It acts as anti-mutagenic and hence protects RNA and DNA integrity to reduce mutations that can lead to drug resistance. With their immunoregulatory potential, production of healthy lymphocytes have shown to be stimulated up to 300% and those of immune system killer cells up to 400%. In a new study, it has been shown that EGCG can attenuate neuronal damage mediated by JAK/STAT1 pathway and protects form cytokine IFN-gamma neurotoxicity and AIDS related dementia (Brian *et al.*, 2006). Also *in vitro* studies have revealed anti viral activity of GT catechins against adenovirus, Epstein-Barr virus herpes simplex virus and influenza virus (Matsumoto *et al.*, 2011; Smeeton, 2011).

Effective in Renal Failures and Diabetes

Decreased kidney function due to aging and kidney failure are a frequent cause of death. Pertaining to renal failure, GT has shown some promising outcomes as indicated in some preliminary studies. *In vivo* studies have shown significant recovery of glomerular filteration rate following the GT consumption in animals with kidney dysfunctions such as streptozotocin (STZ)-induced diabetic nephropathy (Mowafy *et al.*, 2011). It prevents glycogen accumulation in the renal tubules, probably by lowering blood levels of glucose. Similarly, anti-diabetic effects of GT flavonoids have been suggested in several humans and animal-based studies where improvement in glucose tolerance and insulin sensitivity in individuals with diabetes have been reported (Wu *et al.*, 2004; Iso *et al.*, 2006; Wolfram *et al.*, 2006). An aqueous solution of green tea polyphenols (GTP) was found to inhibit lipid peroxidation (LP), scavenge hydroxyl and superoxide radicals *in vitro*. Administration of GTP (500 mg/kg b.wt.) to normal rats increased glucose tolerance significantly (Rani *et al.*, 2013).

Lowering of blood glucose level insulin, triglycerides and free fatty acids upon fasting have been reported following 12 weeks administration of GT in diabetic rats and the ability of their adipocytes to respond to insulin and absorption of blood sugar greatly increased (Fiorino *et al.*, 2012). Therefore, GT could be a beneficial additional therapy in the management of diabetic nephropathy (Renno *et al.*, 2008).

Cardiovascular Defence

A positive inter-relationship between GT consumption and cardiovascular health has been developed through multiple epidemiological, clinical and experimental studies (Velayutham *et al.*, 2008). In a study conducted in Japan, impacts of GT consumption were evaluated and was found that higher green tea and coffee consumption were inversely associated with risk of cardiovascular diseases and stroke in general population (Yoshihiro *et al.*, 2013). It is assumed that those who drink at least three cups of GT every day, are at a 2% lower risk of suffering a stroke as compared with those who drink less than a cup a day (Lenore *et al.*, 2009).

Regular drinking of GT seems to lower the chance of getting high blood pressure and inhibits atherosclerosis. GT has been shown to effectively lower LDL Cholesterol, triglycerides, lipid peroxides and fibrinogen while improving the ratio of bad good cholesterol i.e. ratio of LDL to HDL cholesterol. The potent antioxidant effect of GT inhibits the oxidation of KDK cholesterol in the arteries which plays a major contributor role in the formation of atherosclerosis. Thromboxane and Angiotension converting enzymes (ACE) are key factors responsible for loss of arterial elasticity leading to high blood pressure and arterial constriction. GT catechin plays an essential role in blocking thromboxane and ACE production and appears to be their natural inhibitor which significantly reduces the blood pressure (Nileeka *et al.*, 2011).

Synergistic Activity with Antibiotics

With an increasing awareness of the side effects of antibiotics and chemical agents, there has been an increase in evaluating the prospects of natural products, plants and their extracts as new sources of antimicrobial agents (Bink *et al.*, 2011). It has been anticipated that the drug discovery and development may not necessarily be limited to developing new molecule entities, however, new avenues for research of carefully and rationally designed, synergistic traditional herbal formulations can also open in herbal remedies (Patwardhan and Mashelkar, 2009). The foreseen advantages of combination therapy encompass broad spectrum, rapid and synergistic potency of drug activity, reduced dosage of toxic components and lowered risk of drug resistance (Chanda and Rakholiya, 2011; Tamma *et al.*, 2012).

Keeping in view the above prospects of combinational drug therapy, the synergistic effects of GT catechins and antibiotics have been explored. A recent investigation reported that the antibacterial activity obtained using boiled water GT extract is enhanced in combination with Penicillin G against *Bacillus subtilis* bacterium (Smeeton, 2011). There is an enhancement in the antimycotic effects of amphotericin B and fluconazole against *Candida albicans* when used in combinations with GTP (Navarro-Martinez *et al.*, 2006). The combined use of GT methanolic, acetone extracts with ampicillin with ampicillin significantly reduced the growth of *Staphylococcus aureus*, *Pseudomonas aeruginosa* and *Escherichia coli* (Anand *et al.*, 2015).

GT extract in combination with probiotics significantly reduces the viable count of *Staphylococcus aureus* and *Streptococcus pyogens* (Ping *et al.*, 2008). Also synergy between GT extract with levofloxacin, and chloramphenicol, has been reported against enterohaemorrhagic *Escherichia coli* (Isogai *et al.*, 2001) and *Shigella dysenteriae* (Tiwari *et al.*, 2005).

Therapeutic Potentials against Neuronal Disorders

One of the recent advancement in investigation of putative positive role of tea for human health is the evidences reported by researchers where GT extract have shown to enhance the cognitive functions i.e., the working memory. The findings suggest promising clinical implications for the treatment of cognitive impairments in psychiatric disorders such as dementia. However, the neural mechanisms underlying this cognitive enhancing effect of GT remained unknown (Schmidt *et al.*, 2014).

Considerable health promoting qualities of GT has been revealed which demonstrates therapeutic potentials of GTP for nerve degenerative diseases such as Parkinson's and Alzheimer's disease. Interestingly, in combinational drug therapy, GT catechins with anti-inflammatory drugs and other immune modulating compounds, might offer an effective strategy for prevention and treatment of the disease. The synergistic effects of GT with anti- Parkinson's drug "rasagiline" were observed where lower doses of the GT and rasagiline restored the activity and replenished level of dopamine, an affected neurotransmitter in Parkinson's disease (Reznichenko *et al.*, 2010). GTPs have shown neuroprotective activity against development of Alzheimer's disease (Okello *et al.*, 2011) and antidepressant-like effects in mouse behavioral models of depression (Wei-Li *et al.*, 2012).

Rheumatoid Arthritis and Osteoarthritis

GTP offers a promising novel and effective strategies for treatment and prevention of Osteoarthritis (OA) and Rheumatoid Arthritis (RA) (Anand *et al.*, 2012). Studies have showed vital role of EGCG in protection of human chondrocytes from IL-Iα induced inflammatory responses (Akhtar *et al.*, 2011).

Although phase-controlled trials are yet to be performed to confirm the efficacy of EGCG or GT extract in human RA or OA, an extensive evaluation of the potential risks or benefits of using EGCG alone or together with anti-rheumatic drugs may open a new area of research wherein EGCG or its synthetic cognates could be developed to enhance its clinical appeal (Ahmed, 2010).

It has been proposed that combinational use of GTPs and currently available anti-rheumatoid arthritis/osteoarthritis drugs can have a synergistic chemo-protective effect and the application can prove beneficial in overcoming the adverse effects of such toxic drugs when used alone (Katiyar and Raman, 2011).

Antiallergy Potentials

EGCG, is believed to be the primary source of GT's beneficial effects (Fujimura *et al.*, 2002). In a recent study, mast cell stabilizing and anti-anaphylactic activity of phenolic and aromatic compounds in aqueous extract of GT have shown positive effects in treatment of asthma and allergic rhinitis (Balaji *et al.*, 2014). However, oolong tea, which is one of the four types of tea have been reported to have higher anti-allergic activity. O-methylated derivative of EGCG (-)-epigallo-catechin-3-O-(3-O-methyl)-gallate isolated from Oolong tea, is reported to have more potent inhibitory effects on type I and IV allergies in mice than EGCG (Tachibana *et al.*, 2000).

In vitro studies suggest immunoregulatory effects of GT on human IgE responses. However, these anti-allergy findings are bases on preliminary investigation, further, animal and human trials are yet to be conducted to further understand the mechanism of inhibiting the IgE response by GT extract. Catechins in GT play a significant role in combating Seasonal allergic rhinitis (SAR). As per the research analysis, GT will deliver promising anti-allergy response, if prior to pollen exposure, GT is consumed consecutively for 1.5 month (Yamamoto *et al.*, 2009).

Anti-Carcinogenic Activity

In the past decade, GT's cancer-preventive effects have been widely supported by epidemiological, cell culture, animal and clinical studies. Abundant experimental and epidemiologic evidence have been accumulated from *in vitro* and *in vivo* studies which depicts chemo-preventive role of GTP in skin, lung, oral cavity, oesophagus, stomach, intestine, colon, liver, pancreas, bladder, mammary gland, and prostate cancers (Yang and Wang, 2010; Henning *et al.*, 2011). EGCG can inhibit tumorigenesis during the initiation, promotion and progression stages in animal models of carcinogenesis (Lambert and Elias, 2010). Recent human pilot studies have shown preventive effect of GT against lung cancer (Yuan, 2011). The evidence suggests that green tea catechins modulate breast cell carcinogenesis and need further investigation in the clinical setting of chemoprevention of high-risk women (Yiannakopoulou, 2014).

As per the analysis, role of GT in breast cancer development in humans is still unclear due to relatively small number of epidemiological studies on GT and breast cancer (Wu and Butler, 2011). Randomized clinical trials have demonstrated efficacy of GT catechins on treatment of cervical lesions and external genital warts. It has been found that women who are at frequent intake of GT, are at 37% less risk of developing colorectal cancer (Butler and Wu, 2011; Yang *et al.*, 2011). The genotoxic effect associated with some anticancer drugs such as Trenbolone and Docetaxel have been reportedly countered GT extracts (Guptaa *et al.*, 2013).

As more and more people are getting aware of multipotentials and health attributes of tea, especially GT, there has been a tremendous increase in its production, combined with relatively constant consumption. With, increased media attention, health consciousness among the people and greater inclination of people for natural ingredients, market for GT extracts is rapidly growing worldwide (San, 2010).

However, this global enhancement in tea production and consumption is associated with their adverse environmental impact in conservation of forest habitat and soil conservation. To overcome the huge global demand, every year great area of land would be required which accompanies massive alteration of habitats for farming tea. The practice will ultimately means some plant and animal species native to that area to suffer which would be a serious concern for the sustainability of ecosystem (Clay, 2004).

According to Global Industry Analysts (GIA), the global market for hot beverages like GT is forecasted to reach US$ 69.77 billion in value and 10.57 million tons in volume terms by the year 2015 (GIA, 2011) and to meet the huge demand of tea in global market, strategies must be adopted to meet up the challenges in the coming years (Basu *et al.*, 2010).

There are some alarming environmental impacts of cultivation of tea plants that need urgent attention to be analysed and rectified.

Environmental Impact of Tea Cultivation

Destruction of Habitat

The tea plants are mainly cultivated in tropical and sub-tropical humid climates, which is a favourable habitat for tropical and sub-tropical forest ecosystems rich in biodiversity. With the increasing demand for tea, more and more land is frequently getting bared and converted into tea plantations. Such occupancies are also the habitat to rich biodiversity which is adversely affected since tea growing regions practice monoculture production system.

One such exploitation of huge biodiversity was reported from tea growing regions in North East India where the densely covered land covered with variety of grasslands, marshes and a host to diverse range of flora and fauna (tigers and rhinos), have been converted to tea plantations. Despite the opposition and strict prohibition from the concern authorities, forests are still being cleared to make way for new plantations. In year, 2011, a tract of Ethiopian rainforest was sold to grow tea, despite opposition from Ethiopia's President and environmental authorities (William, 2011).

Nelliampathy, the second biggest abode of the most endangered lion-tailed macaque after the famous Silent Valley National Park in Kerala, India, is facing destruction of its habitat due to unregulated plantation activities, fragmentation and conversion of forest land. This significant loss of habitat associated with tea plantations has led to the decline of the lion-tailed macaque in India (Palakkad, 2011). In a yet similar case it has been reported that, one of the world's rarest primates the Horton Plains Slender Loris found only in Srilanka, have been driven to the brink of extinction due to habitat destruction.

Experts believe the prime reason for its rarity was due to loss of its natural forest habitat largely destroyed by the drive to create tea plantations (Smith, 2010). Now, both Lion Tailed Macaque and, the Horton Plains Slender Loris are on IUCN's Red List of endangered species (William, 2011).

Tea Plantation

Tea cultivation which provides good ground cover with deep rooted tea implants reduces the risk of soil erosion. However, area where tea seedlings are replanted, vulnerability to erosion is more prominent. The monoculture practices don't support ecosystem functions and led to decrease in water retention and increase in water runoff which considerably favours soil erosion. Consequently, there is indiscriminate removal of wetland habitats and frequent deforestation which causes destructions of natural habitat that supports biodiversity.

In a research conducted on soil degradation of Rwandan tea growing area of Africa, processes of soil degradation and chemical element concentration in tea-growing regions was assessed. It was concluded that accelerated soil degradation due to erosion is caused not only by topography but also by human activities which involves both the physical loss and reduction in the amount of topsoil associated with nutrient decline (Mupenzi *et al.*, 2011).

The soil conditions under intensive tea plantations are not conducive to high soil fauna populations and their activity. With intensive tea farming, productivity of soil reduces and the land is rarely left uncultivated. Thus, it doesn't get an opportunity to rest and replenish, leading to nutrient-sparse soils that are easily degraded and washed or blown away. Also, with frequent human induced trampling of the soil during tea harvesting, much of soil macrofauna populations (particularly their biomass) get further reduced (Senapati *et al.*, 1994).

Over 18% of all tea growing regions in India lie within the Western Ghats which are biodiversity hot spot. Large scale intensive agriculture especially tea cultivation has become one of the reasons for the wide-spread decline in amphibians population in these regions (Ranjit, 2003). Likewise, a tremendous decrease in biodiversity has been observed in the Tanzanian Usumbara Mountains, a hotspot of unique species, in streams near tea plantations in Tanzania, Africa (Williams, 2011).

The soil degradation associated with tea plantation is a cycling process which feeds upon itself and increases the environmental degradation associated with it. This degradation process compel farmers to rely on chemical inputs to maintain the productivity which further add on to the soil degradation leading to decrease productivity.

Indiscriminate Agrochemicals Input

Monocultures practices provide suitable environmental conditions for growth and survival of pests that compel the farmers for indiscriminate use of toxic pesticides. These harmful agrochemicals have long lasting hazardous impact upon soil quality as well as on local wildlife (Van der Wal, 2008). Also, the monocrop production and its associated chemical inputs reduce the soil biodiversity and soil organic matter. The subsequent leaching of polluted soil carrying toxic pesticides into the neighbouring rivers is associated with killing of natural fish flora, and causing harm to the animals and people dependent on these polluted rivers (Michael, 2013).

India hold 11% of World tea exports and is the leading supplier of tea to countries like Russia, U.S., UK and Germany and however, despite the implementation of alternative ecological pest control practices, use of hazardous agrochemicals is still practiced.

Pesticides poisoning is frequent in India and carries a high mortality and morbidity (Goel and Aggarwal, 2007). Of the total burden of acute pesticide poisoning, the majority of deaths occurs due to self-poisoning with organophosphorus pesticides (OP), aluminium phosphide paraquat, glyphosate, pyrthoids and other pesticides. Of these, highest case fatality rates have been reported with poisoning due to aluminium phosphide, endosulphan and paraquat which are considered as serious pollutants (Nagami *et al.*, 2005; Srinivas *et al.*, 2005; Van *et al.*, 2005) (Nagami *et al.*, 2005; Srinivas *et al.*, 2005;Van *et al.*, 2005).

Pyrethroids possess health risks to immediate environment and can also be quite toxic to fish, downstream organisms, and certain beneficial insects such as bees. With its frequent application, it can even deplete the ozone layer (Fareed, 1996). On the other hand, Glyphosate is a selective herbicide whose intensive input into fields has caused new weed problems, and rendering plant species resistant to the chemical. Paraquat, which is commonly known as Gramoxone, is a broad-spectrum herbicide and an alternative for glyphosate. However, paraquat too possesses serious environmental implications which have raised a serious threat for ecosystem. Due to insignificant and unmarked application of this deadly chemical, about 22 different species of weeds in 13 countries have become resistant to paraquat (Khosya and Gothwal, 2012).

Currently, Paraquat which is described by US Environmental Protection Agency as "extremely biologically active and toxic to plants and animals" and by the Environmental Risk Management Authority of New Zealand as "very ecotoxic to the aquatic environment" (Watts, 2011), has been banned in 32 countries (including the countries of the European Union) mainly because of its associated health disabilities.

Shortness of breath due to lung damages, damage to kidney, liver, oesophagus and skin and birth defects in rodents and frogs are some of the side effect which have been reported after consumption of paraquat. Despite these alarming, incidences, Indian tea industry until recently has used these pesticides that had been banned in developed countries (Clay, 2004).

In a report from Assam, four elephants were reported dead in Kaziranga National Park, India, after they wandered into a tea plantation and ate contaminated grass which had been sprayed with pesticides. In 2011, deaths of cows and vultures in the Assam region of India has also been blamed on pesticides and has led to renewed calls for its use to be banned (William, 2011).

Processing

Energy is a critical input for tea industries and its costs constitute 30 % of the total tea processing cost. It is consistently required for operations such as withering, processing (rolling/cutting), fermentation, drying, sifting and packing. All these operations make use of different forms of energy i.e. electrical, thermal and human. More than 80% of the energy is occupied by thermal energy which is required to remove moisture from tea leaves during withering and drying (Rudramoorthy *et al.*, 2000), whereas electrical energy is required at almost all stages of unit operations.

As per United Nations Environment Programme (UNEP), about 8kWh of energy is required to process one kilogram of finished tea, compared with 6.3 kWh for the same amount of processed steel (Van der Wal, 2008). The drying process of tea leaves is an essential step required to improve the taste, flavour and nature of the tea leaves. It also reduces moisture which prevents moulding and increases shelf life of the tea.

Drying of tea requires a lot of energy (Brouder, 2013) and can be accomplished by four different ways namely steaming, baking, roasting and sun-drying in which various fuels (wood or gas) are used. Tea industries are some of the most important fuel wood consumers in rural areas. In particular, logging for firewood as a part of tea processing has provoked extensive deforestation in countries such as Kenya, Sri Lanka, Malawi and India. As per the information from WWF, in Sri Lanka it takes between 1.5 and 2.5 kWh/kg of tea (Michael, 2013). The total specific thermal energy consumption in Sri Lanka and India varies between 4.45 and 6.84 kWh/kg made tea while it is about 10 kWh/kg made tea in Vietnam (Baruah *et al.*, 2012). This high energy use can have devastating environmental impacts, which varies among different countries. For example, in India, use of firewood in the drying process has led to severe deforestation. In parts of East Africa, where power is expensive and unreliable, many tea factories have had to install polluting standby diesel generators to meet their needs (Van der Wal, 2008). Considering how the wood is harvested, there are large implications for its environmental impacts since most of wood that is used for drying tea comes from harvesting in natural forests.

Packaging

World-wide, loose tea leaves are preferred, with only small percentage of tea is available as in form of tea bags. Generally, tea bags are chosen for its ease in utility and disposability. However, tea bag instantly looses their freshness and thus remains fresh under ideal conditions for only six months. Whereas, loose tea leaves can last their freshness and potentials for up to two years. Loose tea usually comes in a container with a liner keeping the environmental constrains. But, the environmental impacts of packaging in case of tea bags are alarming and they are not considered too much extent.

Traditionally, a special, high quality paper derived from Abaca (a type of banana tree) were used to make tea bags, but in last few years, a major rush by tea companies has been observed which are now using nylon, and PET that has caused some concerns regarding chemical leaching (Michael, 2013). Even paper tea bags have some health concerns and they may be even worse than the plastic tea bags because many of them are made of polymer known as epichlorohydrin which is a potent carcinogen as considered by National Institute for Occupational Safety and Health, NIOSH (Mercola, 2013).

The paper tea bags are manufactured from a blend of wood and vegetables (hemp) fibres, pulp which are usually chlorine- bleached (as a part of their processing). Chlorine dioxide (CLO_2), dioxin [2, 3, 7, 8-Tetrachlorodibenzo-p-dioxin (TCDD)] are some common chemicals which are used in pulp bleaching instead of elementary chlorine. Bleached filters or tea bags can leach toxic dioxins and epichlorohydrin into milk, coffee, and other foods with which they come in contact which could be linked to cancer, hormone disorders and developmental problems (Fredpereira, 2012).

Preventive Measure following Better Management Practices (BMPs)

Several preventive measures have been recognized which can have convincing effects on elucidating environmental impacts of tea production. The guidelines of better management practices or BMPs allow tea manufacturers to improvise their plantation practices and encourage them to increase biodiversities within their plantation area since many of these biodiversity helps in reducing the input of pesticides into the fields. The BMPs involve following attributes:

Biodiversity Conservation

This is one of principle of BMP's which focuses on conserving the biodiversity before any new plantings are undertaken. Its prime emphasis is on improving yields rather than increasing planting and habitat conservation. Abandoning the non- profitable tea zones like steep slopes, shallow soils, alkaline soils and poorly drained lanes, which are home to the precious flora and fauna, much of the biodiversity residing within these zones existing tea estates can be enhanced. It will often results in higher net producer profits and help in preventing biodiversity loss.

It has been reported from China that ancient tea gardens were bestowed with more enriched biodiversity as compared to the present tea estates. On comparison their life form, different types of rare and endangered life forms have been indicated in ancient tea gardens indicated that ancient tea garden as compared to secondary forest and normal tea garden where protected species were not found (Qi *et al.*, 2013). Indiscriminate deforestation within the tea gardens premises has removed such endangered plants from the tea gardens. Thus, it has become very much obligatory for the responsible authorities to check the unlawful practice within such plantation zones such that the healthy existence of biodiversity can be protected from their extinction.

Tea plantations accompanying rare and endangered species of shade trees should be encouraged which will support the native species with their frugivores activity to colonize the regions. It will help in seed dispersal which will provide suitable conditions for the colonization and abundance of several other plant species to flourish at the tea gardens (Chetana and Ganesh, 2012).

Tea is a long lived plant, and preserving small patches of tea bushes particularly old varieties in an area that is targeted for replantation would a beneficial strategy in conservation of biodiversity. This practice also promotes conservation of genetic material that may be useful for tea propagation and developing disease resistance varieties which might become a threat for tea conservation in future.

Reduction of Soil Erosion and Degradation

Tea cultivation is often practiced in areas which receives high levels of rainfall and have considerable slopes. Circumstantial erosion occurring most extensively during periods of planting or replanting in such vulnerable location follows reduction in soil fertility and the tea crop yield. The massive soil destruction also affects locations away from the actual site of erosion known as off site effects where damaging effects can be apparent including direct damage to neighbouring tea lands, siltation of irrigation canals, reservoir sedimentation which consequently reduces affected reservoirs life span and pollutes the water ways (Jayanth, 1998). Also, there are some indirect impacts on environment which is followed by increase in fertilizer application, energy consumption, and change of adaptability of land uses (Yan *et al.*, 2003).

To overcome sever loss of soil and its devastating side effects under tea cultivated land, it is crucial to have plantation ground covered all the time. The practice of ground cover prevents erosion by protecting the soil from splash erosion and increases the infiltration from over land flow. Different cover crop species such as butterfly peas (*Centrosema pubescens*), wild ground nut (*Calopogonium mucunoides*) and silverleaf (*Desmodium uncinatum*) can be used as ground covers which are not grown for harvesting. These crops grown underneath the tea plants in a compact sheet cover the soil and prevent it from degradation (Emanuelsson and Rasmusson, 2012).

Tea pruning is yet another way which can be implemented to cover bare soils in order to prevent soil erosion. If an area is to be replanted, then the explored ground should be mulched to conserve the moisture and improving the fertility of soil. Replantation should be takes done as early as possible.

Contour planting is another practice which can also be implemented to prevent soil erosion in slopes (Ekanayake, 2010). Single band of grass such as napier or elephant grass (*Pennisetum purpureum*) can be established every five to ten rows of tea to supplement contour planting which will contribute in ground covering for soil all the year around. The grass vetiver, *Chrysopogon zizanioides* can be grown in tea croplands to protect the soil from getting washed away. Due to its solid root system it stabilizes the soil and prevents erosion caused by water runoff. Also, it does not demand frequent maintenance and is very persistent against disturbances like flooding, draught and diseases.

Performing Organic Farming

The bioorganic fertilization technique and the principles of biological management of soil fertility with soil biota and organic matter have great potential for widespread application, particularly in agro-forestry systems. Switching to organic tea farming would protect habitats and wildlife, safe-guards the productivity of the land, and reduces the need for pesticides.

Because agro-chemicals could not sustain increase in tea production and their input cannot restore the soil fertility, therefore, it is essential to search for such alternatives which can support in recovering soil's original characteristics (as in the forest), i.e., its biological, physical and chemical properties, before it gets degraded (Senapati *et al.*, 2002). The biofertilizers acting as a source of nitrogen and phosphorus will instantaneously increase the efficiency of shoot and root system and will support the healthy growth of tea plants. For quality seedling production of tea plant, due consideration should be given to the biofertilizers such as VAM, Azospirillum and PSB rather than chemical inputs. This technology could be adopted in tea growing areas to improve the rate of growth, soil structure, enhance productivity in cultivable field and also helps reducing the usage of chemical fertilizers which subsequently reduces the cost of seedling production (Nepolean *et al.*, 2012).

Biofertilizer can be also manufactured by composting tea pruning and high quality organic matters. Instead of using harmful pesticide in tea plants, verm-composting practice should be encouraged to prevent degradation of soil and damage to the crop. Replacing toxic herbicides, efficient plant extracts can be applied into the fields to prevent the tea plantation from debilitating effects of herbicides (Bakry, 2013).

Owing to the improvement of tea plant growth and productivity, KDHP VermidermaTM has been designed which is a unique bioformulation prepared with high quality vermicompost manufactured scientifically from pruned tea leaves enriched with *Trichoderma viride*. It is a proven non-pathogenic and ecofriendly microbe which can control many pathogen from causing damping off, root rot, rhizome rot, stem rot and wilt in tea crop species. This extraordinary compost enriched with multiple nutrients like N, P, K, Ca, Mg and Zn, improves tea growth and the presence of the biofungi restricts the prevalent diseases of tea crop.

Alternative Energy Source

For tea estates situated at high altitude, receiving high annual rainfall and all season river flows, switching to renewable energy could prove as benevolence to them for their tea manufacturing and this practice can make them favourable sites for hydropower projects (Van der Wal, 2008). Mini hydel schemes can be successful in tea estates which are located in hilly regions with mountainous terrains. South Indian tea plantations are best suited for the use of hydel power plants as they are blessed with streams and ravines to provide them with required water sources for energy production. If solar energy is used for tea processing, there would be around 50% of saving on fuel as compared with fuel woods for drying (Rudramoorthy *et al.*, 2000). Biomass energy can also be generated from wood wastes like briquettes which are made from sawdust or charcoal dust and can be better choice against wood (Githiomi and Oduor, 2012).

Conclusion

Tea is one of the most consumable drinks after water and with the increase in its popularity pertaining to its wide health attributes in particular of green tea, the consumption of tea is expected to increase moderately in few decades. Plethora's of scientific evidences have depicted health benefits of green tea and have advocated its role as a powerful nutraceutical and health supplement. Researchers have shown antimicrobial potential of green tea and attempts are being made to utilize GT as an impressive candidate in herbal drug therapy. All these affirmations relating to health benefits have consequently accelerated the absolute belief of its consumers and henceforth, commercially there is tremendous hike in consumer demand for this magical beverage.

To meet consumers demand, concerned authorities are increasing production and processing of tea. However, most of their actions are operating with negligence to the affiliated detrimental effects on environment. The present status in prospects of protection and conservation of rich biodiversity and ecosystem is alarming pertaining to environmental sustainability in tea farms. It is indeed a high time that the tea manufacturing units must implement the possible means of saving rich biodiversity and environmental sustainability of nature. As far as the production is concern, it is inhuman to sacrifice our precious biodiversity and accommodating increase production by desertification and converting rich biodiversified land into tea plantations. Adoption of the guidelines under GMPs can greatly enhance the production of tea crop. Authorities should have check on indiscriminate and improper use of hazardous pesticides by farmers. It should be ascertained that tea farming should be practiced at sites which does not scarifies eco-balance. At the same time, value addition and diversification for a wide range of tea products need to be encouraged in order to maintain balance between supply demand chains.

References

Ahmad A., Senapati S., Khan M.I., Kumar R., and Sastry M. (2003) Extracellular biosynthesis of monodisperse gold nanoparticles by a novel extremophilic actinomycete *Thermomonospora* sp. *Langmuir*. 19(8): 3350-3553.

Ahmed S. (2010) Green tea polyphenol epigallocatechin 3-gallate in arthritis: progress and promise. *Arthritis. Res. Ther.* 12 (2):208.

Akhtar N., and Haqqi T.M. (2011) Epigallocatechin-3-gallate suppresses the global interleukin-1beta-induced inflammatory response in human chondrocytes. *Arthritis. Res. Ther.* 13(3): R93.

Amin J., Simamora E.L.P., Anwar E. and Djajadisastra J. (2014) Green tea (*Camellia sinensis*, l.) ethanolic extract as hair tonic in nutraceutical: physical stability, hair growth activity on rats, and safety test. *Int.J. Pharm. Pharma Sci.* 6 (5): 94-99.

Anand J., Gautam P., and Rai N. (2015) Comparative Study of antibacterial and anti-proliferative potential of green tea from different geographical locations in India. *Asian J. Pharm. Clin. Res.* 8(1): 253-258.

Anand J., Rai N., Kumar N. and Gautam P. (2012) Green tea: A magical herb with miraculous outcomes. *Int. Res. J. Pharm.* 3(5): 139-147.

Anand J., Upadhyaya B., Rawat P., and Rai N. (2014) Biochemical characterization and pharmacognostic evaluation of purified catechins in green tea (Camellia sinensis) cultivars of India. 3 Biotech. DOI 10.1007/s13205-014-0230-0.

Anand J., Gautam P., and Rai N.(2015) Comparative study of antibacterial and anti-proliferative potential of green tea from different geographical locations in India. *Asian. J. Pharm. Clin Res.* 8(1): 253-258.

Aoyama N., and Akiyama N. (2013) Tea consumption increasing, beating coffee. Asahi Shimbun Globe (Article). Retrieved from http://ajw.asahi.com/article/globe/feature/tea/AJ201302100024

Archana S ., and Jayanthi A. (2011) Comparative analysis of antimicrobial activity of leaf extract from fresh green tea, commercial green tea and black tea on pathogens. *J. App. Pharma. Sci.* 1(8): 149-152.

Armoskaite V., Ramanauskiene K., Razukas A., Dagilyte A., Baranauskas A., and Briedis V. (2011) The analysis of quality and antioxidant activity of green tea extracts. *J. Med. Plant. Res.* 5(5):811-816.

Bakry F.A., Eleiwa M.E., Taha S.A., and Ismil S.M. (2013). Comparative toxicity of Paraquat herbicide and some plant extracts in *Lymnaea natalensis* snails. Toxicol. Ind. Health. [Epub ahead of print]

Balaji G, Chalamaiah M., Hanumanna P., Vamsikrishna B., Jagadeesh K.D.J., and Babu V.V. *(2014).* Mast cell stabilizing and anti-anaphylactic activity of aqueous extract*of green tea* (*Camellia sinensis*). *Int. J. Vet. Sci. Med.* 2(1):89-94.

Barras A., Mezzetti A., Richard A., Lazzaroni S., Roux S., Melnyk P., Betbeder D., and Monfi lliette-Dupont N. (2009) Formulation and characterization of polyphenolloaded lipid nanocapsules. *Int. J. Pharm*. 379 (2):270-277.

Baruah B.P., Khare P., and Raoa P.G. (2012). The energy utilization pattern in tea industries of NE India and environmental issues. *Two and Buds*. 59(2):9-13.

Basu M A., Bera B., and Rajan A. (2010) Tea Statistics: Global Scenario. *Inc. J. Tea Sci.* 8 (1): 121-124.

Bhattacharya D., and Rajinder G. (2005) Nanotechnology and potential of microorganisms. Crit. Rev. Biotechnol. 25 (4):199–204.

Bink A., Pellens K., Cammue B.P.A., and Thevissen K. (2011) Anti-biofilm strategies: How to eradicate Candida biofilms? *Open. Mycol. J.* 511(29): 29-38.

Brian G., Demian O., Hauyan H., Jin Z., Nan S., Veljko N., Jared E., Douglas S., Francisco F., and Jun T. (2006) EGCG mitigates neurotoxicity mediated by HIV-1 proteins gp120 and Tat in the presence of IFN-ã: Role of JAK/STAT1 signalling and implications for HIV-associated dementia. *Brain. Res.* 1123(1):216-225.

Brouder A.M. (2013). 19 factors driving the future of the tea industry. Forum for the future. Retrieved from http://www.forumforthefuture.org/blog/19-factors-driving-future-tea-industry

Bureenok S., Tamaki M., Kawamoto Y., and Nakada T. (2007) Additive effects of green tea on fermented juice of epiphytic lactic acid bacteria (FJLB) and the fermentative quality of rhodesgrass silage. Asian-Aust. *J. Anim. Sci.* 20(6):920-924.

Butler L.M., and Wu A.H. (2011) Green and black tea in relation to gynecologic cancers. *Mol. Nutr. Food. Res.* 55(6):931-40.

Chanda S., and Rakholiya K. (2011) Combination therapy: Synergism between natural plant extracts and antibiotics against infectious diseases, In: Mendez-Vilas A. ed. Science against microbial pathogens: communicating current research and technological advances, FORMATEX.520-529.

Carmen C., Artacho R., and Gimenez R. (2006) Beneficial Effects of Green Tea—A Review. *J Am. Coll. Nutr*. 25(2) 79-99.

Chandrasekharan N., and Kamat P.V. (2000) Improving the photo-electrochemical performance of nanostructured TiO_2 films by adsorption of gold nanoparticles. *J. Phys. Chem.* B. 104 (46):10851-10857.

Charles E.I., Guang Y.W., Weimin X., Jun H.J., Lisa R., Christopher C., Vincenzo D.M., Edmund C.J., and Sharon H. (2008) Epigallocatechin Gallate Inactivates Clinical Isolates of Herpes Simplex Virus. Antimicrob. *Agent. Chemother.* 52(3): 962–970.

Chatterjee A., Saluja M., Agarwal G., and Alam M. (2012) Green tea: A boon for periodontal and general health. *J. Indian. Soc. Periodontol*. 16(2):161-167.

Chetana H.C., and Ganesh T. (2012). Importance of shade trees (*Grevillea robusta*) in the dispersal of forest tree species in managed tea plantations of southern Western Ghats, India. *J. Trop. Ecol*, 28: 187–197.

Chu K.O., Chan K.P., Wang C.C., Chu C.Y., Li W.Y., Choy K.W., Rogers M.S., and Pang C.P.(2010) Green tea catechins and their oxidative protection in the Rat Eye. *J. Agric. Food. Chem.* 58 (3): 1523-34.

Clay J. (2004). World Agriculture and the Environment: A Commodity-By-Commodity Guide to Impacts and Practices. Island Press, Business & Economics: 570.

Doddanna S.J., Patel S., Sundarrao M.A., and Veerabhadrappa R.S. (2013) Antimicrobial activity of plant extracts on Candida albicans: An in vitro study. Indian. J. Dent. Res. 24(4):401-405.

Ekanayake P.B. (2010). Agronomic and Cultural Practices. Caring tea. Retrieved from http://caringtea.blogspot.in/2010/07/agronomic-and-cultural-practices.html

Emanuelsson C., and Rasmusson E. (2012). The effects of soil erosion on nutrient content in smallholding tea lands in Matara district, Sri Lanka. Thesis. Bachelor degree thesis, 15 credits in Physical geography and ecosystem analysis Department of Physical Geography and Ecosystems Science, Lund University. Page 5-6.

FAO (Food and Agriculture Organization of the United Nations). (2001) Medium term outlook for tea. Committee on Commodity problems. 14th session of the Intergovernmental group on Tea. New Delhi, India. Available at http://www.fao.org/docrep/meeting/003/Y1419e.htm

Fareed M. (1996). Tea and Environmental Pollution. *Tea and Coffee Trade Journal*. 168 (12).

Fiorino P., Evangelista F.S., Santos F., Magri F.M.M., Delorenzi J.C.M., Milton G.M., and Farah V. (2012) The Effects of Green Tea Consumption on Cardiometabolic Alterations Induced by Experimental Diabetes. Exp. Diabetes. Res. Article ID 309231: 7.

FOASTAT database. (2008) retrieved from http://faostat.fao.org/site/339/default.aspx

FOASTAT database. (2011) retrieved from http://faostat.fao.org/site/339/default.aspx

Fredpereira (2012). Tea: The disadvantages of a TEA BAG. Fredpereira's blog. Retrieved from https://fredpereira.wordpress.com/2012/04/26/tea-the-disadvantages-of-a-tea-bag/

Fujimura Y., Tachibana H., Maeda-Yamamoto M., Miyase T., Sano M., and Yamada K. (2002) Antiallergic tea catechin, (")-epigallocatechin-3-O-(3-O-methyl)-gallate, suppresses Fc-epsilonRI expression in human basophilic KU812 cells. *J.Agric.Food.Chem.* 50(20):5729–5734.

Gardea-Torresdey J.L., Gomez E., Peralta-Videa J., Parsons J.G., Troiani H.E., and Jose-Yacaman M. (2003) Alfalfa sprouts: a natural source for the synthesis of silver nanoparticles, *Langmuir*. 19(4): 357- 1361.

Garg P., Tyagi S.P., Sinha D.J., Malik V., and Maccune E.R. (2014) Comparison of antimicrobial efficacy of propolis, Morinda citrifolia, Azadirachta indica, triphala, green tea polyphenols and 5.25% sodium hypochlorite against Enterococcus fecalis biofilm. *Saudi. Endod. J.* 4(3): 122-7.

GIA. (2011). Global Industry Analysts, Inc. Report on Global Hot Beverages (Coffee and Tea) Market. Available from: http://www.strategyr.com/Hot_Beverages_

Githiomi J.K. and Oduor N. (2012). Strategies for Sustainable Wood fuel Production in Kenya. *Int. J. App. Sci. Tech.* 2(10):21-25.

Godic A., Poljsak B., Adamic M.,and Dahmane R. (2014) The Role of Antioxidants in Skin Cancer Prevention and Treatment. Oxidative Medicine and Cellular Longevity. 2014 (Article ID 860479) 6 pages.

Goel A., and Aggarwal P. (2007). Pesticide poisoning. *Nat. Med. J. India*. 20(4):182: 192.

Gupta S.K., Selvan V.K., Agrawal S.S., and Saxena R. (2009) Advances in pharmacological strategies for the prevention of cataract development. *Indian. J .Ophthalmol*. 57(3): 175–183.

Guptaa J., Siddiqueb Y.H., and Afzalc M. (2013). Protective role of green tea extract against genotoxic damage induced by anticancer drugs in cultured human lymphocytes. *Food. Biol.* 2(1):08-13

Hamilton M., and Shah S. (1999) Activity of the tea component gallate and analogues against methicillin resistant Staphylococcus aureus. *J. Antimicrob. Chemother.* 46 (5): 852-853.

Henning S.M., Wang P., and Heber D. (2011) Chemopreventive effects of tea in prostate cancer: green tea versus black tea. *Mol. Nutr. Food. Res.* 55(6):905-20.

Huang J., Wang Y., Xie Z., Zhou Y., Zhang Y., and Wan X. (2014) The anti-obesity effects of green tea in human intervention and basic molecular studies. *European J. Clin. Nutrition*. 68, 1075-1087.

Iso H., Date C., Wakai K., Fukui M., Tamakoshi A., and JACC study group. (2006) The relationship between green tea and total caffeine intake and risk for self reported type 2 diabetes among Japanese adults. *Ann. Intern. Med.* 144(8):550-562.

Isogai E., Isogai H., Hirose K., Hayashi S., and Oguma K. (2001) In vivo synergy between green tea extract and levofloxacin against enterohemorrhagic Esherichia coli 0157 infection. *Curr. Microbiol.* 42(4):248-51.

Jayanth A. (1998). Soil Erosion Damage Function for Smallholder Tea in Sri Lanka: An Empirical Estimation. Paper presented at 1st World Congress of Environmental and Resource Economists, Isola di San Giorgio, Venice, Italy.

Katiyar S.K. (2011) Green tea prevents non-melanoma skin cancer by enhancing DNA repair. *Arch. Biochem. Biophys*. 508(2): 152–158.

Katiyar S.K., and Raman C. (2011) Green tea: a new option for the prevention or control of osteoarthritis. *Arthritis. Res. Ther.* 13(3):121.

Khosya S., and Gothwal S. (2012). Two Cases of Paraquat Poisoning from Kota, Rajasthan, India.Case Reports in Critical Care. Hindawi Publishing Corporation. Article ID 652146, 3 pages

Kim A., Chiu A., Barone M.K., Avino D., Wang F., Coleman C.I., and Phung O. J. (2011) Green tea catechins decrease total and low-density lipoprotein cholesterol: a systematic review and meta-analysis. *J. Am. Diet. Assoc.* 111(11):1720-9.

Kim K.Y., Davidson P.M., and Chung H.T. (2001) Antibacterial activity in extracts of *Camellia japonica* L. Petals and its application to a model food system. *J. Food. Prot.* 64(8):1255-60.

Kopjar M., Tadic M., and Pilizota V. (2015). Phenol content and antioxidant activity of green, yellow and black tea leaves. *Chem. Biol.Technol. Agri.* 2(1).

Krolikowska A., Kudelski A., Michota A., and Bukowska J. (2003) SERS studies on the structure of thioglycolic acid monolayers on silver and gold. In: Proceedings of the 7th International Conference on Nanometre-Scale Science and Technology and the 21st European Conference on Surface Science, *Surf. Sci.* 532: 227-232.

Kushiyama M., Shimazaki Y., Murakami M., and Yamashita Y. (2009) Relationship between Intake of Green Tea and Periodontal Disease. *J. Periodontol.* 80(3): 372-377.

Lambert J.D., and Elias R.J. (2010) The antioxidant and pro-oxidant activities of green tea polyphenols: A role in cancer prevention. *Arch. Biochem. Biophys.* 501(1): 65-72.

Lee M.J., Lambert J.D., Prabhu S., Meng X.F., Lu H., Maliakal P., Ho C.T., and Yang C.S. (2004) Delivery of tea polyphenols to the oral cavity by green tea levels and black tea extract. *Cancer. Epidemiol. Biomarkers. Prev.* 13(1):132–137.

Lenore A., Weiqing L., and David E. (2009) Green and Black Tea Consumption and Risk of Stroke: A Meta-Analysis. *Stroke*. 40(5):1786-1792.

Li G., Zhang Y., Mbuagbaw L., Holbrook A., Mitchell A.H.L., and Thabane L.(2014) Effect of green tea supplementation on blood pressure among overweight and obese adults: a protocol for a systematic review. BMJ Open. 4: e004971. doi:10.1136/

Liao S, Kao Y.H., and Hiipakka R.A. (2001) Green tea: biochemical and biological basis for health benefits. *Vitam. Horm.* 62: 1-94.

Lin S.M., Lin F.Q., Guo H.Q., Zhang Z.H., and Wang Z.G. (2000) Surface states induced photoluminescence from Mn2+ doped CdS nano—particles. Solid. State. *Commum.* 115(11):615–618.

Maki K.C., Reeves M.S., Farmer M., Yasunaga K., Matsuo N., Katsuragi Y., Komikado M., Tokimitsu I., Wilder D., Jones F., Blumberg J.B., and Cartwright Y. (2009) Green Tea Catechin Consumption Enhances Exercise-Induced Abdominal Fat Loss in Overweight and Obese Adults". *J. Nutr.* 139(2):264-270.

Markova Z., Novak P., Kaslik J., Plachtova P., Brazdova M., Jancula D., Siskova K.M., Machala L., Marsalek B., Zboril R. and Varma R. (2014) Iron(II,III)–Polyphenol Complex Nanoparticles Derived from Green Tea with Remarkable Ecotoxicological Impact. ACS Sustainable. *Chem. Eng.* 2 (7): 1674–1680.

Matsumoto K., Yamada H., Takuma N., Niino H., and Sagesaka Y.M. (2011) Effects of green tea catechins and theanine on preventing influenza infection among healthcare workers: a randomized controlled trial. BMC. Complement. *Altern. Med.* 11:15.

McKay D.L., and Blumberg J.B. (2002) The role of tea in human health: An update. *J. Am. Coll. Nutr.* 21(1):1–13.

McNaught J.G. (1906) On the action of cold or lukewarm tea on Bacillus typhosus. *J. Royal. Army. Med. Corp.* 7:372-373.

Mercola J. (2013). Plastic and Cancerous Compounds in Tea Bags— A Surprising Source of Potential Toxins.

Michael B. (2013). The environment impact of tea. Green living tips. Retrieved from http://www.greenlivingtips.com/articles/tea-and-the-environment.html

Moghbel A., Farjzadeh A., Aghel N., Aghel H., Raisi N. (2011) The Effect of Green Tea on Prevention of Mouth Bacterial Infection, Halitosis, and Plaque Formation on Teeth. Iranian. *J. Toxicol.* 5(14):502-515.

Mowafy A.M., Salem H.A., Gayyar M.M., Mesery M.E., and Azab M.F. (2011) Evaluation of renal protective effects of the green-tea (EGCG) and red grape resveratrol: role of oxidative stress and inflammatory cytokines. *Nat. Prod. Res.* 25(8):850-6.

Mu-Hsin C., Chia-Fang T., Yu-Wen H., Fung-Jou L. (2014) Epigallocatechin gallate eye drops protect against ultraviolet B– induced corneal oxidative damage in mice. *Molecular Vision.* 20:153-162

Mupenzi J.D.P., Li L., Ge J., Varenyam A., Habiyaremye G., Theoneste N., and Emmanuel K. (2011) Assessment of soil degradation and chemical compositions in Rwandan tea-growing areas. *Geoscience Frontiers.* 2(4): 599-607.

Nagami H., Nishigaki Y., Matsushima S., Matsushita T., Asanuma S., Yajima N., Usuda M., and Hirosawa M. (2005). Hospital-based survey of pesticide poisoning in Japan, 1998–2002. Int. J. Occup. Environ. *Health.* 11(2):180–4.

Navarro-Peran E., Cabezas-Herrera J., Campo L.S., and Rodriguez-Lopez J.N. (2007) Effect of folate cycle distruption by the green tea polyphenol epigallocatechin-3 –gallate. *Int. J. Biochem. Cell. Biol.* 39(12): 215-25.

Nepolean P., Jayanthi R., Vidhya P.R., Balamurugan A., Kuberan T., Beulah T., and Premkumar R. (2012). Role of biofertilizers in increasing tea productivity. *Asian. Pacific. J. Trop. Biomed.* 2(3): S1443-S1445.

Neturi R.S., Srinivas R., Simha V.B., Sree S.Y., Shekar T.C., and Kumar P.S. (2014) Effects Of Green Tea On Streptococcus Mutans Counts – A Randomised Control Trail. *J. Clin. Diagnostic. Res.* 8(11): ZC128-ZC130.

Nileeka B.B.W., and Vasantha R.H.P.(2011) Plant flavonoids as angiotensin converting enzyme inhibitors in regulation of hypertension. *Functional Foods Health Dis.* 5:172-188.

Okamoto M., Sugimoto A., Legun K.P., Nakayama K., Kamaguchi A., and Maeda N. (2004) Inhibitory effect of green tea catechins on cysteine proteinases in Porphyromonas gingivalis. Oral. Microbiol. *Immunol.* 19(2):118–120.

Okello E.K., McDougall G.K., Kumar S., and Seal C.J. (2011) In vitro protective effects of colon-available extract of Camellia sinensis (tea) against hydrogen peroxide and beta-amyloid (Aβ (1–42)) induced cytotoxicity in differentiated PC12 cells. *Phytomedicine* 18(8–9): 691–696.

Pajonk F., Riedisser A., Henke M., McBride W.H., and Fiebich B. (2006) The effects of tea extracts on proinflammatory signaling. *BMC Medicine.* 4: 28.

Palakkad. (2011). The lion-tailed macaque faces habitat destruction. Mathrubhumi Yearbook 2014. Retrieved from http://www.thehindu.com/sci-tech/energy-and-environment/the-liontailed-macaque-faces-habitat-destruction/article2427486.ece

Parley M., Bansal N., and Bansal S. (2011) Is life- span under our control? *Int. Res. J. Phar.* 2 (1):40-48.

Patil S.M., Sapkale G.N., Surwase U.S., and Bhombe B.T. (2010) Herbal medicines as an effective therapy in hair loss – A review. *Res. J. Pharma. Biol. Chem. Sci.* 1 (2): 773- 81.

Patwardhan B., and Mashelkar R.A. (2009) Traditional medicine-inspired approaches to drug discovery: can Ayurveda show the way forward? *Drug. Discovery. Today.* 14(15-16): 804-811.

Ping S., Anders H., Christina N., and Hazel M. (2008). Synergistic effect of green tea extract and probiotics on the pathogenic bacteria, *Staphylococcus aureus* and *Streptococcus pyogens.* J. Microbiol. *Biotechnol.* 24(9): 1837-42.

Qi D.H., Guo H.J., and Sheng C.Y. (2013). Assessment of plant species diversity of ancient tea garden communities in Yunnan, Southwest of China. *Agrofor. Syst.* 87(2): 465–474.

Rani R., Nagpal D., Gullaiya S., Madan S., Agrawal S.S. (2013) Phytochemical, Pharmacological and Beneficial Effects of Green Tea. Int. J. Pharmacog. *Phytochem.Res.* 6(3):420-426.

Ranjit R.J.D. (2003) Impact of tea cultivation on anurans in the Western Ghats. Curr Sci. 85(10).

Ravichandran, M., Hettiarachchy N.S., Ganesh V., Ricke S.C., and Singh S. (2011) Enhancement of antimicrobial activities of naturally occurring phenolic compounds by nanoscale Delivery against *Listeria monocytogenes*, *Escherichia coli* O157:H7 and *Salmonella* typhimurium in broth and chicken meat system. *J. Food Safety*. 31(4): 462–471.

Renno W.M., Abdeen S., Alkhalaf M., and Asfar S. (2008) Effect of green tea on kidney tubules of diabetic rats. *Br. J. Nutr.* 100(3):652-9.

Retrieved from http://articles.mercola.com/sites/articles/archive/2013/04/24/tea-bags.aspx

Reznichenko L., Kalfon L., Amit T., Youdim M.B., and Mandel S.A. (2010) Low dosage of rasagiline and epigallocatechin gallate synergistically restored the nigrostriatal axis in MPTP-induced Parkinsonism. *Neurodegener. Dis.* 7(4):219-31.Format

Roh Y., Mc Millan A.D., Zhang C.L., Rawn C.J., Lauf R.J., Bai J., and Phelps T.J. (2001) Microbiol synthesis and the characterization of metal substituted magnetites. *Solid State Commun.* 118:529-534.

Rudramoorthy R., Kumar S.C.P. Velavan R., and Sivasubramaniam S. (2000). Innovative measures for energy management in tea industry. Proceedings of the 42nd National Convention of Indian Institute of Industrial Engineering, organised by IIIE, Coimbatore. Page 163-167.

San J. (2010) Green tea: A Global Strategic Business Report. Global Industry Analysts, Inc. (408): 528-9966. http://www.prweb.com/releases/green_tea/green_tea_extracts/prweb3515094.htm

Schmidt A., Hammann F., Wolnerhanssen B., Meyer-Gerspach A.C., Drewe J., Beglinger C., and Borgwardt S. (2014) Green tea extract enhances parieto-frontal connectivity during working memory processing. *Psychopharmacol.* 231(19): 3879-3888.

Schramm J. (2013) Going Green: The Role of the Green Tea Component EGCG in Chemoprevention. J. Carcinog. *Mutagen*. 4(142): 1000142.

Senapati B.K., Lavelle P., Panigrahi P.K., Giri. S., and Brown G.G. (2002). Restoring soil fertility and enhancing productivity in Indian tea plantations with earthworm and organic fertilizers. International Technical Workshop on Biological Management of Soil Ecosystems for Sustainable Agriculture. Brazilian Agriculture Research Corporation, *Brazil*. 182: 172-190.

Senapati B.K., Panigrahi P.K., and Lavelle P. (1994). Macrofaunal status and restoration strategy in degraded soil under intensive tea cultivation in India. In Transactions of the 15th World Congress of Soil Science. 4: 64-75.

Sharma R.K., Gulati S., and Mehta S. (2012) Preparation of Gold Nanoparticles Using Tea: A Green Chemistry Experiment. *J. Chem. Educ*. 89 (10): 1316–1318.

Shinn M. (2014) Report Finds 34 Pesticides in Tea From India. Greenpeace. Retrieved from http://ecowatch.com/2014/08/11/clean-chai-demand-pesticide-free-tea/

Simpson A., Shaw L., and Smith A.J. (2001) The bio-availability of fluoride from black tea. *J. Dent*. 29(1):15–21.

Smeeton B. (2011) The synergy of green tea and penicillin G against *Bacillus subtilis*. *J. App. Pharma.* 2 (3): 197-200.

Smith L. (2010). Found: Sri Lankan primate thought to be extinct for 60 years. The Guardian. Article. Retrieved from http://www.theguardian.com/world/2010/jul/19/horton-plains-slender-loris-found

Srinivas R.C., Venkateswarlu V., Surender T., Eddleston M., and Buckley N.A. (2005). Pesticide poisoning in south India: Opportunities for prevention and improved medical management. *Trop. Med. Int. Health*. 10(6):581–8.

Tachibana H., Sunada Y., Miyase T., Sano M., Maeda-Yamamoto M., and Yamada K. (2000) Identification of a methylated tea catechin as an inhibitor of degranulation in human basophilic KU812 cells. *Biosci. Biotechnol. Biochem*. 64(2):452–454.

Tahir A., and Moeen R. (2011) Comparison of antibacterial activity of water and ethanol extracts of *Camellia sinensis* (L.) Kuntze against dental caries and detection of antibacterial components. *J Med Plants Res* . 5(18): 4504-4510.

Takabayashi F., Harada N., Yamada M., Murohisa B., and Oguni I. (2004) Inhibitory effect of green tea catechins in combination with sucralfate on *Helicobacter pylori* infection in Mongolian gerbils. *J. Gastroenterol.* 39(1):61–63.

Tamma P.D., Cosgrove S.E., and Maragakis L.L (2012) Combination therapy for treatment of infections with gram-negative bacteria. *Clin. Microbiol. Rev.* 25(3): 450-470.

Tanigawa T. , Kanazawa S., Ichibori R., FujiwaraT., Magome T., Shingaki K., Miyata S., Hata Y., Tomita K., Matsuda K., Kubo T., Tohyama M., Yano K., and Hosokawa K. (2014) (+)-Catechin protects dermal fibroblasts against oxidative stress-induced apoptosis. *BMC Complementary and Alternative Medicine*. 14:133

Tariq M., Naveed A., and Barkat A.K. (2010) The morphology, characteristics, and medicinal properties of *Camellia sinensis*' tea. *J. Med. Plants. Res*. 4(19): 2028-2033.

Tiwari R.P., Bharti S.K., Kaur H.D., Dikshit R.P., and Hoondal G.S. (2005) Synergistic antimicrobial activity of tea and antibiotics. *Indian J. Med. Res*. 122(1): 80-84.

Van D.H.W., and Konradsen F. (2005). Risk factors for acute pesticide poisoning in Sri Lanka. *Trop. Med. Int. Health*. 10(6):589–596.

Van der Wal S. (2008). Sustainability issues in the tea sector: A comparative analysis of six leading producing countries, SOMO (2008): http://bit.ly/19Jal5a

Velayutham P., Babu A., and Liu D. (2008). Green Tea Catechins and Cardiovascular Health: An Update. *Curr. Med. Chem.* 15(18): 1840–1850.

Venkateswara B., Sirisha K., and Chava V.K. (2011) Green tea extract for periodontal health. *J Indian. Soc. Peridontol,* 15(1): 18-22.

Watts M. (2011). Paraquat. Pesticide action network Asia and the pacific. Retrieve from http://www.academia.edu/5540550/8-Paraquat

Wei-Li Z., Hai-Shui S., Yi-Ming W., Shen-Jun W., Cheng-Yu S., Zeng-Bo D., and Lin L. (2012) Green tea polyphenols produce antidepressant-like effects in adult mice. *Pharmacol. Res.* 65(1): 74-80.

Weng X., Huang L., Chen Z., Megharaj M. and Naidu R. (2013) Synthesis of iron-based nanoparticles by green tea extract and their degradation of malachite. *Industrial Crops and Products*. 51:342-347.

William M. (2011). Special report on Environmental damage and human rights abuses blight global tea sector. Resurgence and ecologist. Retrieved from: http://www.theecologist.org/News/news_analysis/847970/environmental_damage_and_human_rights_abuses_blight_global_tea_sector.html .

Wolfram S., Raederstorff D., Preller M., Wang Y., Teixeira S.R., Riegger C., and Weber P. (2006). Epigallocatechin gallate supplementation alleviates diabetes in rodents. *J. Nuts.* 136(10):3512-3518.

Wu A.H., and Butler L.M. (2011) Green tea and breast cancer. *Mol. Nutr. Food. Res.* 55(6):921-30.

Wu C.D., and Wei G.X. (2002) Tea as a functional food for oral health. *Nutrition.* 18(5): 443–444.

Wu L.Y., Juan C.C., Ho L.T., Hsu Y.P., and Hwang L.S. (2004) Effect of green tea supplementation on insulin sensitivity in Sprague- Dawley rats. *J. Agric. Food. Chem.* 52(3):643-648.

Yamamoto M.M., Ema K., Monobe M., Shibuichi I., Shinoda Y., Yamamoto T., and Fujisawa T. (2009) The efficacy of early treatment of seasonal allergic rhinitis with Benifuuki green tea containing o-methylated catechin before pollen exposure: An open randomized study. *Allergol Int*. 58 (3):437-444

Yan Z., Hong Z., Bu-zhuo P., and HaoY. (2003) Soil erosion and its impacts on environment in Yixing tea plantation of Jiangsu Province. *Chinese. Geo. Sci.* 13(2): 142-148.

Yang C.J., Yang Y.C., and Uuganbayar D. (2003) Effect of feeding diets containing green tea by-products on laying performance and egg quality in hens. Kor. J. Poult. Sci. 30(3):183-189.

Yang C.S., and Wang X. (2010) Green Tea and Cancer Prevention. *Nutr. Cancer*. 62(7): 931-937.

Yang G., Zheng W., Xiang Y.B., Gao J., Li H.L., Zhang X., Gao Y.T., and Shu X.O. (2011) Green tea consumption and colorectal cancer risk: a report from the Shanghai Men's Health Study. *Carcinogenesis*. 32(11):1684-8.

Yang J., and Liu R.H. (2013) The phenolic profiles and antioxidant activity in different types of tea. *Int. J. Food. Sci. Technol*. 48(1):163-171.

Yang S.W., Lee B.R., and Koh J.W. (2007) Protective effects of epigallocatechin gallate after UV irradiation in cultured human retinal pigment epithelial cells. Korean. *J. Ophthalmol.* 21(4):232-7.

Yiannakopoulou E.C. (2014) Effect of green tea catechins on breast carcinogenesis: a systematic review of in-vitro and in-vivo experimental studies. *Eur J. Cancer. Prev.* 23(2):84-9.

Yoshihiro K., Hiroyasu I., Isao S., Kazumasa Y., Hiroshi Y., Junko I., Manami I., and Shoichiro T (2013) The Impact of Green Tea and Coffee Consumption on the Reduced Risk of Stroke Incidence in Japanese Population. Stroke. DOI: 10.1161/STROKEAHA.111.677500

Yuan J.M. (2011) Green tea and prevention of esophageal and lung cancers. *Mol. Nutr. Food. Res.* 55(6):886-904.

Zhang B., and Osborne N.N. (2006) Oxidative induced retinal degeneration is attenuated by epigallocatechin gallate. *Brain. Res*. 1124(1): 176-87.

Tea: Technological Initiatives, pp. 301-327
New India Publishing Agency, New Delhi, India
Edited by Niladri Bag, Arundhati Bag and L.M.S. Palni

10

Biotechnology of Tea

Mainaak Mukhopadhyay and Tapan Kumar Mondal

Abstract

Tea [*Camellia sinensis* (L.) O. Kuntze], *of family Theaceae, is an important cash crop worldwide. It is an evergreen shrub cultivated in humid and sub-humid tropical, sub-tropical, and temperate regions of the world, and grows mainly on acidic soils. Due to limitation of expansion of tea cultivation in new land, import of low cost high quality tea, increasing vertical productivity is the best option of Indian tea Industry. The primary need of the tea industry today is to develop high yielding clones with better quality and stress tolerance potential. Conventional tea research has contributed appreciably to the genetic improvement of tea over the past several decades, but it is time-consuming and labor intensive. Micropropagation and somatic embryogenesis, both are considered the most capable system of tea concerning efficient in vitro plant regeneration, its manipulation via transgenic approaches and their rapid multiplication. Due to the absence of genome sequence, generation of genomic resources are essential for development in marker-assisted breeding, association mapping, cloning of genes, and mapping of quantitative trait loci, which is extremely essential to develop elite genotypes. Till today satisfactory progrees has been made by various works which are discussed here. Apart from that, functional genomics study of tea plants has made significant progress. The work done under this domain are the construction of cDNA libraries, generation and annotation of ESTs, analysis of gene expression profile, construction and use of cDNA microarrays, development and utilization of expressed sequence tags-simple sequence repeat and short tandem repeat markers, cloning and expression of genes involved in secondary metabolism, disease and pest resistance and stress tolerance which are elaborated here.*

Introduction

Camellia sinensis (L.) O. Kuntze, of family Theaceae, is an evergreen shrub cultivated in humid and sub-humid tropical, sub-tropical, and temperate regions of the world, and grows mainly on acidic soils (Mukhopadhyay and Mondal, 2014). The primary center of origin of tea was South-East Asia, explicitly at the point of intersection between the 29°N (latitude) and 98°E (longitude) near the source of the river Irrawaddy at the confluence of North-East India, North Burma, South-West China and Tibet provinces. Tea flourishes well within the latitudinal ranges between 45°N to 34°S that traverse about 52 countries (Deka *et al.*, 2006; Mondal *et al.*, 2004). This genus has more than 325 different species (Mondal 2002a; Mondal *et al.*, 2003) and over 600 popular genotypes of tea are cul-tivated worldwide. It occupies about 2.7 million hectares of cultivated land of the world with an annual production of about 2.2 million tons (Mondal 2014; Mondal *et al.*, 2004, Mondal 2011) and occupying 16.4% of the total tea growing areas of the world. Apart from tea, *C. japonica,* wild relative of tea, is valued due to its excellent floricultural splendor. Quite a few other wild species such as *C. reticulata*, *C. sasanqua*, and *C. saluensis* are also popular owing to their ornamental value. Species such as *C. oleifera*, *C. semiserrata* and *C. chekiangolomy* produce oil from mature seeds destined to pharmaceutical industry, albeit to a limited scale (Mukhopadhyay and Mondal, 2014).

Leaf is the key criterion by which three types are distinguished. They are; Assam type with biggest leaves, China type with smallest leaves and Cambod type whose leaf size are found to be inbetween Assam and China type. *C. sinensis* L., *C. assamica* and *C. assamica* ssp. *Lasiocalyx* are the three main *Camellia* species (Mukhopadhyay 2012; Mondal 2002b, 2008) which are the important sources of foreign exchange for almost all the tea-producing countries in the world.

Commercially, tender leaves are plucked, processed and used as a drink that give the necessary 'pep' and the 'just required' stimulus for doing more of both mental as well as physical work. Alkaloids, caffeine (1,3,5-trimethylxanthine), polyphenols (e.g. catechin) and two minor isomeric dimethylxanthines namely theobromine and theophyline are responsible for the mildly stimulant effects of tea. Briskness and other taste characteristics of made tea qualities are attributed to caffeine (Dosumu *et al.*, 2010). In ancient times, Chinese used tea as a beverage and have been doing so for the past 2000 years (Eden, 1958), and today more than half of the world's population consumes tea as morning drink.

Need of Biotechnology in Tea

An increase in tea production to meet the increasing demand requires advances in research and technological development. Hence, the preliminary step is to develop high yielding clones with better quality and enhanced stress tolerance capability. Due to its monoculture and perennial nature, tea plants often, encounter various biotic and abiotic stresses (Mukhopadhyay *et al.*, 2012, 2013a, 2013b). Apart from these, reduction of winter dormancy, interflush dormancy, better response to mechanical harvesting are some of the emerging needs of Indian tea industry. Incidentally, exploitation of DNA based techniques for molecular or genetic characterization of various tea clones is an effective way. Simultaneously, the complex life cycle and out-breeding nature of tea poses several limitations for its genetic improvement through conventional breeding. Due to heterogeneous nature of tea, discrimination between the three tea varieties is difficult. Furthermore, morphological characteristics are unable to reflect the inherent genetic variation within the crop, which actually shows high plasticity with respect to biochemical and physiochemical descriptors. Being perennial in nature, conventional tea breeding is slow due to several reasons. They are long gestation periods, self-incompatibility, unavailability of distinct mutant of diverse biotic and abiotic stresses, low success rate of hand pollination, short flowering time and long duration for seed maturation. Therefore, use of molecular tools (Mondal *et al.*, 2000a) is extremely important to reveal the unexplored genetic variation in tea (Sharma *et al.*, 2009a). Importantly, in the Indian context, tea industry is on the threshold of decline in yield and due to non-availability of land, there is limitation in area expansion. Therefore, continuous effort to widen tea gene pool has become a significant need (Jain and Newton, 1990) and hence biotechnological advancement in this field has become necessary to develop and maintain superior clones.

Majority of the agronomic traits of tea are quantitative, hence they are not amenable to easy manipulations in breeding programs without elaborate and long-term field testing in at least more than one environment in order to determine their inheritance, adaptability, and stability. However, limitation of conventional breeding and lack of distinct mutation further limit the discovery of quantitative trait loci (QTLs) (Kamunya *et al.*, 2010). These along with the fact that slow vegetative propagation, availability of low genomic resources further remind the need of biotechnological intervention for varital improvement of tea.

Tissue Culture

Tissue and cell culture is the most primitive tool of plant biotechnology. The need to improve *in vitro* multiplication rate depends upon factors related to efficacy of micropropagation such as; explants type, media composition, initiation, proliferation, *in vitro* as well as *ex vitro* rooting which were started in the early 1990s, standardized, improved further in tea and its wild relatives. However, the critical phase of micropropagation is the institution of *in vitro* plantlets to greenhouse and achievement of uniform growth vis-a-vis high survival rate, which call for superior greenhouse provisions, and alteration of microclimate to resemble the local environment.

The potential of tissue and organ culture for rapid clonal propagation of elite tea clones and the use of genetic engineering methods for bringing about improvement in tea quality and production were well recognized, and many laboratories had successfully developed *in vitro* propagation for tea including induction of somatic embryogenesis (Palni *et al.*, 1999). The overall success of any micropropagation technology depends upon efficient shoot proliferation, formation of a well-developed root system in micropropagated shoots, successful acclimatization of plantlets and final establishment in the field. Application of tissue culture technology for mass propagation of tea has been demonstrated. Factors, other than hormones, include mineral nutrients, carbon source, pH of the medium, gelling agents, light intensity, duration and temperature etc has been found to be important. However, to improve the *ex vitro* rooting, several new techniques have been developed. Grafting of tissue culture raised shootlets over young seedling known as micrografting was developed by Prakash *et al.*, 1999; Vyas *et al.*, 1999 and later commercialized by Parathi *et al.*, (2004). *Bacillus subtilis* and *Pseudomonas corrugata*, isolated from tea rhizosphere and screened for their antifungal activity, was used for hardening and survival of tea plants which is known as biological hardening (Pandey *et al.*, 2000). Induction of hairy roots in micropropagated shootlets of tea using *Agrobacterium rhizogenes* had also been reported as a means of hardening (Zehra *et at.*, 1996).

Apart from micropropagation, several other *in vitro* cell culture techniques were attempted in tea (Mondal, 2014). Somatic hybrids or cybrids production has enormous importance in tea but enough success has not been achieved. The accomplishment of this technique depends upon successful development of, 'cell suspension culture', which was first initiated by Bagratishvili *et al.*, 1979. On the other hand, tea being heterozygous and heterogeneous, production of homozygous diploids is of immense importance for improvement. However, except few reports on anther/pollen culture (Chen and Liao,1983; Hazarika and Chaturvedi, 2013), no successful plantlet have been developed. Somatic hybridization also has potential for the production of caffeine-free tea through the fusion of protoplasts of caffeine-free but aroma rich *C. luetacense* or *C. irrawadiensis* with that of cultivated tea (Mondal *et al.*, 2004). Although Katsuo (1969) as well as Okano and Fuchinone (1970) initiated anther culture of tea that produced roots from anther derived callus, till today fully grown plant has not been developed.

Earlier reports indicated that, different researchers made several endeavors for viable protoplast culture but success has always eluded them. Nevertheless, Protoplast culture has tremendous potential in tea crop improvement. It is recognized that many of the wild relatives of tea have agronomically important biotic and abiotic stress resistant characters, which can be incorporated into the cultivated variety of tea (Mondal *et al.*, 2004).

Tea is treasured for pleasant taste and aroma, owing to the presence of alkaloids, caffeine and other methyl xanthenes such as theobromine and theophylline. These alkaloids have also been used as therapeutic and drugs. The exploration of production of secondary metabolites in tea is very important and several useful attempts have been made (Orihara and Furuya 1990; Matsuura *et al.*, 1991).

Somatic Embryogenesis

Somatic embryogenesis is considered the most capable system of tea concerning efficient *in vitro* plant regeneration system (Jain and Newton, 1990) and its efficacy depends on multiplication efficiency and conversion rate. Development of adventitious embryos from explants barring an intervening callus phase facilitates genetic fidelity and thus it has an excellent potential in clonal propagation (Mondal *et al.*, 2001a: Mondal *et al.*, 2000b) through genetic transformation (Mondal *et al.*, 1999). In artificial seed production immature somatic embryos were rescued and cultured before abortion. It produced resistant and haploid tea plants (Chen and Liao, 1982). Somatic embryogenesis is the solitary alternative to conventional micropropagation due to little success on formation of regeneration of protoplast or cell suspension cultures. There are various factors that govern somatic embryogenesis of tea such as explants, its physiological stage, genotypic variation; induction media formulation, maturation and proliferation along with growth regulators and adjuvant (Mondal *et al.*, 2004). Various researchers have undertaken somatic embryogenesis, in tea; a tabular representation, in summarized form, is given in Table 1.

Table 1: Somatic embryogenesis in *Camellia sinensis*

Species/cultivar	Explant	Medium				Reference
		Induction	Maturation	Germination	Multiplication	
		MS + BA (0.5) +IBA (0.5)	-	MS + BA (0.5) + NAA (5)	-	Abraham & Raman 1986
C. sinensis TRI-2025	Nodal segment	MS + BA (0.5) + GA_3 (3)	-	-	MS0 +1/2 macro salts	Akula & Dood, 1998
C. sinensis	Mature seed	½ MS macro + Full micro MS+ AHS (100)+ Gln (100)	½ MS macro+ Full micro MS + AHS (100) + Gln (100)	½ MS macro + Full micro MS+ AHS (100)+ Gln (100)	-	Akula *et al.*, 2000
C. sinensis (L.) O. Kuntze.	Cotyledon	Modified MS + BA (10) +YE (2)	-	Modified MS + BA (10) with 2% sucrose	-	Arulpragasam *et al.*, 1988
C. sinensis	Cotyledon	MS + BA (1.13) + NAA (22.7)	MS + BA (1.13) +NAA (22.7)	MS + BA (2.26) + NAA (22.7) + GA_3 (0.12)	-	Bag *et al.*, 1997
C. sinensis	Mature cotyledon, leaf	MS + BA (2) + 2,4-D (5)	-	MS + BA (2) + NAA (3)	-	Balasubramanian *et al.*, 2000
Theasinensis (L.), B-61	Cotyledon	MS + Kin (0.05) + 2,4-D(0.5)	½ MS + Kin (0.05)+ AC (0.2%) + glucose (1.5%)	½ MS + Kin (0.05) - + AC (0.2%) + glucose (1.5%)		Bano *et al.*,1991
C. sinensis	Immature Cotyledon	Modified MS + BA (3) + NAA (2)	-	MS + BA (5)	-	Haridas *et al.*, 2000
C. sinensis, T-78	Cotyledon	MS + BA (10) + IBA (0.5) + A (80)	MS+BA (10)+ IBA (0.5) +A (80)	B_5+ BA (3)+IAA (2)	-	Jha *et al.*,1992
C. sinensis, Yabukita	Cotyledon	MS + BA (4) + IBA (2)	-	MS + BA (10) + IBA (0.5)	-	Kato,1982; Kato,1986

C. sinensis, Yabukita, Sayamamidori, Benikaori, Akane.	Immature leaves of *in vitro* grown shoots	Liquid MS + 2,4 - D (0.5)	MS + BA (10) + IBA (0.5) or MS + BA (4)+ IBA (2)	MS + BA (10) + IBA (0.5) or MS + BA (4) + IBA (2)	-	Kato, 1996
C. sinensis	Mature de-embryonated cotyledon	MS + BA (2) + IBA (0.2)	Modified MS + BA (2) + IBA (0.2) + Gln (1) + K_2SO_4	Modified MS + BA (2) + IBA (0.2) + Gln (1)	-	Mondal *et al.*,2001a
C. sinensis	Cotyledonary segment	NN	NN +N-Z-amino type-A(0.1%)	NN + N-Z amino type-A (2.6 %)	-	Nakamura, 1985
C. sinensis with 13 cultivars, *C. japonica 3 cultivars.*, *C. sasanqua, C. brevistela, C. nokoensis, C. japonica* (cv. Kosyougatu)*x C. granthamiana.*	Half sliced cotyledon	MS+BA (1-5)	-	-	-	Nakamura, 1988
C. sinensis cv Shan Chat Tien	Somatic embryo	MS + BAP (8)	-	MS + BAP (3)	BAP (3) and IBA (0.3)*C*	Nguyen V T (2012)
C. sinensis & *C. assamica*	De-embryonated Cotyledon segments	B_5 + NAA (0.1) + BA (2)	B_5 + NAA (0.1) + BA (2)	B_5 + NAA (0.1) + BA (2)	-	Paratasilpin, 1990
C. sinensis, UPASI-10	De-embryo-nated immature Cotyledonary segment	MS + PBOA (1µM) + BA (0.1)	MS + Brassin (1µM)	MS + Brassin+ (1µM)	-	Ponsamuel *et al.*,1996
C. sinensis, UPASI-10	Cotyledon with/ without	Modified MS + BA (0.25)	MS + BA(3-5) + CW (10%) + GA_3	MS + BA (3-5) + CW (10%) +	-	Rajkumar & Ayyappan, 1992

	embryogenic axis		(0.25-1.0)	GA_3(0.25 -1.0)		
C. sinensis, TRI-2024.	Nodal segment & leaf	VW + CW (15%) MS + IBA (0.1) + BA (1) for nodal segment; MS + 2, 4-D (1) + Kin (1) for leaf	-	-	Callus, embryo-like structure	Sarathchandra *et al.*,1988
C. sinensis	Embryogenic leaf callus	MS + BA (3 mg) + NAA (0.1 mg)	BA (1.0 mg) + NAA (0.1)	-	-	Seren *et al.*, 2006
C. sinensis	Leaf from filed grown plant	MS + BAP (2) + NAA (3) for callus induction and BAP (1.0 mg/L) + NAA (0.1 mg/L) for embryo induction	-	-	-	Seren *et al.*, 2007
C. sinensis, Kangra Jat	Cotyledon slice	½ MS + BA (2) + IAA (0.2)	-	½ MS + BA (2) + IAA (0.2) + GA_3 (0.2)	-	Sood *et al.*, 1993
C. sinensis	Cotyledon slice	Modified MS	Modified MS	Modified MS+ Kin (1.8) or BA (1)	-	Wachira & Ogado, 1995
C. sinensis, Chyi-Men & Pyng-Shoei.	Mature cotyledon	MS+ Kin (10) + IAA (1)	-	-	-	Wu ,1976 & Wu *et al.*,1981
C. sinensis	Mature cotyledon	-	-	-	-	Yan & Ping, 1983

Figures in parenthesis denote concentrations in mg/l (unless otherwise stated).

A : adenine; ABA : abscissic acid; AC : activated charcoal; AHS : adenine hemisulphate; AM : Anderson (1984) basal medium; CW ;coconut water; DTT : dithiothreitol; Gln : glutamine; MS0 : MS (Murashige and Skoog, 1962) basal medium without added growth regulators; NN : Nitsch and Nitsch (1969) medium; PBOA : phenylboronic acid; PVP : polyvinyl pyrrolidone; vit : vitamins; VM : Vacin & Went (1949) medium; YE : yeast extract.

Exploitation of embryogenesis through biotechnological approach induces genetic transformation, because foreign genes can be introduced into primary embryos that multiply subsequently (Mondal *et al.*, 2001b). Two different growth patterns for secondary embryogenesis have been reported in tea like, repetitive embryogenesis and callus-to-somatic embryo, which requires callus sub-culturing (Vieitez, 1994). Bioreactor for secondary embryogenesis has also been established and induction of *in vivo* embryogenesis could be achieved excluding conventional tissue culture media (Mondal *et al.*, 2001c).

Genetic Transformation

Agrobacterium-mediated genetic transformation is still the most widely used method of producing transgenic tea plants (Mondal *et al.*, 2001b, c). Besides being cheaper and simpler than most direct gene transfer methods, it allows little rearrangement of transgenes, and efficient integration of the transgene into the plant genome (Sandal *et al.*, 2007). Successful genetic transformation of tea leaf explants requires fulfillment of certain conditions such as; the nonappearance of bactericidal effect of leaf polyphenols, proper expression of *Agrobacterium vir* genes, and persistence of growth potential of the transformed leaf explants for shoot regeneration. So far, the commonly used antioxidants or polyphenol adsorbents have not only been unable to negate the bactericidal effect of leaf polyphenols, but also inhibited the regeneration process (Sandal *et al.*, 2007). Somatic embryos derived from cotyledon explants have been transformed with *Agrobacterium* (Mondal *et al.*, 2002a; Mondal *et al.*, 1999, Mondal *et al.*, 2001c, d: Mondal *et al.*, 1999, Lopez *et al.*, 2004); however, several commercially important tea cultivars have remained largely recalcitrant to transformation. This is mainly due to the exudation of high contents of bactericidal polyphenols (Kumar *et al.*, 2004) and lack of suitable regeneration systems (Mondal *et al.*, 2004). L-glutamine facilitates the *Agrobacterium* mediated gene transfer into tea leaves, because, unlike the commonly used polyphenol adsorbents and antioxidants, preserves the ability of *Agrobacterium* to employ its virulence system, i.e., *vir* genes, for T-DNA transfer and does not have adverse effects on the plant regeneration process and L-glutamine counteracts the bactericidal effects of polyphenols that are exuded from tea leaves when used as explants (Sandal *et al.*, 2005).

Production of transgenic tea [*C. sinensis* (L.) O. Kuntze cv. Kangra Jat] was developed via *Agrobacterium*-mediated genetic transformation of somatic embryos. Two disarmed *A. tumefaciens* strains (EHA 105 and LBA 4404) carrying the binary plasmid with the *npt*II gene and *gus*-intron were evaluated as vector systems. During pre-culture, wounding and acetosyringone treatment were inhibitory, but the bacterial growth phase (optical density, 600 = 0.6), cell density (109/ml), co-cultivation period (5 days) and pH of the co-cultivation medium (5.6) had positive effects on transformation. Following co-cultivation, globular somatic embryos were treated with kanamycin in multiplication medium. Further selection took place in the maturation and germination medium at an elevated kanamycin level. An average of 40% transient expression was evident based on the GUS histochemical assay. Kanamycin-resistant, GUS-positive embryos were germinated, and the resulting microshoots were multiplied *in vitro.* Integration of the transgenes into the tea nuclear genome was confirmed by PCR analysis using *npt*II- and *gus*-specific primers and by Southern hybridization using an *npt*II-specific probe. The transgenic shoots were micrografted onto seed-grown rootstocks of cv. Kangra Jat (Mondal *et al.*, 2001b). Stable integration of transgene had been carried out after molecular characterization through southern hybridization. Five (out of 12) independent kanamycin resistant, GUS positive lines yielded PCR amplification products of 693 bp and 650 bp with *npt-II* and *gus* gene specific primers respectively indicating the linking of marker gene *npt-II* with that of *gus* as a single 'T-DNA strand' in the genomic DNA of the transformed plant. Southern blot analysis of *Pst I* digest of genomic DNA from each of the five putative transgenic lines generated an internal transgene fragment of 1.6 kb that hybridized to the *npt-II* probe. Additional shorter fragments produced in some transgenic lines further indicated a deletion of the T-DNA containing *npt-II.* The deletion may perhaps occur during transformation or consequent regeneration. Different banding pattern observed in the southern hybridization could be due to multiple insertion, rearrangement or deletions of the integrated transgenes in the regenerated plants. Although 80–90% transgenic plant survival was achieved under greenhouse conditions, the stability of the transgene remains to be elucidated as, tea plants take years to flower and set seeds (Mondal *et al.*, 2004).

Agrobacterium mediated transformation has been exploited to reduce caffeine by suppressing glutathione synthetase (Mohanpuria *et al.*, 2008) or to produce low caffeine containing tea by silencing caffeine synthase gene (Mohanpuria *et al.*, 2011). Although several attempts have been made to standardize the protocol but transgenic tea plant remain elusive. Recently Bhattacharya *et al.*, (2014) produced transgenic tea plant containing Osmotin gene aiming to produce better low moisture stress plant. Improved tolerance of polyethylene glycol-induced water stress and faster recovery from stress were evident in transgenic lines compared with the normal phenotype. Significant improvements in growth under in-vitro conditions were also observed Besides enhanced reactive oxygen species-scavenging enzyme activity, the transgenic lines contained significantly higher levels of flavan-3-ols and caffeine, key compounds that govern quality and commercial yield of the beverage. The selected transgenic lines have the potential to meet the demands of the tea industry for stress-tolerant plants with higher yield and quality.

Molecular Markers

Molecular markers, also called DNA markers, can be considered as useful tools to locate the site of desirable genetic traits or indicate specific genetic differences. Based on detection techniques, they can be broadly classified as; hybridization based, polymerase chain reaction based and DNA sequence based markers. They are selectively neutral, unlimited in number and not affected by environmental factors or developmental stage of the plant (Bandyopadhyay, 2011). Morphological characteristics of tea are not capable to replicate the genetic variation because it shows high plasticity with respect to biochemical and physiochemical descriptors. Therefore, identification of highly reliable molecular tools such as microsatellite or SSR markers is extremely important to reveal the unexplored genetic variation in tea which is the key for developing the improved cultivars or identification of genes/ QTLs. Currently tea plantation is dominated by few superior high yielding genotypes, and widespread cultivation of clonal tea could diminish genetic diversity. Hence, germplasm characterization based on molecular markers will create the base material for improvement of deserved agronomical trait, identification of cultivar, prevention of duplicate entry in gene pool, efficient selection, and taxonomic classification of tea.

In tea, Randomly Amplified Polymorphic DNA (RAPD) was the first molecular marker that was used by several researchers (Wachira *et al.*, 1995 and 1997; Kaundan *et al.*, 2000; Chen and Yamaguchi, 2002; Young Goo *et al.*, 2002; Chen *et al.*, 2005; and Mewan *et al.*, 2005) mainly to study the genetic diversity of local genotypes. For an example, 23 primers that generated 157 polymorphic bands and maximum polymorphism of 20 bands were detected characterized thirty-eight different cultivars of Kenyan tea. The amplified fragments and similarity matrix ranged from 0.3 to 3 kb and 43% to 96%, respectively, among the clones and based on linkage cluster analysis, the constructed dendrogram clearly discriminated Assam, Cambod and China types (Wachira *et al.*, 1995). 10-mer and 12-mer primers detected variation among Korean, Japanese, Chinese, Indian and Vietnamese tea. It was concluded that Korean tea undergone little genetic diversification after introduction from China but Japanese tea had a nearer relationship with Chinese and Indian counterpart revealing the fact that tea in Japan came from China as well as India (Tanaka *et al.*, 1995). RAPD markers (Mondal, 2000; Mondal *et al.*, 2000a) characterized twenty-five Indian tea cultivars including 2 wild species. Eleven, 10-mer primers generated polymorphic banding pattern, which were ranged from 7 to 21 per genotype. One hundred thirty eight bands, among 154, were polymorphic resulting 95.2%

genetic variability. Further an average of 57% within and 43% between populations variability was also recorded. Dendrogram separated the population in to three clusters as per their morphological classification. They are China, Assam and Ornamental type. The results showed that RAPD markers could be used effectively to distinguish and characterize Indian tea germplasm (Mondal, 2000). RAPD with 50 primers were applied to investigate genetic variability of *in vitro* raised tea plants (T-78) and among them, 39 primers developed 197 monomorphic bands in all the concerned plants. It was concluded that variation occurred because of mutation during micropropagation. Interestingly, the marker profiles among these 4 plants were identical with all 11 primers suggesting a complete homogeneity among them (Mondal and Chand, 2002). RAPD was also employed to identify the true-crossing progenies in tea breeding programme and to determine relationship between parents and their hybrids. PCR amplification of DNA from 10 processed made tea demonstrated their originality and identification of cultivars indicating that it could be a suitable technique to identify the adulteration of made tea (Singh *et al.*, 1999). Interestingly, Tanaka and Taniguchi (2002) improved the RAPD technique by adding nucleotides to the 3' end to make more clearer band in tea. RAPD markers have also been used to test the genetic fidelity of vegetatively propagated tea plant (Singh *et al.*, 2004).

Inter Simple Sequence Repeats (ISSR), for genetic characterization (Mondal, 2002b), has been used in tea and they show greater repeatability and stability of map position in the genome while comparing closely related individual. There are several examples of using ISSR markers to discriminate plant genetic resources, evaluation and identification of germplasm. The analysis also provides an efficient method for the discrimination of germplasm at the inter-specific level (Liu *et al.*, 2012). Twenty-five diverse tea cultivars were analyzed using the ISSR-PCR of which 84% were polymorphic. Dendogram revealed three distinct clusters of Cambod, Assam and China type, which coincides with the known taxonomical classification. Therefore, ISSR-PCR is a potential tool for genetic fingerprinting and taxonomic classification of tea genotypes (Mondal, 2002c). In tea, Yao *et al.* (2005) also developed ISSR. Off-late efforts were made to discriminate tea germplasm at the inter-specific level using specific ISSR markers. One thirty four tea germplasms were classified at variety level using the band pattern combination by 42 bands generated with four ISSR primers. These putative variety-specific markers could be transformed to sequence characterized amplification regions after sequencing and designing primer pairs to develop specific markers for tea germplasm. Therefore, ISSR analysis, on one hand reveals high genetic polymorphism among tea plants apart from presenting a practical as well as effective approach to differentiate tea germplasm at the inter-specific level (Liu *et al.*, 2012).

Restriction Fragment Length Polymorphism (RFLP) has been used to investigate genetic diversity in cultivated plants. In tea, Matsumoto *et al.* (1994) cloned the PAL gene; to study genetic variation of Japanese green tea using RFLP and classified Japanese green tea cultivars into five groups having multiple routes of origin and Assam hybrids was distinguished from Japanese green tea cultivars based on grouping. The inheritance of PAL gene indicated a single copy per haploid genome that inherited as a single gene. In order to prevent adulteration, Japanese tea 'Yabukita' was studied by employing STS-RFLP using the sequence information of 3 genes (i.e. phenyl-ammonia lyase, chalcone synthase and dihydroflavonol 4-reductase genes) involved in secondary metabolism in tea (Kaundun and Matsumoto, 2003).

Simple Sequence Repeats or SSRs, *i.e.*, microsatellites, are 2-5 bp long tandemly repeated DNA sequence motifs with high polymorphism in plant genomes and are well recognized due to hypervariability, simplicity of scoring, co-dominance, vis-a-vis high reproducibility (Fig 1). SSRs are the most powerful genetic markers for genetic linkage analysis, diversity study and marker assisted selection. The microsatellite markers are mainly used in plant genetic analysis due to the reducing cost of DNA sequencing and increasing availability of EST sequence data. In tea several works have been used SSR markers primarily to access the genetic diversity of tea (Freeman *et al.*, 2004; Zhao *et al.*, 2007 and Hung *et al.*, 2008).

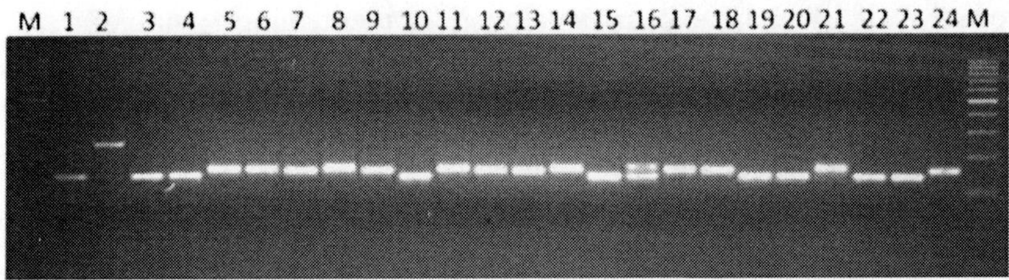

Fig. 1: Representative picture of SSR profile with TKM 21-3 primer. M= 100 bp marker, 1 to 24 represent different tea genotypes (*Source*: unpublished data, Tapan Kumar Mondal).

Expressed sequenced tags are potential source of developing SSR markers. EST-SSR markers are potential candidates for gene tagging and comparative studies as they are gene specific. Studies have been reported regarding the use of EST-SSR markers in a large number of commercially important plants. Around 49403 EST data in tea have been generated till date. Study has been performed by detecting SSRs from the EST sequences of *Camellia sinensis* available in the public domains and functional annotation of the SSR containing ESTs has also been undertaken. The annotation assisted to identify the putative functions of the ESTs and to find out the important functional domain markers (FDM) related to the SSR-ESTs leading to gene ontology study. Gene ontology includes biological process containing operations or sets of molecular events with a defined beginning and end, functioning of integrated living units like, cells, tissues, organs, and organisms, cellular components, the parts of a cell or its extracellular environment and molecular function and the elemental activities of a gene product at the molecular level, such as binding or catalysis. In tea, the mined EST-SSR markers would probably help in further study of variability, mapping, evolutionary relationship etc. Keeping an eye on the above-mentioned intent, 12,851 EST sequences of *Camellia sinensis*, downloaded from National Center for Biotechnology Information (NCBI) were mined for the development of Microsatellites. 6148 (4779 singletons and 1369 contigs) non redundant EST sequences were found after preprocessing and assembly of these sequences using various computational tools. Out of total 3822.68 kb sequence examined, 1636 (26.61%) EST sequences containing 2371 SSRs were detected with a density of one SSR/1.61 kb leading to development of 245 primer pairs. These types of studies would, in all probability, enhance the cross species applications to develop conserved orthologous marker sets (Sahu *et al.*, 2012).

CAPS (cleaved amplified polymorphic sequence) utilizes amplified DNA fragments digested with a restriction endonuclease to display restriction site polymorphisms (Konieczny and Ausubel , 1993). Kaundan and Matsumoto (2003) were frist to developed and utilized this marker for varietal; classification of Japanese green tea cultivars. Later 37 CAPS markers derived from cytoplasmic genome and ESTs of tea have been developed for identifying 12 prevailing tea cultivars in Taiwan. Further analysis such as principal coordinate analysis and cluster analysis group them in 3 group which support their morphological classification (Hu *et al.*, 2014).

Biodiversity conservation is essential for tea and the assessment of genetic variability existing in the field gene banks is important for maintenance of genetic resources and broadening of the genetic base. Twenty-one genomic and genic microsatellite markers were used to evaluate genetic diversity and DNA fingerprinting of 15 popular tea accessions. Each accession had a unique marker profile, indicating that microsatellite markers were useful in differentiation studies among the tea collections. 127 polymorphic alleles were scored with an allele frequency of 6.05 per primer. The polymorphism information content ranged from 0.2 to 0.60, with an average of 0.359. SSR markers analysis detected a high level of heterozygosity in tea. The Jaccard's similarity coefficients ranged from 0.15 to 0.56 with an average similarity index (ASI) of 0.234. The first two coordinate explained 54.33% of the total variance. Further analysis indicated that the populations formed two major groups with exclusive China and China hybrids (I) and Assam types (II). The collections from western Himalayan possessed a moderate to high level of genetic diversity, which could provide valid guidelines for genetic improvement of tea (Bhardwaj *et al.*, 2014).

Amplified fragment length polymorphism (AFLP), a persistent DNA marker, detects polymorphisms better than RFLPs or RAPDs and offer meticulous genetic studies in closely interrelated population (Meksen *et al.*, 1995). It is highly reliable for the assessment of genetic variation among and within populations. The advantage of AFLP is its capability to produce multilocus fingerprints in a single analysis, significantly reducing the cost of analysis and increasing the possibility of detecting polymorphisms (Vos *et al.*, 1995). In tea, Paul *et al.* (1997) first employed AFLP in 32 clones belong to Indian and Kenyan origin and five enzyme-primer combinations that revealed 73 unambiguous polymorphic bands. Genetic diversity indicated the Chinary types are inconsistent than the others. The similarity matrix co-efficient varied from 35% to 96% and the dendogram was largely similar with the existing biosystematics of tea. Assam clones from India and Kenya clustered closely indicating a common ancestry. AFLP analysis of 42 tea clones, which includes UPASI-23, 17 popular South Indian clones, and two Kenyan tea clones were done and it was found that 90% of the UPASI clones are inbred and inappropriate for commercial cultivation and it was concluded that AFLP could be utilized for developing markers related to resistance to blister blight (Rajasekaran, 1997). AFLP markers with eight primer pairs for genetic characterization of 29 Darjeeling tea cultivars produced 677 bands, of which 469 were polymorphic. Subsequently, using UPGMA method, a similarity matrix was constructed that clearly divided the clones into three groups (Assam, China and Cambod type) and it was concluded that two (T-246 and T-135, which are Assam types) clones had originated by extensive cross breeding between species within the genus *Camellia* (Misra and Sen-Mandi, 2001). Apart from these, Wachira *et al.* (2001), Balasarvanan *et al.* (2003), Mishra and Sen-Mandi (2004a, b), and Sharma *et al.* (2009b) contributed in the progress of AFLP. Huge genetic variation was found within tea plants of *C. assamica*, but the genetic diversity among cloned tea plants was lower than that among the seed produced and wild plants. The cloned tea plant had a low genetic diversity (13.77%), while the wild was high (85.02%). The genetic diversity of native wild tea was higher than that of cultivated cloned tea. It was also found that variation of morphological, biochemical characters was found within and among tea plant germplasm of China tea variety. Compared with new cultivars, the relatively primitive tea tree contain greater amounts of compounds ECG and EC and lower amounts of compounds EGCG and EGC. Unlike most of the commercial tea cultivars, the wild teas are resistant to cold and common diseases affecting the tea species. Taken together, these studies indicate that wild tea is genetically highly variable and a large proportion of valuable tea germplasm may have been lost during the commercial planting of vegetatively propagated plants (Ji *et al*., 2012). Attempt was also made to explore the genetic variation of clones of Darjeeling by employing AFLP and high degree of polymorphism was observed in 29 tested Darjeeling tea clones. The extent of genetic relatedness among these clones was found to be 70% using different primer pair combinations. They had a very high similarity index (0.697-0.869), which suggested that tea clones have a narrow gene pool (Mishra *et al*., 2009). AFLP profiling of popular 49 tea cultivars of southern India were generated. The study indicated that a large percentage of valuable tea germplasm might have been already lost through the continuous removal of older plants, especially seedlings for commercial planting of vegetatively propagated plants. However, AFLP analysis specified that, further extensive planting of a few clones would cause analogous erosion of genetic diversity. The Assam cultivars, which have been extensively exploited for higher yield, in particular, showed a narrow genetic diversity. From the AFLP analysis, it was noticed that, in order to avoid further degradation of germplasm resources, existing populations must be preserved and further crosses should be made with genetically distant varieties or genotypes of diverse origin (Balasarvanan *et al*., 2003). Sharma *et al*., (2010) analyzed 123 commercially important tea accessions representing major populations in India. The overall genetic similarity recorded was 51%. No significant differences were recorded in average genetic similarity among tea populations cultivated in various geographic regions (northwest 0.60, northeast and south both 0.59). UPGMA cluster analysis grouped the tea accessions according to geographic locations, with a bias toward China or Assam/Cambod types. Cluster analysis results were congruent with principal component analysis. Further, analysis of molecular variance detected a high level of genetic variation (85%) within and limited genetic variation (15%) among the populations, suggesting their origin from a similar genetic pool.

Genetic maps are essential tools for implementing quantitative trait loci (QTL) analysis and marker-assisted selection (MAS) breeding. A few genetic maps, based on dominant markers such as RAPD or AFLP, have been reported for *C. sinensis*, but do not permit easy comparisons of maps at interspecific or intergeneric levels. Taniguchi *et al*. (2012) constructed a high-density linkage map of tea containing 441 SSRs, 674 of RAPDs and other markers. However, next generation sequencing (NGS) methods, including Illumina and 454, can produce millions of sequences at a relatively low cost. Recent studies have offered substantial increase in DNA sequence availability with NGS. Thus, obtaining gene expression information will not only expand species genetic resources, but also enable the exploration of molecular mechanisms behind flowering and pollination, such as self-incompatibility and various fruit-bearing rates in different cross-parent combinations (Tan *et al*., 2013).

Narrow genetic base of tea cultivars is a serious obstacle to sustain and improve productivity due to rapid vulnerability of genetically uniform cultivars due to various biotic and abiotic stresses. Molecular markers are rapidly adopted for crop improvement as an effective and appropriate tool for basic and applied studies addressing biological components in tea production. Hence, study of genetic diversity of the newly improved clones under exploitation is necessary to avoid narrowing of genetic pool (Leonida *et al.*, 2013).

Functional Genomics

Due to rapid increase in DNA sequence information and the complete genome sequences for several plant species, plant research started focusing on analysis of gene function. Functional genomics studies principally categorize gene function and gene expression in the spatial, temporal, cell-dependent and regulatory mechanisms. Once the *eilte* gene with desirable trait is discovered, the same is used to develop the new plant type through transgenic approaches.

Tea has a genome of 4.0 Gb with a basic chromosome number of $n = 15$. The genome size is 10 times larger that of rice (389 Mb) . An efficient first step for the analysis of the large-genome species such as tea is to survey the expressed genes. Expressed sequence tag analysis, in which partial sequences of a large number of cDNA clones are isolated, is a useful approach to reveal expressed sequences in the genome and it enables the identification of many genes responsible for important traits. In addition, ESTs can be used as a resource for functional genomics experiments, such as gene expression analysis using microarrays (Taniguchi *et al.*, 2012).

Functional genomics study of *C. sinensis* has made significant progress (Mondal and Sutoh, 2013). The fundamental steps in this approach involves, the construction of cDNA libraries, generation and annotation of ESTs, analysis of gene expression profiling, construction and use of cDNA microarrays, development and utilization of expressed sequence tags-simple sequence repeat (EST-SSR) and short tandem repeat (STR) markers, cloning and expression of genes involved in secondary metabolism, disease and pest resistance and stress tolerance (Chen *et al.*, 2009). In tea, application of functional genomics was started with the isolation of chalcone synthase (Mukhopadhyay *et al.*, 2013c) gene from a Japanese green tea cultivar 'Yabukita', which revealed its organ specific and sugar-responsive expression. Since then, considerable amount of work has been done in the functional genomic research of tea. Fundamentally two types of efforts have been made, such as i) cloning of individual gene associated with particular trait and ii) differential gene expression, which leads to identify group of genes that are associated to a particular trait. Interestingly, majority of the genes cloned so far belongs to quality followed by yield as these two parameters have major demand in the tea industry (Mondal, 2014).

Recent efforts have demonstrated that analysis of ESTs is an appropriate strategy for identifying genes involved in specific biological functions in model plants and even in non-model plants in which genome has not been sequenced. The suppression subtractive hybridization (SSH) technique enables specific cloning of ESTs representing genes that are differentially expressed in different mRNA populations and isolates genes without prior knowledge of their sequence or identity. Moreover, the availability of databases of known genes and proteins and gene ontology (GO) annotation provides an opportunity to predict the functions of newly isolated putative gene sequences (Das *et al.*, 2012).

The cost-effective and ultra-high-throughput DNA sequencing technology, RNA-seq, is a revolutionary advance in genome-scale sequencing. This method is fast and simple because it does not require bacterial cloning of the cDNAs. Direct sequencing of these cDNAs can generate short reads at an extraordinary depth. Following sequencing, the resulting reads can be assembled into a genome-scale transcription profile. It is a more comprehensive and efficient way to measure transcriptome composition, obtain RNA expression patterns, and discover new genes. Using high-throughput Illumina RNA-seq, the transcriptome from poly $(A)^+$ RNA of *C. sinensis* was sequenced which generated 34.5 million reads. that produced 127,094 unigenes (with an average length of 355 bp and an N50 of 506 bp) which consisted of 788 contig clusters and 126,306 singletons. Sequence similarity analyses against public databases like; Uniprot, NR and COGs at NCBI, Pfam, InterPro and KEGG found 55,088 unigenes could be annotated with gene descriptions, conserved protein domains, or gene ontology terms. Some of the unigenes were assigned to putative metabolic pathways. Targeted searches using these annotations identified the majority of genes associated with several primary metabolic pathways and natural product pathways that are important to tea quality, such as flavonoid, theanine and caffeine biosynthesis pathways. Novel candidate genes of these secondary pathways were discovered. Comparisons with four previously prepared cDNA libraries revealed that this transcriptome dataset had both a high degree of consistency with previous EST data and an approximate 20 times increase in coverage. Thirteen unigenes related to theanine and flavonoid synthesis were validated. Their expression patterns in different organs of the tea plant were analyzed by RT-PCR and quantitative real time PCR (Shi *et al.*, 2011).

Low temperature related stress is important environmental factors that tea plants experience during their life cycle. Cold acclimation is a process by which some plants can enhance their freezing tolerance after exposure to low but non-freezing temperatures for a certain period. RNA-Seq and digital gene expression (DGE) technologies were employed to the study the genome wide expression profiles during cold acclimation in tea plants. Using the Illumina sequencing platform, 57.35 million reads were obtained from cold stressed leaf tissue of tea. They were assembled into 216,831 transcripts from which 1,770 differentially expressed transcripts were identified, of which 1,168 were up-regulated and 602 down-regulated. These included signal transduction genes, cold-responsive transcription factor genes, plasma membrane stabilization related genes, osmosensing-responsive genes and genes for detoxification enzymes. Pathway analysis indicated that the "carbohydrate metabolism pathway" and the "calcium signaling pathway" probably play a vital role during cold stress response. Thus, a global survey of transcriptome profiles of tea plants in response to low; non-freezing temperatures was evident. Furthermore, insights into the molecular mechanisms of tea plants during the cold acclimation process were also obtained (Wang *et al.*, 2013).

In order to understand an outline of the mRNA expression profile in tender leaves and apical buds of tea plant, a *C. sinensis*, cDNA library was constructed and 210 cDNA clones were sequenced and analyzed. Among them, 84 high quality ESTs were generated which were classified into putative cellular roles, like; transcription (14.2%), protein synthesis (14.2%), cell growth and division (8.6%), cell structure (5.7%), signal transduction (5.7%), transporters (2.9%), disease and defenses (2.9%), secondary metabolism (2.9%) and gene regulation (2.9%) (Phukon *et al.*, 2012).

Seven cDNA libraries from various organs were used to generate 17,458 ESTs which were assembled into 5,262 unigenes. About 50% of the unigenes were assigned annotations by Gene Ontology. Some were homologous to genes involved in important biological processes, such as nitrogen assimilation, aluminium response, and biosynthesis of caffeine and catechins. Digital northern analysis showed that 67 unigenes were expressed differentially among the seven organs. Simple sequence repeat (SSR) motif searches among the unigenes identified 1,835 unigenes (34.9%) harboring SSR motifs of more than six repeat units. A subset of 100 EST-SSR primer sets was tested for amplification and polymorphism in 16 tea accessions. Seventy-one primer sets successfully amplified EST-SSRs and 70 EST-SSR loci were polymorphic. Furthermore, these 70 EST-SSR markers were transferable to 14 other *Camellia* species. The ESTs and EST-SSR markers are therefore important genomic resources to study important traits of tea plants (Taniguchi *et al.*, 2012).

MicroRNAs (miRNAs) are approximately 19~21 nucleotide noncoding RNAs produced by Dicer-catalyzed excision from stem-loop precursors. Many plant miRNAs have critical functions in development, nutrient homeostasis, abiotic stress responses, and pathogen responses via interaction with specific target mRNAs. With the invention of microRNA by Das and Mondal (2010) several other have been discovered in tea and other *Camellia* species in *in silico* (Prabhu and Mandal, 2010; Mohanpuria and Yadav (2012); Zhu and Luo, 2013). Zhang *et al.*, (2014) discovered 31 up-regulated miRNAs and 43 down-regulated miRNAs in 'Yingshuang', and 46 up-regulated miRNA and 45 down-regulated miRNAs in 'Baiye' in response to cold stress, respectively. A total of 763 related target genes were detected by degradome sequencing. The RLM-5'RACE procedure was successfully used to map the cleavage sites in six target genes of *C. sinensis*. These findings reveal important information about the regulatory mechanism of miRNAs in *C. sinensis*, and promote the understanding of miRNA functions during the cold response.

Tea is well known as an aluminium (Al) accumulating plant that grows well in very acidic soils containing high levels of Al^{3+}; this is of interest because Al toxicity limits the growth of many other species in acidic soils and the Al in the xylem sap of tea is complexed with citrate (Mukhopadhyay *et al.*, 2012). Three unigenes potentially related to Al response were detected; one citrate synthetase and two Al-response proteins. Hence, further analyses, such as expression analysis of the responses of tea to Al, will probably reveal whether those genes have roles in Al resistance or tolerance (Taniguchi *et al.*, 2012).

Future Prospects

Till date significant progress has been made in tea biotechnology (Mondal *et al.*, 1998, 2000b, 2001a; Mondal, 2002b) and other area of molecular biology such as; construction of cDNA libraries, generation and annotation of ESTs, analysis of gene expression profiling, construction and use of cDNA microarrays, development and utilization of simple sequence repeat, microsatellite and short tandem repeat markers and the cloning and expression of genes involved in secondary metabolism, disease and pest resistance (Chen *et al.*, 2009).

Though markers have been developed and used mainly to study the genetic diversity yet, none of them have been utilized for mapping of agronomically important QTLs, marker assisted breeding (MAB) for gene introgression which has a tremendous role in genetic improvement of tea. In the genus *Camellia*, developing biparental mapping population is difficult and hence association-mapping approaches could be a better alternative. Unfortunately, such an approach has remained hitherto unexplored for identifying the agronomically important genes. Although several aspects of tea molecular biology work can be intended, yet priority should be given to, (i) undertake a *de novo* whole genome sequencing for development of reference genome, and (ii) re-sequence popular trait-specific cultivars for generating large scale SNPs markers and their utilization for association mapping, gene introgression, high density linkage map constructions etc.

References

Abraham G.C. and Raman K. (1986) Somatic embryogenesis in tissue culture of immature cotyledons of tea (*Camellia sinensis*). In: Somers DA, Gengenbach BG, Biesboor DD, Hackett WP, Green CE, editors. 6th International Congress on Plant Tissue and Cell Culture. Univ Minnesota, Minneapolis; p. 294.

Agrobacterium-mediated genetic transformation of tea (*Camellia sinensis*). *J. of Plantation Crops,* 29: 45-48.

Akula A. and Dodd W.A. (1998) Direct somatic embryogenesis in a selected tea clone, 'TRI-2025' (*Camellia sinensis* (L).O. Kuntze) from nodal segment. *Plant Cell. Rep.* 17: 804–809.

Akula A., Akula C. and Bateson M. (2000) Betaine, a novel candidate for rapid induction of somatic embryogenesis in tea (*Camelia sinensis* (L.) O.Kuntze). *Plant. Growth. Regu.* 30: 241-246.

Arulpragasam P.V., Latiff R. and Seneviratne P. (1988) Studies on the tissue culture of tea (*Camellia sinensis* (L.) O. Kuntze). 3. Regeneration of plants from cotyledon callus. *Sri Lank J. Tea Sci.* 57: 20-23.

Bag N., Palni L.M.S. and Nandi S.K. (1997) Mass propagation of tea using tissue culture methods. Physiol. *Mol. Biol. Plants* 3: 99-103.

Bagratishvili D.G., Zaprometov M.N. and Butenko R.G. (1979) Obtaining a cell suspension culture from the tea plant. *Fiziol. Rast*. 26: 449-451

Balasaravanan T., Pius P.K., Kumar R.R., Muraleedharan N. and Shasany A.K. (2003) Genetic diversity among south Indian tea germplasm (*Camellia sinensis, C. assamica* and *C. assamica* spp. Lasiocalyx) using AFLP markers. *Plant Sci* 165: 365-372.

Balasubramanian S., Marimuthu S., Rajkumar R. and Balasaravanan T. (2000) Isolation, culture and fusion of protoplast in tea. In: Muraleedharan N, Rajkumar R, editors. Recent advance in plants crops research. India: Allied Publishers Ltd; p. 3-9.

Bandyopadhyay T. (2011) Molecular marker technology in genetic improvement of tea. *Int. J. Plant Breed. Genet*. 5: 23-33.

Bano Z., Rajaratnam S. and Mohanty B.D. (1991) Somatic embryogenesis in cotyledon culture of tea (*Thea sinensis* L.) *J. Hort. Sci.* 66: 465-470.

Bhardwaj P., Sharma R.K., Kumar R., Sharma H. and Ahuja P.S. (2014) SSR marker based DNA fingerprinting and diversity assessment in superior tea germplasm cultivated in western Himalaya. *Proc. Indian Natn. Sci. Acad.* 80: 157-162.

Bhattacharya A., Saini U., Joshi R., Kaur D., Pal A.K., Kumar N., Gulati A., Mohanpuria P., Yadav S.K., Kumar S. and Ahuja P.S. (2014) Osmotin-expressing transgenic tea plants have improved stress tolerance and are of higher quality. *Transgenic Res* 23: 211–223.

Chen L. and Yamaguchi L. (2002) Genetic diversity and phylogeny of tea plant (*Camellia sinensis*) and its related species and varieties in the section Thea genus *Camellia* determined by randomly amplified polymorphic DNA analysis. *J. Hort. Sci. Bio.* 77: 729-732.

Chen L., Gao Q.K, Chen D.M. and Xu C.J. (2005) The use of RAPD markers for detecting genetic diversity, relationship and molecular identification of Chinese elite tea genetic resources [*Camellia sinensis* (L.) O. Kuntze] preserved in a tea germplasm repository. *Biodivers Conserv* 14: 1433-1444.

Chen L., Zhao L.P., Ma C.L., Zhang Y.L., Liu Z., Qiao X.Y., Yao M.Z. and Wang X.C. (2009) Recent progress in the molecular biology of tea (*Camellia sinensis*) based on the expressed sequence tag strategy: a review. *J. Hort. Sci. Biotech.* 84: 476–482.

Chen Z. and Liao H. (1982) Obtaining plantlet through anther culture of tea plants. Zhongguo chaye. 4: 6-7.

Das A. and Mondal T.K. (2010) *In silico* analysis of miRNA and their targets in tea. *American Journal of Plant Science* 1: 77-86.

Das A., Das S. and Mondal T.K. (2012) Identification of differentially expressed gene profiles in young roots of tea [*Camellia sinensis* (L.) O. Kuntze] subjected to drought stress using suppression subtractive hybridization. *Plant Mol. Biol. Rep.* 30: 1088-1101

Deka A., Deka P.C. and Mondal T.K. (2006) Tea. In: Parthasarathy, V.A., Chattopadhyay, P.K., Bose, T.K. (eds), *Plantation Crops-I*, Calcutta: Naya Udyog, pp. 1-148.

Dosumu O.O., Oluwaniyi O.O., Awolola V.G. and Ogunkunle O.A. (2010) Toxicity assessment of some tea labels from supermarkets in Ilorin, Nigeria using brine shrimp (*Artemia salina*) lethality assay. *Afr. J. Food Sci.* 4: 282-285.

Eden T. (1958) The development of tea culture. In: Eden T. (Ed) Tea. Longman, London, pp. 1-4.

Freeman S.J., West C.J., Lea V. and Mayes S. (2004) Isolation and characterization of highly polymorphic microsatellites in tea (*Camellia sinensis*). *Mol. Ecol. Notes* 4: 324-326.

Haridas V., Balasaravanan T., Rajkumar R. and Marimuthu S. (2000) Factor influencing somatic embryogenesis in *Camellia sinensis* (L.) O. Kuntze. In: Muraleedharan N, RajKumar R, editors. Recent Advances in Plantation Crops Research. India: Allied Publishers Ltd; pp. 31-35.

Hazarika R.R. and Chaturvedi R. (2013) Establishment of dedifferentiated callus of haploid origin from unfertil- ized ovaries of tea (*Camellia sinensis* (L.) O. Kuntze) as a potential source of total phenolics and antioxidant activity. *In Vitro Cell Dev Biol-Plant* 49: 60–69.

Hu C.Y., Tsai Y.Z. and Lin S.F. (2014) Development of STS and CAPS markers for variety identification and genetic diversity analysis of tea germplasm in Taiwan. *Botanical Studies*, 55: 12-27.

Hung C.Y., Wang K.H., Huang C.C., Gong X., Ge X.J. and Chiang T.Y. (2008) Isolation and characterization of 11 microsatellite loci from *Camellia sinensis* in Taiwan using PCR-based isolation of microsatellite arrays (PIMA). *Conserv. Genet.* 9: 779-781.

Jain S.M. and Newton R.J. (1990) Prospects of biotechnology for tea improvement. *Proc. Indian Natn. Sci. Acad.* 56: 441-448.

Jha T.B., Jha S. and Sen S.K. (1992) Somatic embryogenesis from immature cotyledon of an elite Darjeeling tea clone. *Plant Sci.* 84: 209-213.

Ji P.Z., Gao J., Liang M.Z. and Huang X.Q. (2012) AFLP analysis of genetic variation among cloned, seed produced and wild *Camellia sinensis* var. *assamica* tea plant in Yunnan, China. *Pak. J. Bot,* 44(6): 1989-1992.

Kamunya S.M., Wachira F.N., Pathak R.S., Korir R., Sharma V., Kumar R., Bhardwaj P., Chalo R., Ahuja P.S. and Sharma R.K. (2010) Genomic mapping and testing for quantitative trait loci in tea (*Camellia sinensis* (L.) O. Kuntze). Tree Genet Genome DOI 10.1007/s11295-010-0301-2.

Kato M. (1982) Results of organ culture on *Camellia japonica* and *C. sinensis*. Jpn. *J. Breed.* 32: 267-277.

Kato M. (1986) Micropropagation through cotyledon culture in *Camellia japonica* L. and *Camellia sinensis* L. Jpn. *J. Breed.* 36: 31-38.

Kato M. (1996) Somatic embryogenesis from immature leaves of *in vitro* grown tea shoots. *Plant. Cell Rep.* 15: 920-923.

Katsuo K. (1969) Anther culture in tea plant (A Preliminary Report) Study of Tea. 4: 31

Kaundun S.S. and Matsumoto S. (2003) Identification of processed Japanese green tea based on polymorphism generated by STS-RFLP analysis. *J. Food and Chem.* 51: 1765-1770.

Kaundun S.S., Zhyvoloup A. and Park Y.G. (2000) Evaluation of genetic diversity among elite tea (*Camellia sinensis* var. *sinensis*) accessions using RAPD markers. *Euphytica* 115: 7-16.

Konieczny A. and Ausubel F.M. (1993) A procedure for mapping *Arabidopsis* mutations using co-dominant ecotype-specific PCR-based markers. *Plant J.* 4(2): 403–410.

Kumar N., Pandey S., Bhattacharya A. and Ahuja P.S. (2004) Do leaf surface characteristics affect *Agrobacterium* infection in tea (*Camellia sinensis* (L.) O. Kuntze? J. Biosci. 29: 309-317.

Leonida C., Kamunya S.M., Alakonya A., Msomba S.W., Uwimanna M.A. and okinda P.O. (2013) Characterization of 20 clones of tea (*Camellia sinensis* (L.) O. Kuntze) using ISSR and SSR markers. *Agric. Sci. Res. J.* 3: 292-302.

Liu B.Y., Cheng H., Li Y.Y., Wang L.Y., He W. and Wang P.S. (2012) Fingerprinting for discriminating tea germplasm using Inter Simple Sequence Repeats (ISSR) markers. *Pak. J. Bot.* 44(4): 1247-1260.

Lopez S.J., Kumar R.R., Pius P.K. and Muraleedharan N. (2004) *Agrobacterium tumefaciens*-mediated genetic transformation in tea (*Camellia sinensis* [L.] O. Kuntze). *Plant Mol. Biol. Rep.* 22: 201-210.

Matsumoto S., Takeuchi A., Hayastsu M. and Kondo S. (1994) Molecular cloning of phenylalanine ammonia-lyase cDNA and classification of varieties and cultivars of tea plants (*Camellia sinensis*) using the tea PAL cDNA probes. *Theor. Appl. Genet.* 89: 671-675.

Matsuura T., Kakuda T., Kinoshita T., Takeuchi N. and Sasaki K (1991) Production of theanine by callus culture of tea. In: Proceedings of the International Symposium on Tea Science, Shizuka, Japan, pp. 432-43.

Meksen K., Leister D., Peleman J., Zabeau M., Salamini F. and Gebhardt C. (1995) A high resolution map of the vicinity of the R1 locus on chomosome V of potato based on RFLP and AFLP markers. *Mol. Gen. Genet.* 249: 74-81.

Mewan K.M., Liyanage A.C., Everard J.M., Gunasekare M.T.K. and Karunanayaka E. (2005) Studying genetic relationship among tea accessions in Sri Lanka using RAPD. *Sri Lanka J. Tea Sci.* 70: 42-53.

Mishra R.K. and Sen-Mandi S. (2001) DNA fingerprinting and genetic relationship study of tea plants using amplified Fragment Length Polymorphism (AFLP) technique. *Indian J. Plant Genet. Reso.* 14: 148-149.

Mishra R.K. and Sen-Mandi S. (2004b) Molecular profiling and development of DNA marker associated with drought tolerance in Darjeeling. *Curr. Sci.* 86: 60-66.

Mishra R.K., and Sen-Mandi S. (2004a) Genetic diversity estimates for Darjeeling tea clones based on AFLP markers. J. Tea Sci. 24: 86-92.

Mishra R.K., Chaudhury S., Ahmad A., Pradhan M. and Siddiqui T.O. (2009) Molecular analysis of tea clones (*Camellia sinensis*) using AFLP markers. *Int. J. Integr. Biol.* 5: 130-135.

Mohanpuria P. and Yadav S.K. (2012) Characterization of novel small RNAs from tea (*Camellia sinensis L.). Mol. Biol. Rep.* 39: 3977-3986.

Mohanpuria P., Kumar V., Ahuja P.S. and Yadav S.K. (2011) Agrobacterium-mediated silencing of caffeine synthesis through root transformation in *Camellia sinensis* L. *Mol Biotechnol* 48: 235-43.

Mohanpuria P., Rana N.K. and Yadav S.K. (2008) Transient RNAi based gene silencing of glutathione synthetase reduces glutathione content in *Camellia sinensis* (L.) O. Kuntze somatic embryos. *Biol. Plantarum* 52: 381-384.

Mondal T.K. (2000) Studies on RAPD marker for detection of genetic diversity, *in vitro* regeneration and *Agrobacterium*-mediated genetic transformation of tea (*Camellia sinensis*). Ph.D Thesis, Utkal University, India.

Mondal T.K. (2002a) Micropropagation of tea (*Camellia sinensis*). In: Jain, S.M. (ed.), Micropropagation of Woody Plants, Kluwer, Dordrecht, pp. 671-720.

Mondal T.K. (2002b) *Camellia* biotechnology: A bibliographic search. *Int. J. of Tea Sci.*, 1: 28-37.

Mondal T.K. (2002c) Detection of genetic diversity among the Indian tea (*Camellia sinensis)* germplasm by Inter-simple sequence repeats (ISSR). *Euphytica* 128: 307-315.

Mondal T.K. (2008) Tea. In: Kole, C., Hall, T.C. (eds), *Compendium of Transgenic Crop Plants: Transgenic Plantation Crops, Ornamentals and Turf Grasses*, Blackwell. London, pp. 99–115.

Mondal T.K. (2011) Camellia. In: Kole, C. (ed.) Wild Crop Relatives: Genomics and Breeding Resources Plantation and Ornamental Crops, Springer, Berlin, p. 15-40.

Mondal T.K. (2014) In: Breeding and Biotechnoly of tea and wild species, Springer, New Delhi, pp. 50-57

Mondal T.K. and Chand P.K. (2002) Detection of genetic instability among the miocropropagated tea (*Camellia sinensis)* Plants. In Vitro Cell Dev. Biol. *Plant.* 37: 1-5.

Mondal T.K. and Sutoh K. (2013) Application of next generation sequencing for abiotic stress tolerance of plant. In: Barh D., Zambare V. and Azevedo V. (Eds.) Applications in Biomedical, Agricultural, and Environmental Sciences. CRC Press, pp. 347-365.

Mondal T.K., Ahuja P.S. and Chand P.K. (2000a) Molecular characterization of biodiversity in tea (*Camellia sinensis* (L.) O. Kuntze) germplasm and ex situ conservation through *in vitro* culture. Biodiversity conservation for environment protection, Utkal University, Bhubaneswar, india, pp. 9-10.

Mondal T.K., Bhattacharya A. and Ahuja P.S. (2001a) Induction of synchronous secondary embryogenesis of Tea (*Camellia sinensis*). *J. Plant Physiol.* 158: 945-951.

Mondal T.K., Bhattacharya A. and Ahuja P.S. (2001c) Development of a selection system for

Mondal T.K., Bhattacharya A., Ahuja P.S. and Chand P.K. (2001b) Factor effecting *Agrobacterium tumefaciens* mediated transformation of tea (*Camellia sinensis* (L). O.Kuntze). *Plant Cell Rep.* 20: 712-720.

Mondal T.K., Bhattacharya A., Ahuja P.S. and Chand P.K. (2002) Transgenic tea (*Camellia sinensis* (L.) O. Kuntze cv. Kangra Jat) plants obtained by *Agrobacterium*-mediated transformation of somatic embryos. *Plant Cell Rep.* 20: 712–720.

Mondal T.K., Bhattacharya A., Laxmikumaran M. and Ahuja P.S. (2004) Recent advances of tea (*Camellia sinensis*) biotechnology. *Plant Cell Tissue Organ Cult.* 76: 195–254.

Mondal T.K., Bhattacharya A., Sharma M. and Ahuja P.S. (2001d) Induction of *in vivo* somatic embryo-genesis in tea (*Camellia sinensis*) cotyledons. *Curr. Sci.* 81: 101-104.

Mondal T.K., Bhattacharya A., Sood A. and Ahuja P.S. (1998) Micropropagation of tea using thidia-zuran. *Plant Growth Regulation* 26: 57-61.

Mondal T.K., Bhattacharya A., Sood A. and Ahuja P.S. (2000b) Factor effecting induction and storage of encapsulated tea (*Camellia sinensis* L. O.Kuntze) somatic embryos. *Tea* 21: 92-100.

Mondal T.K., Bhattachraya A., Sood A. and Ahuja P.S. (1999) An efficient protocol for somatic embryogenesis and its use in developing transgenic tea (*Camellia sinensis (L)* O. Kuntze) for field transfer. In: Altman A., Ziv M. and Izhar S. (eds.) Plant Biotechnology and *In Vitro* Biology in 21st Century. Kluwer Academic Publishers, Dordrecht, The Netherlands, pp.101-104.

Mondal T.K., Bhattachrya A., Sood A. and Ahuja P.S. (2002) Factors affecting germination and conver-sion frequency of somatic embryos of tea. *J. of Plant Physiol.* 159: 1317-1321.

Mondal T.K., Satya P., Medda P.S. (2003) India needs national tea germplasm repository. In: *International Conference on Global Advances in Tea Science*, 20–22nd November, Calcutta, India, p. 58.

Mukhopadhyay M. (2012) Biochemical, physiological and molecular changes in tea under Zn, B and Al stresses. Ph.D. Thesis, University of Kalyani, West Bengal, India.

Mukhopadhyay M. and Mondal T.K. (2014) The physio-chemical responses of *Camellia* plants to abiotic stresses. *J. of Plant Sci. and Res.* 1: 105.

Mukhopadhyay M., Das A., Subba P., Bantawa P., Sarkar B., Ghosh P.D. and Mondal T.K. (2013a) Structural, physiological and biochemical profiling of tea plantlets (*Camellia sinensis* (L.) O. Kuntze) under zinc stress. *Biol. Plantarum* 57: 474-480.

Mukhopadhyay M., Ghosh P.D. and Mondal T.K. (2013b) Effect of boron deficiency on photosynthesis and antioxidant responses of young tea (*Camellia sinensis* (L.) O. Kuntze) plantlets. *Russ. J. Plant Physiol.* 60: 633-639.

Mukhopadhyay M., Sarkar B. and Mondal T.K. (2013c) Omics Advances in Tea (*Camellia sinensis*). In: Barh D. (Ed) OmicsApproaches In Crop Sciences. CRC Press. pp. 439-465.

Mukhopadyay M., Bantawa P., Das A., Sarkar B., Bera B., Ghosh P.D. and Mondal T.K. (2012) Changes of growth, photosynthesis and alteration of leaf antioxidative defence system of tea (*Camellia sinensis* (L.) O. Kuntze) seedling under aluminum stress. *Biometals* 25: 1141-1154.

Nakamura Y. (1985) Effect of origin of explants on differentiation of root and its varietal difference in tissue culture of tea plant. *Tea Res. J.* 62: 1-8.

Nakamura Y. (1988) Efficient differentiation of adventitious embryos from cotyledon culture of *Camellia sinensis* and other *Camellia* species. *Tea Res. J.* 67: 1-12.

Nguyen V.T. (2012) Regeneration plantlets from somatic embryos of tea plant (*Camellia sinensis* L.) *J. Agri. Tech.* 8: 1821-1827.

Okano N. and Fuchinone Y. (1970) Production of haploid plants by anther culture of tea *in vitro*. Jpn. *J. Breed.* 20: 63-64.

Orihara Y. and Furuya T. (1990) Production of theanine and other ã-glutamine derivatives by Camellia sinensis cultured cells. *Plant Cell Rep*. 9: 1215-122.

Palni L.M.S., Hao C. and Nakamura Y. (1999). In: Jain N.K. (ed.) Global Advances in Tea Science, Aravali Books Intl. (P) Ltd, New Delhi, pp. 449-62.

Pandey A., Palni L.M.S. and Bag N. (2000) Biological hardening of tissue culture raised tea plants through rhizosphere bacteria. Biotech. *Letters* 22: 1087-1091.

Paratasilpin T. (1990) Comparative studies on somatic embryogenesis in *Camellia sinensis* var. *sinensis* and *C. sinensis* var. *assamica* (Mast.) *Pierre. J. Sci. Soc. Thailand.* 16: 23-41.

Paul S., Wachira F.N., Powell W. and Waugh R. (1997) Diversity and genetic differentiation among population of Indian and Kenyan tea (*Camellia sinensis* (L.) O. Kuntze) revealed by AFLP markers. *Theor. Appl. Genet.* 94: 255-263.

Phukon M., Namdev R., Deka D., Modi M.K. and Sen P. (2012) Construction of cDNA library and preliminary analysis of expressed sequence tags from tea plant [*Camellia sinensis* (L) O. Kuntze]. *Gene* 506: 202-206.

Ponsamuel J., Samson N.P., Ganeshan P.S., Satyaprakash V. and Abrahan G.C. (1996) Somatic embryogenesis and plant regeneration from the immature cotyledonary tissues of cultivated tea (*Camellia sinensis* (L.) O. Kuntze). *Plant Cell Rep.* 16: 210-214.

Prabu G.R. and Mandal A.K. (2010). Computational identification of miRNAs and their target genes from expressed sequence tags of tea (*Camellia sinensis*). *Genomics Proteomics Bioinformatics*. 8:113-121.

Prakash O., Sood A., Sharma M. and Ahuja P.S. (1999) Grafting micropropagated tea (*Camellia sinensis* (L.) O. Kuntze) shoots on tea seedling- a new approach to tea propagation. *Plant Cell Rep*. 18: 137-142.

Rajasekaran P. (1997) Development of molecular markers using AFLP in tea. In: Varghese J.P. (Ed.) Molecular Approaches to Crop Improvement. Proceedings of National Seminar on Molecular approaches to Crop Improvement, Kottayam, Kerala, India.

Rajkumar R. and Ayyappan P. (1992) Somatic embryogenesis from cotyledonary explants of *Camellia sinensis* (L.) O. Kuntze. The Planters Chronic, pp. 227-229.

Sahu J., Sarmah R., Dehury B., Sarma K., Sahoo S., Sahu M., Barooah M., Modi M.K. and Sen P. (2012) Mining for SSRs and FDMs from expressed sequence tags of *Camellia sinensis*. *Bioinformation* 8: 260-266.

Sandal I., Kumar A., Bhattacharya A., Sharma M., Shanker A. and Ahuja P.S. (2005) Gradual depletion of 2,4-D in the culture medium for indirect shoot regeneration from leaf explants of *Camellia sinensis* (L.) O. Kuntze. *Plant Growth Regul.* 47: 121-127.

Sandal I., Saini U., Lacroix B., Bhattacharya A., Ahuja P.S. and Citovsky V. (2007) *Agrobacterium*-mediated genetic transformation of tea leaf explants: effects of counteracting bactericidity of leaf polyphenols without loss of bacterial virulence. Plant Cell Rep. 26: 169-176.

Sarathchandra T.M., Upali P.D. and Wijeweardena R.G.A. (1988) Studies on the tissue culture of tea (*Camellia sinensis* (L.) O. Kuntze) 4. Somatic embryogenesis in stem and leaf callus cultures. *Sri Lank J. Tea Sci.* 52: 50–54.

Seran T.H., Hirimburegama K. and Gunasekare M.T.K. (2006) Somatic embryogenesis from embryogenic leaf callus of tea {*Camellia sinensis* (L.) Kuntze)}. *Tropical Agricultural Research* 18: 367- 375.

Seran T.H., Hirimburegama K. and Gunasekare M.T.K. (2007) Production of embryogenic callus from leaf explants of *Camellia sinensis* (L.). J. Nat. Sci. Found Sri Lanka 35: 191-196.

Sharma R.K., Bhardwaj P., Negi R., Mohapatra T. and Ahuja P.S. (2009a) Identification, characterization and utilization of unigene derived microsatellite markers in tea (*Camellia sinensis* L.), *BMC Plant Biol* 9: 53

Sharma R.K., Negi M.S., Sharma S., Bhardwaj P. and Kumar R. (2009b) AFLP based genetic diversity assessment of commercially important tea germplasm in India. *Biochem. Genet.* 48: 549-564.

Sharma R.K., Negi M.S., Sharma S. and *et al.* (2010). AFLP-based genetic diversity assessment of commercially important tea germplasm in India. *Biochem Genet.* 48: 549-564.

Shi C.Y., Yang H. and *et al.* (2011) Deep sequencing of the *Camellia sinensis* transcriptome revealed candidate genes for major metabolic pathways of tea-specific compounds. *BMC Genomics* 12: 131.

Singh M., Bandana and Ahuja P.S. (1999) Isolation and PCR amplification of genomic DNA fropm market samples of Dry tea. Plant Mol. Biol. Rep. 17: 171-178.

Singh M., Saroop J. and Dhiman B. (2004) Detection of intra-clonal genetic variability in vegetatively propagated tea using RAPD markers. *Biol. Plantarum* 48: 113-115.

Sood A., Palni L.M.S., Sharma M., Rao D.V., Chand G. and Jain N.K. (1993) Micropropagation of tea using cotyledon culture and encapsulated somatic embryos. *J. Plant. Crops* 21: 295–300.

Tan L.Q., Wang L.Y., Wei K. and *et al.* (2013) Floral transcriptome sequencing for SSR marker development and linkage map construction in the tea plant (*Camellia sinensis*). PLoS ONE 8(11): e81611. doi:10.1371/journal.pone.0081611.

Tanaka J. and Taniguchi F. (2002) Emphasized-RAPD (e-RAPD): A simple and efficient technique to make RAPD bands clearer. *Breeding Sci*. 52: 225-229.

Tanaka T., Mizutani T., Shibata M., Tanikawa N. and Parks C.R. (2005) Cytogenetic studies on the origin of *Camellia×vernalis*. V. Estimation of the seed parent of *C. × vernalis* that evolved about 400 years ago by cpDNA analysis. *J. Japan Soc. Hort. Sci.* 74: 464-468.

Taniguchi F., Furukawa K., Ota-Metoku S. and *et al.* (2012) Construction of a high-density reference linkage map of tea (*Camellia sinensis*). *Breed. Sci* 62: 263-273.

Vieitez A.M. (1994) Somatic embryogenesis in *Camellia* spp. In: Jain S, Gupta P & Newton R (eds) Somatic Embryogenesis in Woody Plants Kluwer Academic Publishers, Dordrecht, The Netherlands. pp. 235-276.

Vos P.R., Hogers M.B., Reijans M. and *et al.* (1995) AFLP: A new technique for DNA fingerprinting. *Nucleic Acids Res.*, 23: 4407-4414.

Wachira F. and Ogada J. (1995) *In vitro* regeneration of *Camellia sinesis* (L.) O. Kuntze by somatic embryogenesis. *Plant Cell Rep.* 14: 463-466.

Wachira F., Tanaka J. and Takeda Y. (2001) Genetic variation and differentiation in tea *(Camellia sisnensis)* germplasm revealed by RAPD and AFLP variation. *J. Hort. Sci. Biotech.* 76: 557-563.

Wachira F.N., Powell W. and Waugh R. (1997) An assessment of genetic diversity among *Camellia sinensis* L. (cultivated tea) and its wild relatives based on RAPDs and organelle-specific STS. *Heredity* 78: 603-611.

Wachira F.N., Waugh R., Hackett C.A. and Powell W. (1995) Detection of genetic diversity of tea (*Camellia sinensis* L.) using RAPD markers. *Genome* 38: 201-210.

Wang X.C., Zhao Q.Y., Ma C.L. and *et al.* (2013) Global transcriptome profiles of Camellia sinensis during cold acclimation. *BMC Genomics* 14: 415.

Wu C.T., Huang T.K., Chen G.R. and Chen S.Y. (1981) A review on the tissue culture of tea plants and on the utilization of callus derived plantlets In: Rao A.N. (ed.) Tissue culture of Economically Important plants. Proc Costed Symp. Singapore: p. 104-106.

Yan M.Q. and Ping C. (1983) Studies on development of embryoids from the culture cotyledons of *Thea sinensis* L. Sci. Silv. Sin 1983; 19:25–29.

Yao M, Huang H, Yu J, Chen L (2005) Analysis on applicability of ISSR in molecular identification and relationship investigation of tea cultivars. *J Tea Sci.* 25:153-157.

Young-Goo P., Kaundan S.S. and Zhyvoloup A. (2002) Use of bulked genomic DNA based RAPD methodology to assess the genetic diversity among abandoned Korean tea plantations. *Gen. Resourc. Crop. Evol.* 49: 159-165.

Zhang Y., Zhu X., Chen X. and et al. (2014) Identification and characterization of cold-responsive microRNAs in tea plant (*Camellia sinensis*) and their targets using high-throughput sequencing and degradome analysis. *BMC Plant Biology* 14: 271-281.

Zhao L.P., Liu Z., Chen L., Yao M.Z. and Wang X.C. (2007) Generation and characterization of 24 nobel EST derived microsatellites from tea plant (*Camellia sinensis*) and cross species amplification in its closely related species and varieties. *Conserv Genet* 9: 1327-1331.

Zhu Q.W. and Luo Y.P. (2013) Identification of miRNAs and their targets in tea (*Camellia sinensis*) *J Zhejiang Univ Sci B.* 14: 916–923.

Index

A

B

E

F

G

H

N

O

P

Q

R

S

T

U

V

W